Advances in Intelligent Systems and Computing

Volume 854

Series editor

Janusz Kacprzyk, Polish Academy of Sciences, Warsaw, Poland
e-mail: kacprzyk@ibspan.waw.pl

The series "Advances in Intelligent Systems and Computing" contains publications on theory, applications, and design methods of Intelligent Systems and Intelligent Computing. Virtually all disciplines such as engineering, natural sciences, computer and information science, ICT, economics, business, e-commerce, environment, healthcare, life science are covered. The list of topics spans all the areas of modern intelligent systems and computing such as: computational intelligence, soft computing including neural networks, fuzzy systems, evolutionary computing and the fusion of these paradigms, social intelligence, ambient intelligence, computational neuroscience, artificial life, virtual worlds and society, cognitive science and systems, Perception and Vision, DNA and immune based systems, self-organizing and adaptive systems, e-Learning and teaching, human-centered and human-centric computing, recommender systems, intelligent control, robotics and mechatronics including human-machine teaming, knowledge-based paradigms, learning paradigms, machine ethics, intelligent data analysis, knowledge management, intelligent agents, intelligent decision making and support, intelligent network security, trust management, interactive entertainment, Web intelligence and multimedia.

The publications within "Advances in Intelligent Systems and Computing" are primarily proceedings of important conferences, symposia and congresses. They cover significant recent developments in the field, both of a foundational and applicable character. An important characteristic feature of the series is the short publication time and world-wide distribution. This permits a rapid and broad dissemination of research results.

More information about this series at http://www.springer.com/series/11156

Zofia Wilimowska · Leszek Borzemski
Jerzy Świątek
Editors

Information Systems Architecture and Technology: Proceedings of 39th International Conference on Information Systems Architecture and Technology – ISAT 2018

Part III

 Springer

Editors
Zofia Wilimowska
University of Applied Sciences
  in Nysa
Nysa, Poland

Leszek Borzemski
Faculty of Computer Science
  and Management
Wrocław University of Science
  and Technology
Wrocław, Poland

Jerzy Świątek
Faculty of Computer Science
  and Management
Wrocław University of Science
  and Technology
Wrocław, Poland

ISSN 2194-5357   ISSN 2194-5365 (electronic)
Advances in Intelligent Systems and Computing
ISBN 978-3-319-99992-0   ISBN 978-3-319-99993-7 (eBook)
https://doi.org/10.1007/978-3-319-99993-7

Library of Congress Control Number: 2018952643

This Springer imprint is published by the registered company Springer Nature Switzerland AG
The registered company address is: Gewerbestrasse 11, 6330 Cham, Switzerland

Preface

Variability of the environment increases the risk of the business activity. Dynamic development of the IT technologies creates the possibility of using them in the dynamic management process modeling and decision-making processes supporting. In today's information-driven economy, companies uncover the most opportunities. Contemporary organizations seem to be knowledge-based organizations, and in connection with that information becomes the most critical resource. Knowledge management is the process through which organizations generate value from their intellectual and knowledge-based assets. It consists of the scope of strategies and practices used in corporations to explore, represent, and distribute knowledge. It is a management *philosophy*, which combines good practice in purposeful information management with a culture of organizational learning, to improve business performance. An improvement of the decision-making process is possible to be assured by the analytical process supporting. Applying some analytical techniques such as computer simulation, expert systems, genetic algorithms can improve the quality of managerial information. Combining analytical techniques and building computer hybrids give synergic effects—additional functionality—which makes managerial decision process better. Different technologies can help in accomplishing the managerial decision process, but no one is in favor of information technologies, which offer differentiable advantages. Information technologies take place a significant role in this area. A computer is a useful machine in making managers' work more comfortable. However, we have to remember that the computer can become a tool only, but it cannot make the decisions. You can not build computers that replace the human mind. Computers can collect, select information, process it and create statistics, but decisions must be made by managers based on their experience and taking into account computer use. Different technologies can help in accomplishing the managerial decision process, but no one like information technologies, which offer differentiable advantages.

Computer science and computer systems, on the one hand, develop in advance of current applications, and on the other hand, keep up with new areas of application. In today's all-encompassing cyber world, nobody knows who motivates. Hence, there is a need to deal with the world of computers from both points of view.

In our conference, we try to maintain a balance between both ways of development. In particular, we are trying to get a new added value that can flow from the connection of the problems of two worlds: the world of computers and the world of management. Hence, there are two paths in the conference, namely computer science and management science.

This three-volume set of books includes the proceedings of the 2018 39th International Conference Information Systems Architecture and Technology (ISAT), or ISAT 2018 for short, held on September 16–18, 2018, in Nysa, Poland. The conference was organized by the Department of Computer Science and Department of Management Systems, Faculty of Computer Science and Management, Wrocław University of Science and Technology, Poland, and University of Applied Sciences in Nysa, Poland.

The International Conference Information Systems Architecture has been organized by the Wrocław University of Science and Technology from the seventies of the last century. The purpose of the ISAT is to discuss a state-of-the-art of information systems concepts and applications as well as architectures and technologies supporting contemporary information systems. The aim is also to consider an impact of knowledge, information, computing, and communication technologies on managing of the organization scope of functionality as well as on enterprise information systems design, implementation, and maintenance processes taking into account various methodological, technological, and technical aspects. It is also devoted to information systems concepts and applications supporting the exchange of goods and services by using different business models and exploiting opportunities offered by Internet-based electronic business and commerce solutions.

ISAT is a forum for specific disciplinary research, as well as on multi-disciplinary studies to present original contributions and to discuss different subjects of today's information systems planning, designing, development, and implementation.

The event is addressed to the scientific community, people involved in a variety of topics related to information, management, computer and communication systems, and people involved in the development of business information systems and business computer applications. ISAT is also devoted as a forum for the presentation of scientific contributions prepared by MSc. and Ph.D. students. Business, Commercial, and Industry participants are welcome.

This year, we received 213 papers from 34 countries. The papers included in the three proceedings volumes have been subject to a thoroughgoing review process by highly qualified peer reviewers. The final acceptance rate was 49%. Program Chairs selected 105 best papers for oral presentation and publication in the 39th International Conference Information Systems Architecture and Technology 2018 proceedings.

The papers have been grouped into three volumes:

Part I—discoursing about essential topics of information technology including, but not limited to, computer systems security, computer network architectures, distributed computer systems, quality of service, cloud computing and high-performance computing, human–computer interface, multimedia systems, big data, knowledge discovery and data mining, software engineering, e-business systems, web design, optimization and performance, Internet of things, mobile systems and applications.

Part II—addressing topics including, but not limited to, model-based project and decision support, pattern recognition and image processing algorithms, production planning and management systems, big data analysis, knowledge discovery and knowledge-based decision support and artificial intelligence methods and algorithms.

Part III—is gain to address very hot topics in the field of today's various computer-based applications—is devoted to information systems concepts and applications supporting the managerial decisions by using different business models and exploiting opportunities offered by IT systems. It is dealing with topics including, but not limited to, knowledge-based management, modeling of financial and investment decisions, modeling of managerial decisions, organization and management, project management, risk management, small business management, software tools for production, theories, and models of innovation.

We would like to thank the program committee and external reviewers, essential for reviewing the papers to ensure a high standard of the ISAT 2018 conference and the proceedings. We thank the authors, presenters, and participants of ISAT 2018, without them the conference could not have taken place. Finally, we thank the organizing team for the efforts this and previous years in bringing the conference to a successful conclusion.

September 2018   Leszek Borzemski
                                                                          Jerzy Świątek
                                                                          Zofia Wilimowska

ISAT 2018 Conference Organization

General Chair

Zofia Wilimowska, Poland

Program Co-chairs

Leszek Borzemski, Poland
Jerzy Świątek, Poland
Zofia Wilimowska, Poland

Local Organizing Committee

Zofia Wilimowska (Chair)
Leszek Borzemski (Co-chair)
Jerzy Świątek (Co-chair)
Mariusz Fraś (Conference Secretary, Website Support)
Arkadiusz Górski (Technical Editor)
Anna Kamińska (Technical Secretary)
Ziemowit Nowak (Technical Support)
Kamil Nowak (Website Coordinator)
Danuta Seretna-Sałamaj (Technical Secretary)

International Program Committee

Zofia Wilimowska (Chair), Poland
Jerzy Świątek (Co-chair), Poland
Leszek Borzemski (Co-chair), Poland

Witold Abramowicz, Poland
Dhiya Al-Jumeily, UK
Iosif Androulidakis, Greece

Patricia Anthony, New Zealand
Zbigniew Banaszak, Poland
Elena N. Benderskaya, Russia
Janos Botzheim, Japan
Djallel E. Boubiche, Algeria
Patrice Boursier, France
Anna Burduk, Poland
Andrii Buriachenko, Ukraine
Udo Buscher, Germany
Wojciech Cellary, Poland
Haruna Chiroma, Malaysia
Edward Chlebus, Poland
Gloria Cerasela Crisan, Romania
Marilia Curado, Portugal
Czesław Daniłowicz, Poland
Zhaohong Deng, China
Małgorzata Dolińska, Poland
Ewa Dudek-Dyduch, Poland
Milan Edl, Czech Republic
El-Sayed M. El-Alfy, Saudi Arabia
Peter Frankovsky, Slovakia
Mariusz Fraś, Poland
Naoki Fukuta, Japan
Bogdan Gabryś, UK
Piotr Gawkowski, Poland
Arkadiusz Górski, Poland
Manuel Graña, Spain
Wiesław M. Grudewski, Poland
Katsuhiro Honda, Japan
Marian Hopej, Poland
Zbigniew Huzar, Poland
Natthakan Iam-On, Thailand
Biju Issac, UK
Arun Iyengar, USA
Jürgen Jasperneite, Germany
Janusz Kacprzyk, Poland
Henryk Kaproń, Poland
Yury Y. Korolev, Belarus
Yannis L. Karnavas, Greece
Ryszard Knosala, Poland
Zdzisław Kowalczuk, Poland
Lumír Kulhanek, Czech Republic
Binod Kumar, India
Jan Kwiatkowski, Poland

Antonio Latorre, Spain
Radim Lenort, Czech Republic
Gang Li, Australia
José M. Merigó Lindahl, Chile
Jose M. Luna, Spain
Emilio Luque, Spain
Sofian Maabout, France
Lech Madeyski, Poland
Zbigniew Malara, Poland
Zygmunt Mazur, Poland
Elżbieta Mączyńska, Poland
Pedro Medeiros, Portugal
Toshiro Minami, Japan
Marian Molasy, Poland
Zbigniew Nahorski, Poland
Kazumi Nakamatsu, Japan
Peter Nielsen, Denmark
Tadashi Nomoto, Japan
Cezary Orłowski, Poland
Sandeep Pachpande, India
Michele Pagano, Italy
George A. Papakostas, Greece
Zdzisław Papir, Poland
Marek Pawlak, Poland
Jan Platoš, Czech Republic
Tomasz Popławski, Poland
Edward Radosinski, Poland
Wolfgang Renz, Germany
Dolores I. Rexachs, Spain
José S. Reyes, Spain
Małgorzata Rutkowska, Poland
Leszek Rutkowski, Poland
Abdel-Badeeh M. Salem, Egypt
Sebastian Saniuk, Poland
Joanna Santiago, Portugal
Habib Shah, Malaysia
J. N. Shah, India
Jeng Shyang, Taiwan
Anna Sikora, Spain
Marcin Sikorski, Poland
Małgorzata Sterna, Poland
Janusz Stokłosa, Poland
Remo Suppi, Spain
Edward Szczerbicki, Australia

Eugeniusz Toczyłowski, Poland
Elpida Tzafestas, Greece
José R. Villar, Spain
Bay Vo, Vietnam
Hongzhi Wang, China
Leon S. I. Wang, Taiwan
Junzo Watada, Japan
Eduardo A. Durazo Watanabe, India

Jan Werewka, Poland
Thomas Wielicki, USA
Bernd Wolfinger, Germany
Józef Woźniak, Poland
Roman Wyrzykowski, Poland
Yue Xiao-Guang, Hong Kong
Jaroslav Zendulka, Czech Republic
Bernard Ženko, Slovenia

ISAT 2018 Reviewers

Hamid Al-Asadi, Iraq
Patricia Anthony, New Zealand
S. Balakrishnan, India
Zbigniew Antoni Banaszak, Poland
Piotr Bernat, Poland
Agnieszka Bieńkowska, Poland
Krzysztof Billewicz, Poland
Grzegorz Bocewicz, Poland
Leszek Borzemski, Poland
Janos Botzheim, Hungary
Piotr Bródka, Poland
Krzysztof Brzostkowski, Poland
Anna Burduk, Poland
Udo Buscher, Germany
Wojciech Cellary, Poland
Haruna Chiroma, Malaysia
Witold Chmielarz, Poland
Grzegorz Chodak, Poland
Andrzej Chuchmała, Poland
Piotr Chwastyk, Poland
Anela Čolak, Bosnia and Herzegovina
Gloria Cerasela Crisan, Romania
Anna Czarnecka, Poland
Mariusz Czekała, Poland
Y. Daradkeh, Saudi Arabia
Grzegorz Debita, Poland
Anna Dobrowolska, Poland
Maciej Drwal, Poland
Ewa Dudek-Dyduch, Poland
Jarosław Drapała, Poland
Tadeusz Dudycz, Poland
Grzegorz Filcek, Poland

Mariusz Fraś, Poland
Naoki Fukuta, Japan
Piotr Gawkowski, Poland
Dariusz Gąsior, Poland
Arkadiusz Górski, Poland
Jerzy Grobelny, Poland
Krzysztof Grochla, Poland
Bogumila Hnatkowska, Poland
Katsuhiro Honda, Japan
Zbigniew Huzar, Poland
Biju Issac, UK
Jerzy Józefczyk, Poland
Ireneusz Jóźwiak, Poland
Krzysztof Juszczyszyn, Poland
Tetiana Viktorivna Kalashnikova,
 Ukraine
Anna Kamińska-Chuchmała, Poland
Yannis Karnavas, Greece
Adam Kasperski, Poland
Jerzy Klamka, Poland
Agata Klaus-Rosińska, Poland
Piotr Kosiuczenko, Poland
Zdzisław Kowalczyk, Poland
Grzegorz Kołaczek, Poland
Mariusz Kołosowski, Poland
Kamil Krot, Poland
Dorota Kuchta, Poland
Binod Kumar, India
Jan Kwiatkowski, Poland
Antonio LaTorre, Spain
Arkadiusz Liber, Poland
Marek Lubicz, Poland

Emilio Luque, Spain
Sofian Maabout, France
Lech Madeyski, Poland
Jan Magott, Poland
Zbigniew Malara, Poland
Pedro Medeiros, Portugal
Vojtěch Merunka, Czech Republic
Rafał Michalski, Poland
Bożena Mielczarek, Poland
Vishnu N. Mishra, India
Jolanta Mizera-Pietraszko, Poland
Zbigniew Nahorski, Poland
Binh P. Nguyen, Singapore
Peter Nielsen, Denmark
Cezary Orłowski, Poland
Donat Orski, Poland
Michele Pagano, Italy
Zdzisław Papir, Poland
B. D. Parameshachari, India
Agnieszka Parkitna, Poland
Marek Pawlak, Poland
Jan Platoš, Czech Republic
Dolores Rexachs, Spain
Paweł Rola, Poland
Stefano Rovetta, Italy
Jacek, Piotr Rudnicki, Poland
Małgorzata Rutkowska, Poland
Joanna Santiago, Portugal
José Santos, Spain
Danuta Seretna-Sałamaj, Poland

Anna Sikora, Spain
Marcin Sikorski, Poland
Małgorzata Sterna, Poland
Janusz Stokłosa, Poland
Grażyna Suchacka, Poland
Remo Suppi, Spain
Edward Szczerbicki, Australia
Joanna Szczepańska, Poland
Jerzy Świątek, Poland
Paweł Świątek, Poland
Sebastian Tomczak, Poland
Wojciech Turek, Poland
Elpida Tzafestas, Greece
Kamila Urbańska, Poland
José R. Villar, Spain
Bay Vo, Vietnam
Hongzhi Wang, China
Shyue-Liang Wang, Taiwan, China
Krzysztof Waśko, Poland
Jan Werewka, Poland
Łukasz Wiechetek, Poland
Zofia Wilimowska, Poland
Marek Wilimowski, Poland
Bernd Wolfinger, Germany
Józef Woźniak, Poland
Maciej Artur Zaręba, Poland
Krzysztof Zatwarnicki, Poland
Jaroslav Zendulka, Czech Republic
Bernard Ženko, Slovenia
Andrzej Żołnierek, Poland

ISAT 2018 Keynote Speaker

Professor Dr. Abdel-Badeh Mohamed Salem, Faculty of Science, Ain Shams
    University, Cairo, Egypt
Topic: Artificial Intelligence Technology in Intelligent Health Informatics

Contents

Models of Financial and Investment Decisions

Relations Between the Use of Modern Technology and the Financial Results of the Financial Service Sector in Poland

Rafał Szmajser[1]([⊠]), Mariusz Andrzejewski[2], and Marcin Kędzior[2]

[1] Capgemini Poland, Al. Jerozolimskie 96, 00-807 Warsaw, Poland
rafal.szmajser@capgemini.com
[2] Financial Accounting Department, Cracow University of Economics,
ul. Rakowicka 27, 31-510 Kraków, Poland
{mariusz.andrzejewski,kedziorm}@uek.krakow.pl

Abstract. This paper presents technological factors of profitability which were analysed for the outsourcing enterprises in the financial service sector in Poland. The paper describes and presents a discussion of the crucial technological factors determining the profitability of the outsourcing enterprises. The analysis was focused on the following factors: the applications and services available via "cloud", automation and robotization, interactive applications and social media, tools and applications for processing a large volume of data and the artificial intelligence solutions. In view of the empirical verification, the crucial technological factors included robotization and automation, the applications available via "cloud" as well as the tools and applications for processing a large volume of data. The results were verified for a two or three year timeframe. The factors were analysed from the perspective of the positions the respondents held, the line of business of an enterprise and its location. The data were collected on the basis of tailored questionnaires. The statistical results were verified by the ANOVA variance analysis.

Keywords: Technology · Profitability · Financial service sector

1 Introduction

Economic slowdown, globalization and international competition commonly entail risk for a majority of enterprises. One of the ways of diminishing it is searching cost cutting methods by sub-contracting some activities to the external entities. Such an activity consisting in separating certain functions of the organizational structure of an enterprise and sub-contracting them to other entities is called outsourcing [16]. Most often, outsourcing is described as a typical process of a developed economy, where some non-core tasks of a company are entrusted to specialized external entities. The term "outsourcing" originates in English, and is a coinage of *outside resource using* which denotes the use of outside resources [*Ibidem*].

The main merit of outsourcing consists in subcontracting the implementation of a function, processes or tasks to external enterprises which specialize in a certain area. Outsourcing processes could be observed as early or the Middle Ages, yet it was only in

© Springer Nature Switzerland AG 2019
Z. Wilimowska et al. (Eds.): ISAT 2018, AISC 854, pp. 3–15, 2019.
https://doi.org/10.1007/978-3-319-99993-7_1

the 20th century that outsourcing became to be applied by the entrepreneurs more commonly. In its first stage, outsourcing was used mostly as an instrument for cutting the costs of production as well as a method for modifying the *make or buy* concept which allowed either producing on one's own or purchasing from the external entity [1].

It was only in the 1920s that H. Ford asserted that when an enterprise is not able to produce a commodity on its own, at the same time cutting the costs and increasing the quality, then an entrepreneur should resort to the external entities by subcontracting to them a part of the processes involved in the manufacture of a certain commodity. Almost fifty years later, the principle presented by H. Ford started to be practiced on the global market. R. Perrot, the owner of Electronic Data Systems, is considered to be the pioneer of modern outsourcing. Pierrot's company was the first one to commission Fritto-Lay concern to run the IT service which was then called "hardware management" [16]. Soon after, other companies started to offer similar services to other enterprises.

Following the 2007 financial crisis, the companies operating in the financial sector began more and more frequently subcontract selected business processes to the outsourcing centres whilst passing on to them the responsibility for the quality and efficiency of those services. A dynamic growth of the financial sector was one of the traits of globalization, and it allowed companies reaching a competitive advantage and facilitated further development within a competitive environment. The sector of financial services (banking, insurance, financial services – BIFS) comprises the services offered by the banks, insurers and other financial institutions. According to the Association of Business Service Leaders (ABSL) report of 2013, Polish sector of financial services is the largest such sector in Central-Eastern Europe with the total assets of 1400 billion PLN (350 billion EUR) [17]. For the past few years, it has been the fastest growing market in Europe.

One of the most important objectives of activity of Global Business Services (GBS) is to achieve profitability [8, 9]. The articles on this subject include an analysis of the following factors of profitability within Global Business Services (GBS) [15]: an appropriate level of abilities and competences of employees and the level of their seniority in the organization, a greater possibility of scalability of the operating activity, providing service from a number of geographical locations around the world, promptness of implementing changes, developing and procuring talents inside an organization, standardization and transformation of processes, supporting organizations in mergers and takeovers, supporting business by technological solutions, especially via automation and analytical solutions, an appropriate price model, programs of continuous process improvement (e.g. *lean*, *six sigma*). The paper was focused on the technological factors as well as the applications and services available via cloud, automation and robotics, interactive applications and social media, tools and application for the processing of a large volume of data, and finally artificial intelligence solutions.

New technologies significantly contribute to reaching the competitive advantage [11]. Innovation and new technologies facilitate higher productivity of enterprises [3]. It boosts generation of the added value for an enterprise. Innovative companies have a more extensive knowledge of the market, they are customer-oriented, which translates into higher prices. They allow launching high quality products or delivering such service, hence they generate higher margins. Similar views were shared by Spanos, Zaralis and Lioukas [13]. According to Czarnitzki and Kraft, the dependencies between new technologies and profitability are not quite clear [4]. As a rule, research and development programs are costly, and they prove to bring returns over a long period of time. It should be noted that the technology or know-how also can be acquired by the competition. Soininen, Martikainen, Puumalainen and Kyläheiko [12] stress that the outlays on research and development as well as new technologies are encumbered with the risk of not gaining the expected returns, particularly in the area of high technologies which are very quickly replaced by the more advanced ones [12]. This assertion by no means disapproves the fact that the enterprises of low technologies and low innovation level can quickly lose their competitive advantage and in the long term, also profitability [2].

The objective of this article is to verify the technological factors influencing profitability of the GBS centres in the financial service sector operating on the territory of Poland with the special focus on Robotics Process Automation (RPA). The following GBS centres, among others, such as: Hewlett-Packard, Alexander Mann Solutions, Amway, Capgemini, Cisco Systems, Electrolux, Hitachi Data Systems, Accenture Services Ltd., Lufthansa Global Business Services, HSBC Service Delivery Polska Ltd., Zurich Insurance Company, MARS Financial Services Europe, PWC, Tata Consultancy Services, Shell, Sony Pictures, Global Business Services, State Street, UBS and firms being the recipients of the outsourcing services or consulting firms operating in the Global Business Services sector, among others, mBank, Delphi, EY, ING, Staco Polska, EDF, Van Gansewinkel, and WNS Global Services were subjected to the analysis. The data were collected on the basis of tailored questionnaires, and the results were verified by the ANOVA variance analysis.

2 Modern Technological Solutions and Automation of Financial and Accounting Processes Making an Impact on the Profitability of the Sector of Modern Business Services

The functioning of enterprises relies on the mutual flow of information, resources, capital and personnel which are subject to specified business processes. Changes in the legislation and tax laws, technological progress as well as the changing environment demand that the on-going processes should be continually monitored and adjusted to the realities of the market. One of the key challenges faced by many organizations includes identification, analysis and the use of the information they possess, making connection between the information and data found both in various locations as well as within the systems of an enterprise. The enterprises are subjected to both the internal as well as the external pressure due to the implementation of IT solutions which are poised to meet the expectations of customers, shareholders as well as the contractual and/or legal terms on the obligatory disclosure of information on various aspects of an enterprise operation. Typical examples of such information include the quality and the efficiency of work, the degree of risk, the changes and all other information on the strategic, tactical, and strategic levels both within an enterprise as well as within its environment. Proper management is one of the key factors which are conducive to a satisfactory performance. Over the past few years, it has become popular to gain knowledge through analysing massive volumes of information collected over a long time, the so-called *big data*.

Oxford English Dictionary (OED) [17] defines term 'big data' as data of a very large size, typically to the extent that its manipulation and management present significant logistical challenges; (also) the branch of computing involving such data.

A. De Mauro defines big data as "datasets whose size is beyond the ability of typical database software tools to capture, store, manage, and analyse" [5]. The term is used to describe a wide range of concepts: from the technological ability to store, aggregate, and process data, to the cultural shift that is pervasively invading business and society, drowning in information overload.

The ability to use that information which enhances the knowledge of the decision maker allows taking more efficient decisions poised to define the optimal target model of an entity – both in financial, economic, organizational or technological terms. The results of the optimization of business processes include lowering costs, increasing efficiency, reduction of elimination of "bottlenecks". Consequently, an entity or an institution can operate more efficiently, building along a competitive advantage. The analytical potential, both in terms of advanced technology and knowledge, is reached through the applications and services available via cloud.

According to the definition provided by Mell and Grace [10], cloud computing is a model for enabling ubiquitous, convenient, on-demand network access to a shared pool of configurable computing resources (e.g., networks, servers, storage, applications, and services) that can be rapidly provisioned and released with minimal management effort or service provider interaction.

Due to the visualization of important information, the decision makers are provided with the basis for taking preventive steps aimed to lower the costs, improve working capital, lower the volume of uninvoiced deliveries or modify the strategy of recovering the receivables. The customers are increasingly interested in the outsourced analytical services which lead to increasing the revenues of the outsourcing companies. The companies which decide on outsourcing their services expect the providers not only to cut operating costs whilst sustaining the appropriate quality of service, but also to guarantee full transparency in reporting performance and full automation of the reporting procedures so that to eliminate time consuming analysis done by hand.

There are numerous technological analytical solutions available on the market which provide complex analytical tools suspended in "cloud". Amongst them there are products offered by such global companies as Microsoft, Oracle, SAP, IBM, Tableau or the solutions applied by the outsourcing companies jointly with the internal systems built by those companies, e.g. Client Intelligence Centre or Production Monitoring and Planning Tool (PROMPT) built by Capgemini. Business intelligence solutions allow the teams reaching outstanding results due to the self-served access to the appropriate information by all members of an organization. The available tools facilitate taking business decisions by the users, providing them with solid information based on facts, irrespective the place of their storage.

Technologies intended to raise the level of automation and robotization of financial and accounting processes constitute other solutions which significantly improve the efficiency of services within the sector of modern business services. Within the environment of outsourcing services, the concepts of robotization and automation of business processes are described as *robotic process automation* (RPA) and *business process automation* (BPA). Those processes are defined as the automation of complex, recurrent business processes where the conventional, manual processing of data is replaced with specialized applications mimicking the operations earlier done by humans [18].

The tools applied to automation and robotization termed *robotic automation* combine in fact a specific technology and methodology of using a computer and specialized software to duplicate a task which was earlier performed by humans, precisely in the same way as it used to be done. That means that unlike the traditional software, *robotic automation* is a tool or a platform which functions as well as co-ordinates another software with the user's interface. In that sense, it is not integrated with other applications but it manages them from, if you will, a higher level, imitating the work done by humans. The advanced system of *robotic process automation* are the important tools applied by the outsourcing organizations to overcome the inefficiency of their internal processes. According to the forecasts made by the outsourcing companies, *robotic process automation* will dynamically develop in the near future. At the same time, with the application of the elements of automation and artificial intelligence, the process will in time become more complex, autonomous and specialized [18].

In line with the forecasts by the Operational Agility Forum [20], the provider of outsourcing services will be able to set up a "virtual administration office" with the application of robotic automation. That office will be manned by robots programmed to conduct office procedures consistent with certain schemes of operation. Due to the fact that the cost of robots is lower than the cost of labour of traditional staff, it is expected that the application of robots will be conducive to the profitability of contracts drawn within the business environment. On the other hand, the implementation of such solutions will bring a new type of risk for the outsourcing company; instead of commissioning a certain service to the provider of business services, a customer may decide to implement automation on their own and reduce the scope of the service previously outsourced.

The application of virtual robots to performing transactions which are based on certain schemes and routine operations will gradually replace the activities earlier done by humans. This will be accomplished with the use of special algorithms implemented by virtual robots. Such solution not only will increase the efficiency of business processes, but also change the role of a human, freeing him from merely repetitive, routine tasks so that he can focus on the tasks which generate higher added value, demand interaction, decision taking or face to face contact with the customer. Blueprism, one of the leaders of robotic process automation, has listed the advantages of implementing modern automation software in those companies which aim to lower the payroll costs and raise the quality of their services [21]:

- The proportion of the cost of a robot worker to a human is six to one; what is more, the former does not make any mistakes.
- The speed of launching automation or streamlining is reduced to days or weeks in comparison to weeks or months needed to bring in the conventional streamlining techniques.
- Robots are programmed to conduct repetitive business processes and they apply functionalities of the existing applications. In this way, the high costs of the necessary intrasystem integrations and reprogramming the robot are lowered.
- *Robotic process automation* platforms are managed via small, specialized IT entities which makes them safe and easy to audit. Due to the RPA control points, they meet the prescribed standards of confidentiality and security of data, and the speed and correctness of transactions is significantly higher owing to automation software.
- *Robotic process automation* consists in working in a virtual environment. This allows a relatively simple scaling the efficiency of robots via controlling the efficiency of servers, to reflect the varying number of transactions.

With the application of automation, the role of a human is reduced to the supervision of proper functioning of a robot and he is freed from manual, repetitive operations. In this context, a human has become the operator of a robot and he intervenes only when a robot encounters a bug in a process or the lack of script necessitates the decision to be taken by a human. Gradually, in line with expanding the data base with the served exceptions, robot's script is learning and keeps adding new scenarios to its data bank. Social media and interactive applications are another important technology having an impact on shaping profitability of an outsourcing organization. Kaplan and Haenlein have defined social media as "a group of applications based on internet solutions, founded mostly on the ideological and technological base of Web 2.0, which facilitate the generation and exchange of the content provided by the users [7]. Based on this definition, Social Media can be classified into more specific categories as: collaborative projects, content communities, blogs, social networking websites, virtual game worlds and virtual social worlds.

The last but not least category of the analysed technology determinants having impact on the profitability of the hired service organization being subject to this research are solutions based on the Artificial Intelligence. A simple definition of Artificial intelligence is provided by Ginsberg, who defines Artificial Intelligence (AI) as the enterprise of constructing an intelligent artefact [6]. Artificial Intelligence may have various applications starting from deduction, reasoning, and problem-solving applications through natural language generation and social intelligence solutions. Another application of AI may be found in the area of predictive analytics with advanced usage of data mining, statistics and modelling to analyse current data and make predictions about the future. Automated written reporting and/or communications together with voice recognition are still in its early days of adoption.

3   The Empirical Results

The obtained responses (Table 1) afford an assertion that interactive applications and social media as well as artificial intelligence solutions noted the highest rise in importance. In the future, the highest impact will be exerted by the automation and robotization and applications and services available via cloud.

Table 1. Respondents' replies to the question about the impact which selected technologies exert on the current profitability of financial and accounting processes over the next two to three years.

Question no	No impact	Negligible impact	Moderate impact	High impact	Very high impact	Number of responses
(1)	0	2	10	28	19	59
(2)	0	0	9	29	21	59
(3)	4	10	20	18	7	59
(4)	5	1	13	20	20	59
(5)	8	10	12	16	12	58

Explanation: question no: (1) applications and services available via cloud; (2) automation and robotization; (3) interactive applications and social media; (4) tools and applications for processing and analyzing a large volume of data; (5) artificial intelligence solutions. Source: Authors' own elaboration.

A t-test for paired analysis was applied to determine whether there are statistically viable disparities between the impact exerted by various factors on the profitability of outsourcing organizations/centers now and within the 2–3 year perspective. The results are presented in Table 2.

Table 2. The results of paired t – tests in comparisons to evaluations of the determinants affecting an evaluation of the technologies having an impact on current profitability of performance of the financial-accounting services and in 2–3 years.

Question no	Time perspective	Average	Standard deviation	t	p
(1)	Currently	3.033898	0.780398	10.12689	1.94E−14***
	In 2–3 years	4.084746	0.787361		
(2)	Currently	3.033898	0.822689	12.03654	2.08E−17****
	In 2–3 years	4.20339	0.683452		
(3)	Currently	2.627119	0.841675	6.513053	1.92E−08***
	In 2–3 years	3.237288	1.078904		
(4)	Currently	3	0.956689	6.990169	3.04E−09***
	In 2–3 years	3.830508	1.166663		
(5)	At present	2.103448	1.02872	8.579941	6.58E−12***
	In 2–3 years	3.241379	1.330159		

Significant differences at the level of: ***$p < 0.001$; **$p < 0.01$; *$p < 0.05$.
Source: Authors' own elaboration.

By examining the data included in Table 2 it can be concluded that a significant increase in evaluating an influence on profitability of the outsourcing organizations/centres in the perspective of 2–3 years in view of the current situation was noted for all five technologies. The results of the survey confirm that the market for technological solutions including artificial intelligence (AI) technologies is rapidly developing. Although AI is still in its early days of adoption, it is just a matter of time before technology will impact a vast majority of organizations. According to the survey deployed by the National Business Research Institute among respondents whose companies have not yet deployed AI technologies, 41% indicated that doing so is a priority. And, more than half (56%) plan to deploy Artificial Intelligence technologies within the next two years, while nearly a quarter of them (23%) intend to do so within the next 12 months. 38 percent of enterprises are already using Artificial Intelligence, with a predicted growth to 62% by 2018. The results of the survey show that between 2014 and 2015 alone, for example, the number of organizations either deploying or implementing data-driven projects increased by about 125%, where the average enterprise spent $13.8 million on the effort [21].

Global Business Services uses labour on a large scale for projects requiring general knowledge process work, where employees are processing a high-volume of data. A highly transactional process may benefit from reducing costs and time with robotic process automation software and cloud computing tools. The automated process needs to be scalable, reliable, manageable and secure. It means that automation must be developed on a solid technical foundation and IT organization needs to be comfortable with it. Although IT is involved in the RPA process, control of the automated processes should remain with the business owners. This is due to the fact that the process changes frequently which requires business knowledge to adopt the functionality of a robot. Involving IT in every change or making them responsible for the change would slow down the process.

In order to be successful in the implementation of a specific technology, an organization must trust its technological solutions (e.g. software robots to carry out business processes in a correct and compliant way). Therefore to ensure this, an implementation of technological solution (e.g. RPA solution) must provide effective access control and auditing features. This is especially important for the implementation of RPA solutions. A related challenge is to provide the way to audit the operation of robots. Auditability means that a human supervisor can track what changes were made to a robot, along with the evidence who made those changes and when. Typically, RPA solutions maintain log changes to each user's permissions and other data which are security-sensitive. This information is important for tracking the errors and it is also crucial for detecting fraud or other attempts to destabilize a robot's work e.g. to achieve this, RPA solutions typically maintain an audit log for each robot.

By analyzing the presented data, it can be recognized that among the presidents, directors and managers, Robotics Process Automation (RPA) is the highest rated determinant influencing profitability in the perspective of 2–3 years (Table 3). According to the Institute for Robotic Process Automation and Artificial Intelligence (IRPA AI) [22], Robotic Process Automation can bring 25–50% cost savings. Process automation enables 24/7/365 execution at a fraction of the cost of human equivalents.

Table 3. The average and standard deviations of evaluation of technologies influencing profitability of performance of the financial-accounting services in 2–3 years according to a type of a position held at work.

Question no	Characteristics	President	Manager	Director	Total
(1)	Average	4.25	4.185185	3.75	4.084746
	Standard deviation	0.856349	0.786278	0.68313	0.794119
(2)	Average	4.375	4.333333	3.8125	4.20339
	Standard deviation	0.619139	0.620174	0.75	0.689319
(3)	Average	3.5	3.222222	3	3.237288
	Standard deviation	0.966092	1.120897	1.154701	1.088165
(4)	Average	3.8125	4.111111	3.375	3.830508
	Standard deviation	1.223043	0.847319	1.5	1.176678
(5)	Average	3.6875	3.148148	2.9375	3.237288
	Standard deviation	1.302242	1.292097	1.388944	1.33053

Explanation: 5.0 – the highest importance, 1.0 - the lowest importance
Source: Authors' own elaboration.

A software robot can cost as little as one-third the price of an offshore full-time employee (FTE) and as little as one-fifth the price of an onshore FTE.

As it can be noticed on the basis of data in Table 4, the position held at work varies significantly: Robotics Process Automation (RPA) with p-value below 0.05. A different perception of a factor which influences profitability depending on the position may result e.g. from the overall strategy and the technology facilitators existing within the organization as well as different experience in the implementation of RPA technology depending on the position held in the company. For example, during the implementation of RPA, the scripts used in simple RPA solutions frequently are run on their user's desktop. The execution of those types of scripts can be started and stopped by a user, who is often sitting at the computer while they execute the script. This is an effective approach for a manager in case of straightforward automation scenarios. But the senior management and executive group perspective (directors and presidents) may aim at the creation of complex solutions with a virtual workforce of software robots. An organization will not gain much from the automation if each robot has to be individually started, stopped, and monitored. Hence a more scalable approach is required to enable robots run largely on their own. Hence the perspective and expectations may be one of the explanations regarding the differences in the evaluation of the impact of RPA technologies influencing the profitability of the financial-accounting processes in 2–3 years, depending on a type of a position held in the organization (Table 4).

It should be noted that among the group of business service providers and consulting/auditing firms, automation and robotization scored highest within the technological factors making an impact on profitability in two or three years' time. Yet, the recipients of business services preferred the applications and services available via cloud (Table 5)

Table 4. The results of the one-way analysis of variance for evaluating the technologies influencing profitability of performance of the financial-accounting processes in 2–3 years differentiated by a type of the position held.

Question no	SS Effect	DF	MS Effect	SS Error	Df	MS Error	F	p-value
(1)	2.50220	2	1.251099	34.07407	56	0.608466	2.05615	0.137496
(2)	3.37182	2	1.685911	24.18750	56	0.431920	3.90330	0.025884*
(3)	2.01130	2	1.005650	66.66667	56	1.190476	0.84475	0.435068
(4)	5.45092	2	2.725459	74.85417	56	1.336682	2.03897	0.139715
(5)	4.89556	2	2.447779	97.78241	56	1.746114	1.40184	0.254646

Significant differences at the level of: $***p < 0.001$; $**p < 0.01$; $*p < 0.05$.
Source: Authors' own elaboration.

Table 5. Average and standard deviations in valuation of technologies making an impact on profitability of financial and accounting services in two to three years' time by the type of an activity.

Question number	Characteristics	Business service provider	Business service recipient	Consulting and auditing firm	Total
(1)	Average	4.043478	4.428571	4.000000	4.084746
	Stand. deviation	0.842128	0.534522	0.632456	0.794119
(2)	Average	4.152174	4.285714	4.500000	4.203390
	Stand. deviation	0.665579	0.755929	0.83666	0.689319
(3)	Average	3.23913	3.428571	3.00000	3.237288
	Stand. deviation	1.057981	1.272418	1.264911	1.088165
(4)	Average	3.782609	4.285714	3.666667	3.830508
	Stand. deviation	1.190948	0.755929	1.505545	1.176678
(5)	Average	3.195652	3.571429	3.166667	3.237288
	Stand. deviation	1.343620	1.272418	1.471960	1.330530

Source: Authors' own elaboration.

Having analysed the data presented in Table 6, it may be asserted that the type of economic activity has no bearing on any technological factor which makes a significant impact on the profitability of outsourcing organizations/centers within 2–3 year time-frame (p-value above 0.05).

The data presented in Table 7 indicate that the location of a business entity does not make any difference to any technological factor which may make a significant impact on the profitability of outsourcing organizations/centres within 2–3 years. What is more, it may be concluded that amongst the firms operating in Poland, the factor which was ranked highest in terms on its impact on profitability within 2–3 years were the applications and services available via cloud, whilst foreign companies chose automation and robotization.

To conclude, the study results presented in this part indicate that the key determinants of profitability influencing effectiveness of operating of the outsourcing centres

Table 6. The results of a single factor variance analysis for the valuation of technologies making an impact on the profitability of financial and accounting processes within 2–3 year timeframe, by type of activity.

Question number	SS Efekt	DF	MS Efekt	SS Błąd	Df	MS Błąd	F	p-value
(1)	0.94894	2	0.474471	35.6273	56	0.636202	0.745786	0.479013
(2)	0.69597	2	0.347984	26.8634	56	0.479703	0.725416	0.488616
(3)	0.59412	2	0.297058	68.0839	56	1.215783	0.244334	0.784056
(4)	1.71709	2	0.858547	78.5880	56	1.403357	0.611781	0.545969
(5)	0.89122	2	0.445608	101.7867	56	1.817621	0.245160	0.783415

Source: Authors' own elaboration.

Table 7. The results of t-test in comparisons to evaluate the technologies making an impact on the profitability of financial and accounting processes within 2–3 years with a consideration to the location of an enterprise.

Question number	Average-Poland	Average – foreign companies	t	p-value
(1)	4.127660	3.916667	0.81914	0.416116
(2)	4.170213	4.333333	−0.72868	0.469179
(3)	3.148936	3.583333	−1.23996	0.220069
(4)	3.744681	4.166667	−1.11104	0.271216
(5)	3.127660	3.666667	−1.25881	0.213229

Source: Authors' own elaboration.

according to the respondents are: robotization and automation, applications available via cloud, and finally the tools and applications for processing and analysing a large volume of data.

4 Concluding Remarks

In this paper, the empirical verification was focused on the technological factors making an impact on Global Business Services (GBS) in the financial service sector in Poland. The following factors were subject to the analysis: the applications and services available via cloud, automation and robotization, interactive applications and social media, tools and applications for processing and analyzing a large volume of data as well as artificial intelligence solutions. The analysis of the questionnaire results proves the significant impact of robotization and automation, applications available via cloud, and applications for processing and analysing a large volume of data on the profitability of outsourcing centers. That fact that merits reader's attention is within 2–3 year timeframe the discussed factors will exert an increasing impact of the profitability of the common services sector. The highest increase in importance was noted for the interactive applications, social media, and artificial intelligence solutions.

The providers of business services and consulting/auditing firms unanimously claimed that automation and robotization were the technological factors making the highest impact on profitability with 2–3 year horizon, whilst the recipients of business services pointed to the applications and services available via cloud. The location of a business entity has no significant bearing on any of the technological factors making an impact on the profitability of outsourcing organizations/center within a 2–3 year time span. It should be noted that Polish companies rank highest the applications and services available via cloud to make an impact on profitability within 2–3 year time-frame, whilst the foreign companies prefer automation and robotization.

In view of the verification of results, it should be noted that the most significant technological factors include robotization and automation (in the opinion of managers, presidents and directors), then the applications available via cloud as well as the tools and applications used to process and analyse a large volume of data.

Future directions of research on technological factors making an impact on the profitability of Global Business Services should be focused on verifications of such determinants as Robotic Process Automation and evaluating potential and the impact of the fastest developing Artificial Intelligence technologies like Natural Language Generation, Speech Recognition, Virtual Agents, Machine Learning Platforms and others.

References

1. Abt, S., Woźniak, H.: Podstawy logistyki. Wydawnictwo Uniwersytetu Gdańskiego, Gdańsk (1993)
2. Anastassopoulos, G.: Profitability differences between MNE subsidiaries and domestic firms: the case of the food industry in Greece. Agribusiness **20**(1), 45–60 (2016)
3. Berrone, P., Gertel, H., Giuliodori, R., Bernard, L., Meiners, E.: Determinants of performance in microenterprises: preliminary evidence from Argentina. J. Small Bus. Manag. **52**(3), 477–500 (2013)
4. Czarnitzki, D., Kraft, K.: On the profitability of innovative assets. Appl. Econ. **42**(15), 1941–1953 (2010)
5. De Mauro, A., Greco, M., Grimaldi, M.: What is big data? A consensual definition and a review of key research topics. In: Giannakopoulos, G., Sakas, D.P., Kyriaki-Manessi, D. (eds.) AIP Conference Proceedings (2015)
6. Ginsberg, M.: Essentials of Artificial Intelligence. Morgan Kaufmann Publishers, Newnes (2012)
7. Kaplan, A.M., Haenlein, M.: Users of the world, unite! The challenges and opportunities of social media. Bus. Horiz. **53**(1), 59–68 (2010)
8. Maślanka, T.: Płynność finansowa a rentowność przedsiębiorstw. In: Dresler, Z. (ed.) Rentowność przedsiębiorstw w Polsce. Wydawnictwo UEK w Krakowie, Kraków (2014)
9. Maślanka, T.: Wartość czy zysk – rozważania na temat głównego celu zarządzania finansami przedsiębiorstwa. In: Owsiak, S. (ed.) Bankowość w dobie kryzysu finansowego a perspektywy rozwoju regionów. Tom III. Przedsiębiorstwo wobec kryzysu finansowego, WSBiF w Bielsku-Białej, Bielsko-Biała (2009)
10. Mell, P., Grance, T.: The NIST Definition of Cloud Computing. NIST Special Publication (2011)
11. Omil, J.C., Lorenzo, P.C., Liste, A.V.: The power of intangibles in high-profitability firms. Total Qual. Manag. **22**(1), 29–42 (2011)

12. Soininen, J., Martikainen, M., Puumalainen, K., Kyläheiko, K.: Entrepreneurial orientation: growth and profitability of Finnish small-and medium-sized enterprises. Int. J. Prod. Econ. **140**(2), 614–621 (2012)
13. Spanos, Y.E., Zaralis, G., Lioukas, S.: Strategy and industry effects on profitability: evidence from Greece. Strateg. Manag. J. **25**(2), 139–165 (2004)
14. Stevenson, A. (ed.): Oxford Dictionary of English. Oxford University Press, USA (2010)
15. Szmajser, R., Andrzejewski, M., Kędzior, M.: Determining factors of profitability of outsourcing service enterprises, research results from Poland. In: 40th Annual Congress, Valencia (2017)
16. Trocki, M.: Outsourcing. Metoda restrukturyzacji działalności gospodarczej. PWE, Warszawa (2001)

Internet sources

17. OED Online. March 2017. Oxford University Press. http://www.oed.com/view/Entry/18833. Accessed 20 Apr 2017
18. http://www.gartner.com/it-glossary/bpa-business-process-automation. Accessed 20 Apr 2017
19. https://www.capgemini.com/resource-file-access/resource/pdf/bpos_next_wave_of_robotic_process_automation.pdf. Accessible after a log-in. Accessed 20 Apr 2017
20. Blue Prism Product Overview NHS Edition April 2013. http://www.blueprism.com/knowledge/knowledge.php?cat=2. Accessed 20 Apr 2017
21. https://narrativescience.com/Resources/Resource-Library/Article-Detail-Page/outlook-on-artificial-intelligence-in-the-enterprise. Access data 10 April 2017. Accessed 20 Apr 2017
22. http://irpaai.com/definition-and-benefits/. Accessed 20 Apr 2017

Modeling Investment Decisions in the System of Sustainable Financing

Joanna Koczar[1]([⊠]), K. M. Selivanova[2], A. R. Akhmetshina[2],
and V. I. Vagizova[2]

[1] Wroclaw University of Economics, 53-345 Wroclaw, Poland
Joanna.koczar@ue.wroc.pl
[2] Kazan Federal University, Kazan 420012, Russia

Abstract. The article proposes the modernization of the algorithm for assessing the investment potential with the use of new indicators of the qualitative analysis of the long-term investment market: "compressiveness," "depth," and "relaxation" in the system of sustainable financing. This algorithm allows assessing the adequacy of the investment potential, determining the average time for post-crisis relaxation of the investment market, and developing measures to regulate the impact on the market.

Keywords: Modeling investment decisions · Investment market
Compressiveness · Depth · Relaxation

1 Introduction

The issues of structural imbalances regulation in the economic system, the formation of long-term investment policy, the correlation of behavioral business models of the financial and real sectors of the economy, institutional development platforms are among the most relevant ones in modern economic theory and practice. Moreover, they are relevant for the transformational economy, the structural proportions in which have undergone significant changes over the past few decades, which under current conditions actualizes the need to implement the processes of re-industrialization and their investment support.

In the financial sector of the economy at this stage of development there is a large number of instruments for investment in the real sector of the economy in conditions of re-industrialization [1–3, pp. 276–277].

With the application of these investment instruments, it is possible to trace the existence of certain behavioral business models of the financial sector organizations, depending on the urgency of the investments that can be conditionally divided into the following types: (1) Short-term business model; (2) behavioral business model; (3) investment business model; (4) a syncretic business model.

Accordingly, among the financial organizations there are institutions that are not focused on interaction with the real sector of the economy, but which have sufficient investment resources, including long-term ones, to realize the goals of re-industrialization. Based on the results of the study of the life cycle curves of the

© Springer Nature Switzerland AG 2019
Z. Wilimowska et al. (Eds.): ISAT 2018, AISC 854, pp. 16–23, 2019.
https://doi.org/10.1007/978-3-319-99993-7_2

economic sectors, we proved that the problem of interaction is most acute during the period of sharp negative changes in the economic conjuncture of the market. In this regard, we consider it expedient to develop a mechanism for managing sectoral interaction, taking into account the behavioral business models of financial organizations, depending on the cycles of economic development.

2 Methods

Neural network modeling in modern science is a relatively new method of clustering data. Artificial neural networks are based on the work of human biological neural networks, transforming the information obtained and bringing it to human consciousness [6, p. 29]. The neural networks of Kohonen were widely used in scientific research. Within the framework of Kohonen neural networks, the method of self-organizing Kohonen maps has been chosen to solve the task of clustering organizations of the financial sector.

The method of the self-organizing Kohonen map (SOM) is based on non-linear programming with the ability to process large data arrays and self-learning networks. Self-organizing maps Kohonen principally differ from all types of clustering, they are able to learn without a teacher. Teaching with the teacher means having two kinds of data: input and output - the network's task is to recreate the mapping of some into others. In the case of training without a teacher, the network only works with input data, independently determining the structure of the output parameters [7, p. 184]. Self-education allows one to apply other data sets to a trained network, for example, data from other countries can be applied to a trained network for the banks. The Kohonen network algorithm is a method of successive approximations (the iteration method), but the Kohonen model compresses the multidimensional input data space into a two-dimensional output space, while vectors that are close to each other at the input are also on the topological map [8, 9].

To perform a cluster analysis using the chosen method, you must independently determine the required number of clusters. We conducted an analysis of the test error and network training. The constructed model is the most adequate for smaller network error values, since in this case the probability of an error in the distribution of the investigated data into invalid clusters is reduced. A network error is the determining factor when choosing the number of clusters using neural network modeling, since weights and values of data activation are corrected to detect the lower value of the network error for a given number of clusters.

In many models of market performance evaluation, Kayle's approach is used [11], which consists in considering individual, smaller characteristics describing liquidity from different angles of view and when combining giving a fairly complete picture [12, p. 35].

In this approach, three such properties are distinguished, which in the literature are often called aspects (attributes or dimensions) of the long-term investment market [12, p. 35]: "conciseness"; "depth"; "relaxation". These aspects can be adapted directly to indicators of long-term investment support for the real sector of the economy and formation of long-term investment potential, as well as to use as a tool for diagnosing business models of financial organizations in the financial sector.

One of the fundamental models by which the study can be visually illustrated is the Miller-Orr model (Fig. 1), which is based on the assumption that the investment potential of a financial institution changes stochastically. At the same time, there is an upper and lower bounds of fluctuations, after which an analyst of a financial organization decides to reduce/increase the volume of the resource base to an effective level determined subjectively. The decrease in the volume of resources is achieved by selling, while the increase is due to the purchase of assets [12, pp. 34–37].

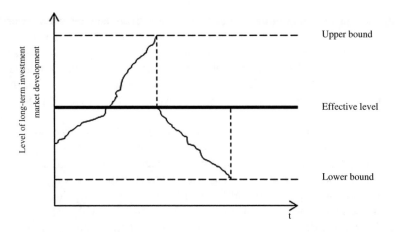

Fig. 1. Miller-Orr model for long-term investment market [12, pp. 34–37]

In our study, this model will be considered from the side of managing the interaction of the financial organization with the real sector of the economy at the macro level, i.e. will determine the upper and lower boundaries of the level of investment support fluctuations in the interaction of the two sectors of the economy, as well as the most acceptable level of this interaction for each of the identified behavioral business models.

In order to exclude the effect of the scale effect on the resulting data, it is proposed to use the relative indicators calculated by dividing the volumes of long-term investments of each financial institution that is part of the behavioral business model into the balance currency.

"Compressiveness" of the market shows how much the deviation of the volume of long-term financial investments in the real sector of the economy from the average level of the market, depending on the economic situation (Fig. 2) [12, pp. 34–37].

The volume of long-term investments at certain points in the life cycle can be overstated or understated relative to optimal investment provision. The understated level indicates a slowdown in investment, hence, the development of the real sector of the economy, the lack of the possibility of introducing new technologies and practices, and a decrease in the activity of the industry.

To calculate the market compressiveness, we take the standard deviation of the SD * 1.96 aggregate, since about 95% of the data falls into this range. This allows us to

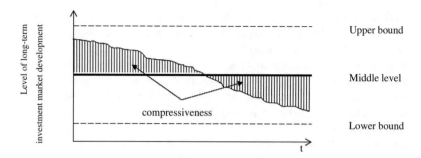

Fig. 2. The level of "compressiveness" of the long-term investment market [12, pp. 37–40]

identify the range of the share of long-term investment in the balance sheet of the average representative of the financial sector with a probability of 95%.

The "depth" of the market is reflected by the minimum and maximum level of long-term investment provision of the real sector of the economy. The depth of the market shows the boundaries of long-term interaction acceptable for the functioning of the financial sector (Fig. 3).

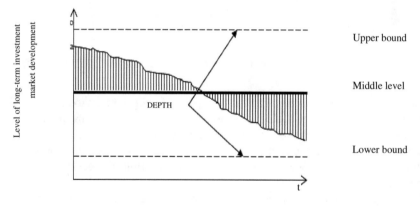

Fig. 3. The level of "depth" of the long-term investment market [12, pp. 41–42]

The depth of the market describes the possibilities of expanding the boundaries of long-term investment in conditions of re-industrialization. Reflects the extreme values of long-term investment security, which can arise with a certain behavioral business model of interaction between the real and financial sectors of the economy [12, pp. 41–42].

To calculate the depth of long-term interaction, we calculate the maximum and minimum of each set. For greater clarity, we also denote the median of the sample, which divides the population into two equal parts, and quartiles with an interquartile distance of 25–75%. Quartiles in this case will mark the position of 50% of the sample financial organizations in each behavioral business model and show the remoteness of the part of the contraction of the market from the maximum and minimum values. The first quartile divides the data array into two parts: 25% of the investigated parameters are less than the first quartile and 75% larger than the first quartile. The third quartile also divides the sample into two parts: 75% of the sample is less than the third quartile and 25% larger than the third quartile.

For all behavioral business models, the volatility of the depth of the market for long-term interaction is likely to be characteristic. The minimum value in all cases will coincide and equal zero. To describe the depth, we will consider the remoteness of the quartiles from the extreme values and the dynamics of the maximum during the period under consideration.

Relaxation of the market is an indicator of the market response to unforeseen circumstances, for example, crisis phenomena. If you find the speed of market relaxation, you can determine how effective the actions of financial institutions, and in some cases, the government and the Central Bank in non-standard conditions (Fig. 4) [12, pp. 43–44].

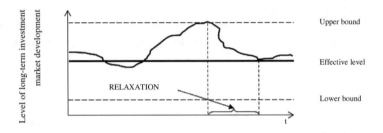

Fig. 4. The level of "relaxation" of the long-term investment market [12, pp. 43–44]

The higher the relaxation rate, i.e. less time period in which the market reaches the optimal level of long-term interaction, the more effective and effective the policy of the financial sector's response to non-standard operating conditions.

The rate of relaxation is defined as the time interval between the moment of deviation of the level of long-term investment potential from the general trend and return to it.

The implementation of the proposed mechanism for managing sectoral interaction taking into account behavioral business models will allow eliminating contradictions in the activity of the real and financial sectors of the economy, providing business entities of the real sector with investment resources for realizing the objectives of re-industrialization on the basis of investment potential of financial organizations.

3   Results

As in today's conditions there is a need to develop a management tool that directly brings together the life cycles of the real and financial sectors of the economy during periods of economic instability, a countercyclical management tool is proposed.

The countercyclical (dynamic) investment buffer is a tool for regulating the investment activity of the financial sector, oriented towards satisfying the interests of the real sector for the purposes of re-industrialization taking into account the cycles of the economic conjuncture of the market

With the help of the countercyclical tool, the share of the necessary long-term investments, calculated from the bank's assets, will be determined, which should be sent by credit institutions to the development of the real sector at a certain point in time.

In order to determine the timing of the introduction of the countercyclical tool, the indicator of the volume of long-term investment in the real sector of the economy is used. On the basis of its dynamics, a trend is constructed, the deviation from which (GAP) is transformed into the value of a countercyclical tool [16, p. 8].

It is assumed that a countercyclical tool will be applied to financial organizations operating within the framework of behavioral business models that are not focused on long-term interaction and have investment potential. For this purpose, based on the retrospective data of the assessment of the contraction and depth of the market for long-term interaction of the real and financial sectors of the economy, the average index of the investment potential for the market as a whole is determined, which is taken as the optimal. The reduction in the level of concentration of long-term investments in the behavioral business model with respect to the optimal one indicates the need to activate the countercyclical tool for financial institutions that adhere to this behavioral business model, given the depth of long-term investment.

Table 1. Value of the indicators of the conciseness and depth of the investment market of behavioral business models in the sectoral interaction management system

Index		Value,%	Deviation
The average value for the whole market	Compression	10	
	Depth	70	
In the first behavioral business model	Compression	5	↓
	Depth	50	
In the second behavioral business model	Compression	5	↓
	Depth	25	
In the third behavioral business model	Compression	40	↑
	Depth	70	
In the fourth behavioral business model	Compression	20	↑
	Depth	70	

Source: compiled by the author

As the real sector develops and profitability grows, the demand for investment resources will decrease, which will allow the growth of own funds directed to its further development. During this period, the countercyclical tool should decrease (down to zero value) to create a competitive basis for real sector investment support.

The activities of economic entities will become more attractive for investors, which will create opportunities for attracting investments on an auction basis.

In order to identify the types of behavioral business models to which a counter-cyclical tool is intended, it is determined that the average contraction index of the investment market is 10%. According to the calculations, the understatement of the level of contraction is observed in relation to the first and second behavioral business model of Russian credit institutions, as well as the high depth of the market in these behavioral business models. However, the coefficient of the countercyclical tool for the first and second behavioral business model should also vary depending on the percentage of market depth that is 50% and 25%, respectively (Table 1).

So, according to the developed mechanism, from November 2009 to November 2014, and also from January 2017, the Central Bank of the Russian Federation needs the use of a countercyclical instrument for credit institutions that use the first and second behavioral business models, since the contraction the investment support of the real sector in them is understated twice in relation to the average market value, while a significant level of depth is determined.

References

1. Vorotnikova, I.V.: Problems of financing innovative activity. RISK: Resources, Information, Supply, Competition, No. 3, pp. 276–278 (2013)
2. Akifieva, S.A.: Development of a comprehensive approach to the definition and classification of investment banking services. Money and Credit, No. 4, pp. 45–49 (2011)
3. Lazareva, E.G.: Investment bank crediting: ways of development. Banking, No. 6, pp. 20–22 (2011)
4. Fomin, E.P., Fomina, N.E., Terentyev, A.V.: Financing of Innovative Activity in Industry: Monograph. Publishing House of Samara State University of Economics, Samara, 144 pp. (2012)
5. Mandel, I.D.: Cluster analysis. Moscow: Finance and Statistics, 176 p. (1988)
6. Osovski, S.: Neural networks for information processing. Trans. with the Polish Rudinsky, I. D. Moscow: Finance and Statistics, 344 p. (2002)
7. Debok, G., Kohonen, T.: Analysis of financial data using self-organizing maps. Debock, G., Kohonen, T., Per. with English. Moscow: Alpina, 317 pp. (2001)
8. StatSoft Russia [Electronic resource]: Electronic textbook on statistics StatSoft - Official website of StatSoft Russia. Access mode: http://statsoft.ru/home/textbook/modules/sttable.html
9. Portal of artificial intelligence [Electronic resource]: Neuron sites of Kohonen - Official site of the portal of artificial intelligence, 2017. Access mode: http://www.aiportal.ru/articles/neural-networks/network-kohonen.html
10. Korhonen, I., Mehrotra, A.: Money demand in post-crisis Russia: dedollarization and remonetization. Emerg. Mark. Finance Trade **46**(2), 5–19 (2010)
11. Kyle, A.S.: Continuous auctions and insider trading. Econometrica **53**(6), 1315–1336 (1985)
12. Naumenko, V.V.: Restructuring of large portfolios of securities in conditions of low market liquidity: the dissertation. A Cand. Econ. Sci. 08.00.10/ Naumenko Vladimir Viktorovich, Moscow, 191 s (2012)
13. StatSoft Russia [Electronic resource]: Software product of the Statistica line—StatSoft Russia official website. Access mode: http://www.statsoft.ru

14. Rodionova, N.D.: Evaluation of the level of interaction between the subjects of the regional innovation system on the basis of the integral approach. Issues of Economics and Law, No. 4, pp. 74–77 (2016)
15. Kuznetsova, V.V.: Macrofinance stability as a priority objective of state regulation. In: Public Administration in the 21st Century: Materials of the 13th International Conference. "KDU", "University Book", Moscow, pp. 50–56 (2016)
16. Popov, K.O., Yudenkov, Yu.N.: We are not in a hurry to foresee how Basel III will respond to us. Internal Control in a Credit Institution, No. 4, pp. 1–10 (2011)
17. The project "Map of Russian clusters" of the Russian cluster observatory ISEHE HSE [Electronic resource]: Map of Russian clusters - Official site of the project "Map of clusters of Russia" (2017). Access mode: http://clusters.monocore.ru

Efficiency of Gradient Boosting Decision Trees Technique in Polish Companies' Bankruptcy Prediction

Joanna Wyrobek[1]([⊠]) and Krzysztof Kluza[2]

[1] Corporate Finance Department, Cracow University of Economics,
ul. Rakowicka 27, 31-510 Kraków, Poland
wyrobekj@uek.krakow.pl
[2] AGH University of Science and Technology, al. A. Mickiewicza 30,
30-059 Krakow, Poland
kluza@agh.edu.pl

Abstract. The goal of the research was to compare the selected traditional bankruptcy prediction models, namely linear discriminant analysis and logit (logistic) models, with the technique called Gradient Boosting. In particular, the paper verifies two research hypotheses (verification was based on the balanced sample of Polish companies): [H1]: Gradient Boosted Decision Trees algorithm is more accurate than traditional bankruptcy prediction models: logit and discriminant analysis; [H2]: Boosted Decision Trees use both: financial ratios and normalized data from financial statements, but the same accuracy one can achieve only with the normalized data and the bigger number of weak learners.

Keywords: Machine learning · Bankruptcy prediction

1 Introduction

1.1 A Subsection Sample

This century is often described as the era of information, where the information processing technology becomes a crucial part of our existence. Nowadays, machine learning methods are used in more and more aspects of human life. Such algorithms are not only used for managing traffic, driving cars, airplanes, or recognizing: images, voice and text, but they are also used in pattern finding for terrorist attacks, teaching, or crime prevention. From the business point of view, however, the most important application of machine learning algorithms remains bankruptcy prediction [1].

The purpose of this paper is to compare the accuracy of the often used bankruptcy prediction tools such as: linear discriminant analysis and logit (logistic) function with the algorithms considered to be the winners of many ML competitions: boosted decision trees [2]. The conducted comparisons use the representative sample of Polish companies (p = 5%) which consists of 1415 bankrupt and 1415 active companies in Poland. The provided data sample covers the years 2008–2017, one year before bankruptcy proceedings were commenced.

© Springer Nature Switzerland AG 2019
Z. Wilimowska et al. (Eds.): ISAT 2018, AISC 854, pp. 24–35, 2019.
https://doi.org/10.1007/978-3-319-99993-7_3

The paper is organized as follows: in Sect. 2, we present the state of the art in the research on the accuracy of traditional (DA and logit) methods and boosted decision trees in bankruptcy prediction for various years and various countries. Section 3 presents the research methodology, i.e. hypotheses to be tested and the outline of the applied methods. Section 4 presents the model accuracy during cross-validation. Verification of research hypotheses, final conclusions and directions for further research are presented in Sect. 5.

2 Related Works

Traditionally, bankruptcy prediction models are based on linear discriminant analysis (DA) or logit (logistic) function. The most famous Altman Z-Score model from 1968 [3, 4] is based on discriminant analysis. It is still used by the practitioners all over the world.

DA and logit models have many advantages such as: the simplicity of application, intelligibility and they are relatively resistant to data not fulfilling the model's assumptions. However, they also have some disadvantages. Linear discriminant analysis, for example, does not deal well with non-linear dependencies. In fact, it fails if the discriminant information can be found in the variance, not in the mean. It is sensitive to overfitting and is difficult to validate. Moreover, it does not deal well with highly unbalanced data sets. Furthermore, logit (logistic) models are susceptible to outliers and require uncorrelated observations. They can appear to have more predictive power as a sampling bias result. Finally, at least in financial applications, it is very rarely that DA or logit models have accuracy exceeding 95% (in terms of cross-validation or separate validation sample).

This is why in recent years bankruptcy or insolvency prediction models started using more and more sophisticated machine learning techniques. One of such methods is boosting. Boosting generally means using the so-called base-learners (models trained on a piece of a training set), but the selection of data to subsequent base-learners is not left to chance. In boosting, each new base-learner is trained on the mistakes of the previous learners. Thus, it can be applied to various base classifiers. In our research, boosting is going to be applied to the decision trees. Although initially boosting was developed in the form of the adaptive boosting method, later on, it was generalized into a gradient boosting approach [5]. Adaptive boosting was proposed by Freund and Schapire [6] who introduced a technique that uses the same training data over and over again. Such method can be applied to any classification learning algorithm, but it is designed particularly for classification [7]. It uses weights which are then applied to all training instances. The classification error can be calculated on the basis of these weights, i.e., weights for misclassified instances are added up and divided by the sum of all weights.

Manipulation of the weights allows giving to the algorithm the incentives which instances (observations) should be classified correctly by the next classifier. The algorithm begins with assigning equal weights to all training observations (instances). Next, at each iteration, a new classifier is built (weak learner), and (1) correctly classified instances receive smaller weights or (2) misclassified instances receive higher weights. At each iteration, some instances may change their status from "easy" to "hard" or the other way round. After each iteration, the weights show how difficult a specific instance was to classify.

The scale of weights alteration depends on the over error "e" of a particular classifier. The correctly classified instances receive new weights according to the following formula: error <- weight * e/(1 − e) [7]. The misclassified instances keep their weight, but eventually, all weights are re-normalized. So, their sum remains the same as it was before. Boosting procedure deletes current classifier and stops if its overall error is 0, equal or higher than 0.5. For predicting, classifiers receive their weights calculated as follows: weight = (−1) * log (e /(1 − e)), where e represents the classifier's overall error. Finally, one does not have to use the weights. In such a case, the training set has to be re-sampled.

It is so because in boosting, instances are chosen for a particular classifier with the probability proportional to their weight. When the training set becomes as large as the original one, it can be used instead of the weighted data [7].

Gradient boosted trees, developed by Friedman [8, 9], constitute one of the most powerful and effective machine learning methods [10]. GBDT work by building trees in a serial manner, that is, each new tree tries to correct the mistakes of the previous one. The algorithm uses strong pre-pruning. Usually, trees are small and shallow, and their depth is in the range of one to five. This reduces the necessary amount of memory. Each tree can produce a good projection only for a small part of data, so the more trees are added, the better is the model.

From the statistical point of view, boosting is the optimization problem in which the objective is to minimize the loss of the model by adding next weak learner using a gradient descent procedure. Gradient boosting extensively uses the model learning rate to see how much each new tree is correcting the mistakes of the previous one. The higher is the learning rate, the more the next tree attempts to correct the mistakes of the previous trees. One cannot exaggerate with the number of the trees, because the higher is the number of the trees (weak learners) the more sophisticated is the model.

Considering the disadvantages of the method, gradient boosted trees are deemed to be a little bit more sensitive to parameters settings than random forests. On the other hand, they can be more accurate than random forests if the parameters are set in a correct manner. What is more, they do not require scaling and work well on binary data mixed with continuous data.

Table 1 shows the cross-validation accuracy of linear discriminant analysis, logit models random forest and boosted decision trees taken from different papers dedicated to bankruptcy prediction. As it can be observed from Table 1, the average testing accuracy for linear discriminant analysis was in the range between 52.18% and 93,5%. For the logit (logistic) models, the average accuracy was between 69.75% and 97,2%, but cross-validation accuracy above 95% was observed only for one model trained on (only) 250 companies. Finally, for Boosted Decision Trees (both gradient boosting and adaptive boosting algorithms), model's accuracy was in the range between 72,3% and 98,1%. Even if the ranges are similar for boosted decision trees and logit models, the distribution of results acts in favor for boosted trees. Since another popular model in machine learning is random forest, we also presented accuracy of this technique as a benchmark. Random Forest accuracy was in the range between 87,06% and 97,4%.

Table 1. Previous non-Polish research on accuracy of linear discriminant analysis (DA), Logit and Boosted Decision Trees (cross-validation average accuracy in [%])

Studies	Country	Years	No of firms	Method	Accu (%)
Barboza et al. [1]	USA + Canada	1985–2013	449 + 449	logit	76.29
Cho et al. [14]	South Korea	2000–2002	500 + 500	Logit	70.58
Hu and Tseng [15]	USA	1975–1982	65 + 65	Logit	88.73
Huang et al. [16]	China	2000–2011	156 + 156	Logit	74.2
Jabeyr and Fahmi [17]	France	2006–2009	400 + 400	Logit	69.75
Min and Lee [18]	South Korea	2000–2002	944 + 944	Logit	79.87
Nagaraj and Sridhar [19]	India	n.a.	107 + 143	Logit	97.2
Sun and Li [20]	China	2000–2005	135 + 135	Logit	84.72
Tseng and Hu [21]	UK	1985–1994	32 + 45	Logit	86.25
Alfaro et al. [22]	Spain	2000–2003	590 + 590	DA	79.66
Anandarajan et al. [23]	USA	1989–1996	265 + 319	DA	52.25
Barboza et al. [1]	USA + Canada	1985–2013	449 + 449	DA	52.18
Cho et al. [24]	South Korea	2000–2002	500 + 500	DA	78.15
Fedorova et al. [25]	Russia	2007–2011	444 + 444	DA	82.00
Ghodselahi and Amirmadhi [26]	Germany	n.a.	300 + 700	DA	65.91
Hu and Tseng [15]	USA	1975–1982	65 + 65	DA	85.42
Huang et al. [16]	China	2000–2011	156 + 156	DA	74.2
Jabeyr and Fahmi [17]	France	2006–2009	400 + 400	DA	93.5
[12] Jardin (2016)	France	2003–2012	8010 + 8010	DA	80.05
Jardin [12]	France	2003–2012	8010 + 8010	DA	82.64
Li and Sun [20]	China	n.a.	135 + 135	DA	83.13
Li and Sun [27]	China	n.a.	135 + 135	DA	88.09
Liao et al. [28]	Taiwan	2005–2011	63 + 2680	DA	92.44
Min and Jeong [13]	South Korea	2001–2004	1271 + 1271	DA	69.1

(*continued*)

Table 1. (*continued*)

Studies	Country	Years	No of firms	Method		Accu (%)
Min and Lee [18]	South Korea	2000–2002	944 + 944	DA		78.81
Pena et al. [29]	UK	1989–2002	140 + 140	DA		86.6
Sun and Li [20]	China	2000–2005	135 + 135	DA		80.68
Alfaro et al. [22]	Spain	2000–2003	590 + 590	Adaboost	(DT)	91.10
Heo and Yang [30]	South Korea	2008–2012	1381 + 1381	Adaboost	(DT)	78.52
Kim and Upneja [31]	USA	1988–2010	21 + 121	Adaboost	(DT)	98.10
Marques et al. [32]	Australia	n.a. –1987	307 + 383	Adaboost	(DT)	82.9
Marques et al. [32]	Germany	n.a. –1994	700 + 300	Adaboost	(DT)	72.3
Marques et al. [32]	Japan	n.a. –1992	296 + 357	Adaboost	(DT)	85.91
Marques et al. [32]	Poland	2007–2013	950 + 50	Adaboost	(DT)	95.1
Marques et al. [32]	Iran	n.a. –2012	128 + 112	Adaboost	(DT)	75.42
Marques et al. [32]	UCSD dataset	n.a.	1836 + 299	Adaboost	(DT)	85.83
Sun et al. [33]	China	2000–2008	346 + 346	Adaboost	(DT)	96.46
Ghodselahi and Amirmadhi [26]	Germany	n.a.	300 + 700	Boosting	(DT)	72.77
Jardin [12]	France	2003–2012	8010 + 8010	Boosting	(DT)	81.86
Kim and Kang [34]	Korea	2002–2005	729 + 729	Boosting	(DT)	75.1
Barboza et al. [1]	USA + Canada	1985–2013	449 + 449	Random Forest	(DT)	87,06
Liao et al. [28]	Taiwan	2005–2011	63 + 2680	Random Forest	(DT)	94,91
Nagaraj and Sridhar [19]	India	n.a.	107 + 143	Random Forest	(DT)	97,4

Table 2 gives an overview of the results of previous papers based on Polish data, which did cross-validation of their models. As it can be seen in Table 2, the accuracy of linear discriminant models was between 86,11% and 96,26%, the accuracy of logit models was between 83,33% and 92,59%. For boosted decision trees, it was a range between 67,1% and 92,9%. Finally, for Random Forest the range was between 78,6% and 88,88%. This could suggest that the most effective models were created on the basis of linear discriminant analysis approach and the worst results can be observed for boosted decision trees, which would be not very promising. Nevertheless, with the exception of Grzybowska and Karwański paper [11], all other papers were not based on a representative sample of Polish companies. Similarly, for non-Polish research, only Jardin [12] and Min and Jeong [13] used a representative sample of bankrupt firms.

Table 2. Previous Polish research on accuracy of linear discriminant analysis (DA), Logit and Boosted Decision Trees (cross-validation average accuracy in [%])

Studies	Country	Years	No of firms	Base classifiers	Accu (%)
Pociecha et al. [35]	Poland	2005–2009	182 + 7147	DA	86.11
Pociecha et al. [35]	Poland	2005–2009	182 + 182	DA	89.58
Korol [36]	Poland	2005–2009	50 + 56	DA	96.29
Pociecha et al. [35]	Poland	2005–2009	182 + 182	Logit	83.33
Pociecha et al. [35]	Poland	2005–2009	182 + 7147	Logit	88.89
Korol [36]	Poland	2005–2009	50 + 56	Logit	92.59
Grzybowska and Karwanski [11]	Poland	n.a.	2294 + 20000	AdaBoost (DT)	92.9
Pawelek and Grochowina [37]	Poland	2013–2015	42 + 42	Boosting (DT)	67.2
Pawelek and Grochowina [37]	Poland	2013–2015	42 + 7181	Boosting (DT)	77.7
Pawelek and Grochowina [37]	Poland	2013–2015	42 + 7181	RF (DT)	78,6
Korol [36]	Poland	2005–2009	50 + 56	RF (DT)	88,88

3   Research Method

For the reasons mentioned above, the purpose of the paper was to analyze the accuracy of the basic bankruptcy prediction models and boosted decision trees algorithm using a representative sample of Polish companies. Data was retrieved from Orbis database and missing data was filled with both: data from the EMIS database. Data was tested whether the sum of assets was equal to the sum of equity and liabilities. Moreover, any suspicious or error records or columns with more than 30% of missing data were removed. Later on, after cross-correlations calculation, we removed variables with correlation above 70% (not all methods required such exclusion, but we wanted to maintain a complete comparability between methods).

Orbis database contains the information about the legal status of each company and this is how we determined which company was bankrupt. As a status change date, for some records we knew the date from the Orbis database, for the rest of companies we assumed that the court's ruling about the insolvency took place in the year when the company had a negative equity for the first time. We assumed that the insolvency petition must have been filed one year earlier and that the model should be able to know about it one year earlier (one year ahead of the insolvency petition). We had the financial information from balance sheets and income statements and calculated several financial ratios (Table 3).

Table 3. Financial ratios definitions

Symbol	Definition
Profitability ratios	
RSHF (roeplbeforetax)	(Profit before tax/Shareholders funds) * 100
RTAS (roaplbeforetax)	(Profit before tax/Total assets) * 100
ROE (roenetincome)	(Net income/Shareholder funds) * 100
ROA (roanetincome)	(Net income/Total Assets) * 100
PRMA (protifmargin)	(Profit before tax/Operating revenue) * 100
GRMA (grossmargin)	(Gross profit/Operating revenue) * 100
RentBrutto	EBIT/Net Sales Revenues
CFOP	(Cash flow/Operating revenue) * 100
RentAkt	Net Profit (Loss)/Total Assets
Operational ratios	
IC (intcover)	Operating profit/Interest paid
STOT (stockturn)	Operating revenue/Stocks
COLL (collperdays)	(Debtors/Operating revenue) * 360
CRPE (credperdays)	(Creditors/Operating revenue) * 360
EXOP	(Exports/Operating revenue) * 100
RotAktTrw	Operating Revenue/Fixed Assets
RotAkt	Operating Revenue/Total Assets
Structure ratios	
CURR (curratio)	Current assets/Current liabilities
LIQR (liqratio)	(Current assets – Stocks)/Current liabilities
SHLQ	Shareholders funds/Non current liabilities
SOLR (solvratioassbased)	(Shareholders funds/Total assets) * 100
NatWym	Short-Term Investments/Trade Liabilities
ZadlAkt	(Total Assets – Shareholder's Funds)/Total Assets
ZadlKapWl	(Total Assets – Shareholder's Funds)/Equity
ZadlDlug	Long-Term Liabilities/Total Assets
PozKoszOp	Operating Costs/Operating Revenue
Per employee ratios	
PPE	Profit before tax/Employees

(*continued*)

Table 3. (*continued*)

Symbol	Definition
TPE	Operating revenue/Employees
ACE	Cost of employees/Employees
SFPE	Shareholders funds/Employees
WCPE	Working capital/Employees
TAPE	Total assets/Employees

Financial data used in our research covered companies operating in different sectors of economic activity – we managed to collect 1415 useful records for bankrupt companies and 1450 records for active companies. The data sample was then divided into a training set including the companies which went bankrupt in years 2008–2013 (and matching active companies) and a testing (evaluation) set including the companies which went bankrupt in years 2014–2017. The selection of bankrupt companies was based on the time order. The selection of active companies was based on the time order as well, but we also took into account the similar type of economic activity. Thus, for every bankrupt company, we drew at random an active company from the same industry type. Moreover, we added a small amount of other randomly chosen firms. As a result of the data preparation, these sets were almost balanced. It was necessary as without such adjustments to the loss function a strongly imbalanced sample (having, for example, twice as many active companies than bankrupt companies) would result with a model having high general accuracy, but with a high I-st type error. Thus, the model would tend to classify a firm as active whereas in reality it went bankrupt.

After checking and processing, the collected data was normalized. We also applied one-hot encoding to discrete data. For processing, we took advantage of the `skleran` Python library. Data sample were divided into 10 parts and for each iteration we used 9 parts for training and 1 part for testing (cross-validation). We trained the following models: linear discriminant analysis, logit model, and gradient boosted decision trees.

4   Model Training Results

Table 4 presents the average accuracy of cross-validation of the trained models. Contrary to many other papers, in our case, the most efficient algorithm proved to be GBDT. Its average validation sample based accuracy was 99,11% (for a logit model it was 87,10% and for linear discriminant model 86,31% and for Random Forest 98,91%). Table 5 shows the importance of various variables in the GBDT model training. As it can be noticed, except for two variables (`rotakttrw` and `natwym`), all other variables represented the "raw" information, i.e. the variables taken directly from the financial statements, not financial ratios. Traditionally, for bankruptcy prediction models, one used only financial ratios. Such ratios were previously tested whether they have the potential to classify companies into bankrupt and non-bankrupt groups. In the case of the GBDT model, the model preferred the raw data input and created its own ratios during training. In the case of DA and logit models, one had to test assumptions of the models, which was particularly crucial for the logit model. Both training and testing samples had to be balanced as otherwise, one would have to adjust the loss function. GBDT was based on 100 decision trees using the Friedman mean squared error with improvement score as a function to estimate the quality of a split. In our case, the learning rate was set to 0.1, and we expected the improvement in the next iteration compared to the previous one as well as to the previous classifier.

Table 4. Testing accuracy of DA, Logit and GBDT algorithms

Sample	GBDT	Logit	Linear discriminant analysis	Random forest
1	99,52	88,84	87,1	99,52
2	99,52	87,11	87,37	99,41
3	99,76	88,23	87,38	99,88
4	99,88	88,43	86,96	99,76
5	99,64	88,52	87,52	99,17
6	98,93	87,12	87,51	98,93
7	99,76	88,91	87,54	99,17
8	99,76	88,31	86,22	99,76
9	99,05	88,24	87,38	99,05
10	99,88	88,11	87,34	99,76
Average	**99,57**	**88,18**	**87,23**	**99,44**
Validation sample	**99,11**	**87,1**	**86,31**	**98,91**

Table 5. Variables important in the GBDT model

lp	Name	Importance [%]	lp	Name	Importance [%]
1	Industry	47,32	11	othercurrliab	1,84
2	Listed	7,25	12	rotakttrw	1,67
3	Cash	4,39	13	conscode U1	1,27
4	Debtors	3,83	14	conscode U2	1,06
5	Loans	3,55	15	pkd	1,00
6	Creditors	3,36	16	zadlakt	0,92
7	Year	3,09	17	natwym	0,90
8	Nace	2,43	18	oprev	0,86
9	Depreciation	2,21	19	capital	0,85
10	Noncurrliab	1,94	20	countrycode	0,72

5 Concluding Remarks

The goal of our research was to verify two research hypotheses:
[H1]   Gradient Boosted Decision Trees algorithm is more accurate than traditional bankruptcy prediction models like logit and discriminant analysis, and

[H2]   Boosted Decision Trees use both: financial ratios and normalized data from
        financial statements, but the same accuracy can be achieved only with the
        normalized "raw" data and a bigger number of weak learners

On the basis of the trained models' accuracy, we did not find any evidence to reject
the hypothesis H1. Gradient Boosted Trees (GBDT) performed much better than the
other analyzed models (they also performed little bit better than Random Forest). As
expected, the GBDT is an ensemble method, and in our case, it used a very versatile
and efficient classifier – decision trees. On top of that, the GBDT trains each next tree
based on the errors made by the previous trees. Linear discriminant model and logit
model did quite well too, although their cross-validation accuracy was below 95%.

The reasons why the GBDT might have performed so well, compared to other
analyzed methods, may include the following factors:

1. the GBDT is an ensemble method, where each additional weak learner is added
   only if it improves previous trees.
2. the decision trees deal very well with missing values and with outliers. Classical
   methods much do worse with these issues.
3. the GBDT automatically detects nonlinear feature interactions and adjusts to it.
   Decision trees deal also very well with non-normally distributed variables, because
   it does not matter for decision trees.

Hypothesis [H2] at the first sight had to be rejected. The GBDT used two financial
ratios. So 2 variables out of 20 important for the training of the model were not raw
values taken from the financial statement, but ratios calculated based on these statements. However, if we removed the financial ratios from the training sample and
increased the maximum number of decision trees to 150, the model accuracy initially
dropped from 99,57% to 99,21%. With the new extra classifiers, it increased to
99,64%. In other words, the model did equally well without calculation of any financial
ratios.

The research can be concluded as follows. Gradient boosted decision trees in cross-
validation did considerably better than the traditional financial models based on linear
discriminant analysis or logit models including a small time shift between training
sample and the validation sample. We believe that it was achieved through a representative sample of companies in the training sample and because the time shift was not
very significant.

If one looks, however, at the variables that GBDT model found important, they
seem to have short-term nature. The type of the industry which was the most important
factor for GBDT seems to have a short-term nature – in a couple of years many
bankruptcies may take place in another type of industry. Thus, we believe that the
model would become inefficient if it was not fed with new, timely data as the time goes
by. Another problem was extrapolation. Decision trees do not have to have linear
nature, and a significant extrapolation may be problematic. We did not test the accuracy
of the model for such extreme cases.

It is also worth noting that GBDT data did not have to be linearly separable as
bankrupt companies usually can have very low and very high values of a certain

variable and active companies can have values in the middle. GBDT deals very well with such issues and it was visible in our research, on the example of the rotation ratios (which express a relation between the inventories, receivables and payables and sales revenues). In other approaches, one often has to remove such situations by removing "outliers" to make the relationship linear.

Training results led us to the question what is the essential feature of the effective bankruptcy prediction model in finance. Let it be the direction for further research. Well-fitted models, such as GBDR or Random Forest, at least in our case used specific short-term variables such as the information in which industry there was the bigger percentage of bankrupt companies. These models were the most precise, but it seemed evident that with time such variables would lose their predictive power.

Another question was how explicit should be the construction of such a model – with ensemble methods applied to decision trees, it is challenging to understand how exactly the models work. And the last question we found interesting- whether there is a trade-off between cross-validation accuracy and predictive power of the model. And if there is such a relationship, how to deal with it.

References

1. Barboza, F., Kimura, H., Altman, E.: Machine learning models and bankruptcy prediction. Expert Syst. Appl. **83**, 405–417 (2017)
2. Johnson, R., Zhang, T.: Learning nonlinear functions using regularized greedy forest. Technical report 29 (2012)
3. Altman, E.: Financial ratios, discriminant analysis and the prediction of corporate bankruptcy. J. Financ. **47**, 589–609 (1968)
4. Altman, E.: Predicting railroad bankruptcies in America. Bell J. Econ. Manag. Sci. **4**, 184–211 (1973)
5. Alpaydin, E.: Introduction to Machine Learning. MIT Press, Cambridge (2010)
6. Freund, Y., Schapire, R.E.: Experiments with a new boosting algorithm. In: Proceedings of the Thirteenth International Conference on Machine Learning, pp. 148–156 (1996)
7. Witten, I., Eibe, F.: Practical Machine Learning Tools and Techniques. Morgan Kaufman, Burlington (2005)
8. Friedman, J., Hastie, T., Tibshirani, R.: Additive logistic regression: a statistical view of boosting. Ann. Stat. **28**, 337–407 (2000)
9. Friedman, J.: Greedy boosting approximation: a gradient boosting machine. Ann. Stat. **29**, 1189–1232 (2001)
10. Mueller, A., Guido, S.: Introduction to Machine Learning with Python for Data Scientists. O'Reilly Media, Newton (2016)
11. Grzybowska, U., Karwański, M.: Szacowanie parametrów ryzyka kredytowego przy użyciu rodzin klasyfikatorów. Zeszyty Naukowe Uniwersytetu Ekonomicznego w Katowicach **248**, 107–120 (2015)
12. Jardin, P.: A two-stage classification technique for bankruptcy prediction. Eur. J. Oper. Res. **254**, 236–252 (2016)
13. Min, J., Jeong, C.: A binary classification method for bankruptcy prediction. Expert Syst. Appl. **36**, 5256–5263 (2009)
14. Cho, S., Hong, H., Ha, B.: A hybrid approach based on the combination of variable selection using decision trees and case-based reasoning using the mahalanobis distance: for bankruptcy prediction. Expert Syst. Appl. **37**, 3482–3488 (2010)

15. Hu, Y.C., Tseng, F.M.: Functional-link net with fuzzy integral for bankruptcy prediction. Neurocomputing **3**, 2959–2968 (2007)
16. Huang, J., Wang, H., Kochenberger, G.: Distressed Chinese firm prediction with discretized data. Manag. Decis. **55**, 786–807 (2017)
17. Jabeur, S., Fahmi, Y.: Forecasting financial distress for french firms: a comparative study. Empir. Econ. **3**, 1–14 (2017)
18. Min, J., Lee, Y.: Bankruptcy prediction using support vector machine with optimal choice of kernel function parameters. Expert Syst. Appl. **28**, 603–614 (2005)
19. Nagaraj, K., Sridhar, A.: A predictive system for detection of bankruptcy using machine learning techniques. Int. J. Data Min. Knowl. Manag. Process **5**, 29–40 (2015)
20. Sun, J., Li, H.: Financial distress prediction based on serial combination of multiple classifiers. Expert Syst. Appl. **18**, 8659–8666 (2009)
21. Tseng, F., Hu, Y.: Comparing four bankruptcy prediction models: logit, quadratic interval logit, neural and fuzzy neural networks. Expert Syst. Appl. **37**, 1846–1853 (2010)
22. Alfaro, E., Garcia, N., Games, M., Elizondo, D.: Bankruptcy forecasting: an empirical comparison of ada boost and neural networks. Decis. Support Syst. **45**, 110–122 (2008)
23. Anandarajan, M., Lee, P., Anandarajan, A.: Bankruptcy prediction of financially stressed firms: an examination of the predictive accuracy of artificial neural networks. Int. J. Intell. Syst. Account. **10**, 69–81 (2001)
24. Cho, S., Kim, J., Bae, J.K.: An integrative model with subject weight based on neural network learning for bankruptcy prediction. Expert Syst. Appl. **10**, 403–410 (2009)
25. Fedorova, E., Gilenko, E., Dovzhenko, S.: Bankruptcy prediction for russian companies: application of combined classifiers. Expert Syst. Appl. **40**, 7285–7293 (2013)
26. Ghodselahi, A., Amirmadhi, A.: Application of artificial intelligence techniques for credit risk evaluation. Int. J. Model. Optim. **1**, 243–249 (2011)
27. Li, H., Sun, J.: Business failure prediction using hybrid2 case-based reasoning. Comput. Oper. Res. **37**, 137–151 (2010)
28. Liao, J.J., Shih, C.H., Chen, T.F., Hsu, M.F.: An ensemble-based model for two-class imbalanced financial problem. Econ. Model. **37**, 175–183 (2014)
29. Pena, T., Martinez, S., Abudu, B.: Bankruptcy prediction: a comparison of some statistical and machine learning techniques. In: SSRN's eLibrary, vol. 18 (2009)
30. Heo, J., Yang, J.Y.: Adaboost based bankruptcy forecasting of korean construction companies. Appl. Soft Comput. **24**, 494–499 (2014)
31. Kim, S.Y., Upneja, A.: Predicting restaurant financial distress using decision tree and AdaBoosted decision tree models. Econ. Model. **36**, 354–362 (2014)
32. Marques, A.I., Garcia, V., Sanchez, J.S.: Exploring the behavior of base classifiers in credit scoring ensembles. Expert Syst. Appl. **39**, 10244–10250 (2012)
33. Sun, J., Jia, M.Y., Li, H.: Adaboost ensemble for financial distress prediction: an empirical comparison with data from chinese listed companies. Expert Syst. Appl. **38**, 9305–9312 (2011)
34. Kim, M.J., Kang, D.K.: Ensemble with neural networks for bankruptcy prediction. Expert Syst. Appl. **37**, 3373–3379 (2010)
35. Pociecha, J., Pawelek, B., Baryla, B.: Statystyczne metody prognozowania bankructwa w zmieniajacej sie koniunkturze gospodarczej. Wydawnictwo UEK (2014)
36. Korol, T.: Systemy ostrzegania przedsiebiorstw przed ryzykiem upadlosci. Oficyna Wolters Kluwer Business (2010)
37. Pawelek, B., Grochowina, D.: Podejscie wielomodelowe w prognozowaniu zagrozenia przedsiebiorstw upadloscia w polsce. Prace Naukowe Uniwersytetu Ekonomicznego we Wroclawiu 171–179 (2017)

Economic Determinants of the Effectiveness of Polish Enterprises

Anna Kamińska[1(✉)], Agnieszka Parkitna[1], Małgorzata Rutkowska[1],
Arkadiusz Górski[1], and Zofia Wilimowska[2]

[1] Faculty of Computer Science and Management,
Wrocław University of Science and Technology,
ul. Ignacego Łukasiewicza 5, 50-371 Wrocław, Poland
{anna.maria.kaminska,agnieszka.parkitna,
malgorzata.rutkowska,arkadiusz.gorski}@pwr.edu.pl
[2] University of Applied Sciences in Nysa,
ul. Armii Krajowej 7, 48-300 Nysa, Poland
zofia.wilimowska@pwsz.nysa.pl

Abstract. This paper deals with the economic effectiveness of the company. The aim is to investigate which external determinants that have an impact on the economic efficiency of an enterprise measured by selected financial indicators. The research was carried out by building econometric models taking into account a number of theoretically indicated external factors that could affect the economy of enterprises, and what is their statistical significance of the relationships between individual factors and selected measures of economic efficiency were examined.

Keywords: Effectiveness · Economic determinants · Effectiveness factors

1 Introduction

The dynamic development of the world around us caused that the concept of efficiency became the subject of reflection in many spheres of economic activity. The term of economic efficiency, management effectiveness, operational efficiency or global efficiency under which there is a constant pursuit of success is popular. The prevalence of this issue and its heterogeneous significance makes the company's striving for efficiency complicated procedure. The desire to efficiently managed resources, reduced costs, or increased profitability of the capital invested requires a detailed analysis of performance indicators related to the in-depth examination of their measurement methods and the determinants that stimulate their level. The authors shows different approach to this issue, what makes it difficult to clearly determine which of the efficiency determinants mentioned in the literature has a significant impact on the level of profitability or liquidity of the business unit. Therefore the purpose of the work is to examine the impact of efficiency determinants on the company's effectiveness.

© Springer Nature Switzerland AG 2019
Z. Wilimowska et al. (Eds.): ISAT 2018, AISC 854, pp. 36–48, 2019.
https://doi.org/10.1007/978-3-319-99993-7_4

2 Determinants of the Company's Efficiency

The concept of "efficiency determinants" is strictly related to the concept of enterprise efficiency, which are divided into several types. Reliable relations between particular types of efficiency schematically approximate Fig. 1 is presenting the general distribution of economic efficiency.

The basic criteria for the division of economic efficiency is the division into cost effectiveness, income and which is the effect of generated profits (resultant). An effective economic entity can be considered when it is technically effective and when at a given level of results and using the minimum outlay or from a given level of inputs, maximum results are obtained (Ćwiąkała-Małys 1). The unit's technical effectiveness can be studied without knowledge about both: market prices of inputs and results. Performance-oriented results are connected with achieving the optimal level of financial results. As for organizational efficiency, Rummler and Brache described it through three levels of organization and influencing factors (Rummler 6):

- Organization level (strategy, organizational objectives and methods for measuring them, organizational structure, way of using resources)
- Process level (development of new products (innovations) supply process, production process, sales process, distribution process, invoicing process, collection process)
- Work level (recruitment and promotion methods, scope of tasks and duties, applicable work standards, feedback, rewards, trainings.

Reliable efficiency management requires formulating goals, designing and directing each of these levels.

The effectiveness of enterprises depends on a number of efficiency determinants, which are different depending on individual types of enterprise effectiveness. Pilecki distinguished internal and external determinants of the company's efficiency. The internal determinants included: the development of marketing, of new products, the brand, a wide range of services, the elateness of operations. As external determinants, he distinguished: changes in production techniques, development of the services sector, global economic growth, fashion and change of tastes, changes in strategy (Pilecki 5).

The determinant of development can be also the factor of the company's functioning, which is competitiveness, indicated by Mitek and Miciuła. Competitiveness is a leading feature of the market economy, reflecting the company's potential, identified with resources, abilities and skills. Proper identification of competition may contribute to gaining advantage over other entities (Mitek 2012). The work focuses on examining the relationship of selected determinants to the economic viability of enterprises. It is about examining which determinants have the biggest impact on the economic efficiency of enterprises. The indicated question requires the selection of effectiveness measures and the study of the relationship between determinants of economic efficiency and selected indicators.

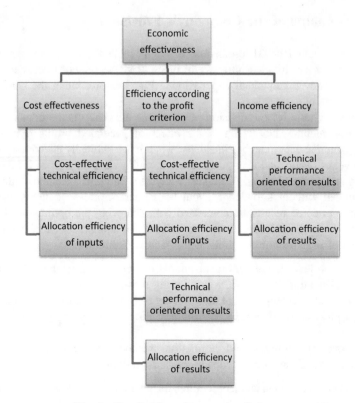

Fig. 1. The division of economic efficiency

3   Measures for Assessing the Economic Effectiveness of Enterprises

The basic measurement of effectiveness in the economic sense can be made by using methods of index analysis, including: accounting, financial and market indicators. The form can be different: absolute (used to measure the value created by an economic unit in a specific period of time or within several consecutive periods), absolute (used to measure the efficiency not only in the context of generated value, but they also as the capital employed to obtain specific effects) or in the form of relative and relative measures. The accounting ratios refer primarily to the relation of revenues and costs, and the most important area of their measurement is profitability. Indicators are based on commonly available financial statements of the company, often confirmed by an independent expert opinion, which makes the data contained therein reliable and reliable. During reviewing the specified measures of economic efficiency of enterprises, they were selected and it was considered that the study of the relationship between the economic efficiency of enterprises and efficiency determinants will be focused on the following indicators: ROE, ROA, ROS, current liquidity ratio and profit per share that are a reflection of the most important economic spheres of the company's operations

4 The Impact of Economic Determinants on the Effectiveness of Enterprises in Poland

The study of the impact of determinants on the economic efficiency of enterprises was conducted on 100 randomly selected Polish enterprises for which financial data was collected from 1997 to 2016. The presented research has been conducted by A. Parkitna and A. Grican in 2017. The most interesting areas were presented, which constitute a cross-section of the team's research work.

In the conducted study, the effectiveness indicators were considered as explanatory variables, while the determinants of efficiency as explanatory variables. Next, based on the tools of the GRETL statistical program, econometric models were made, trying to explain selected bias variables in form of selected measures of the economic efficiency of enterprises through the determinants of economic efficiency. Aim of the research was not to build an econometric model, to calculate the value of selected economic and economic efficiency measures, but through the use of statistical tools to show a relevant relationship between these variables.

To verify the econometric model, it should be determined if the explanatory variables influence the explained variable and verify the model by those factors that don't affect the explained variable. The econometric model is correct if the matching of variables is greater than 60%, the significance of individual regression coefficients occurs, there is no autocorrelation, and the random component is in agreement with the normal distribution and the hypothesis. Then it can be assumed that the variables that are included in the model have a significant impact on the formation of the explained variables.

As part of the conducted research, the following mentioned measures were included as an explanatory variable: ROE, ROA, ROS, current liquidity ratio and earnings per share. With reference to explanatory variables, the focus was on selected determinants. Determinants of economic efficiency are divided into two basic groups:

– internal factors such as: gross profit, net profit, employment in the area of enterprises,
– external factors: the level of unemployment, inflation, money supply, index of income, rediscount rate, population number, befitting natural, tax base.

The results of the study on the impact of economic determinants on ROE and earnings per share are presented in the further part of the study.

4.1 The Impact of Efficiency Determinants on Return on Equity

Research first analyzed the relationship between ROE and selected determinants of the economic efficiency of enterprises. The results of the tests carried out in the GRETL statistical program are presented on the screen shot (Fig. 2).

The developed econometric model, in which the ROE is an explanatory variable, takes into account for 13 different determinants of efficient efficiency of enterprises as clarifying variables. It turned out, however, that some of the determinants do not have a significant relationship, they only slightly explain the selected explanatory variable.

GRETL Model 1; OLS, using observations 1:1-18:1 (T=35)

Dependent Variable : ROE

	coefficient	Std. error	t-ratio	P-value	
Const	-6,13691	9,25569	-0,6630	0,5145	
GBP	0,0216596	0,0104136	2,080	0,0500	**
Reference rate	-0,00627855	0,0112368	-0,5588	0,5822	
PL in EU	0,143071	0,0491407	2,911	0,0083	***
Unemployment	-0,0103403	0,0110441	-0,9363	0,3598	
Inflation	-0,00796746	0,00945903	-0,8423	0,4091	
Money supply	-0,00090141	0,000194216	-4,641	0,0001	***
Emplyment	3,05023e-06	4,57553e-05	0,06666	0,9475	
index of poverty	-0,00030174	0,00624701	-0,04830	0,9619	
discount rate	-0,00173160	0,0076096	-0,2231	0,8256	
population	0,000165031	0,000240219	0,6870	0,4996	
Birthrate	0,215250	0,479281	0,4491	0,6580	
the tax base	3,16856e-06	2,15442e-06	1,471	0,1562	
gross profit	-8,76866e-010	4,61782e-09	-0,1899	0,8512	

Mean dependent	0,080799	S.D. dependent	0,104817	
Sum Squared resid	0,069892	S.E. of regression	0,057690	
R-squared	0,812897	Adjusted R-squared	0,697072	
F(13,21)	7,018291	P-value(F)	0,000050	
Log-likehood	59,11985	Akaike criterion	-90,23970	
Schwarz criterion	-68,46483	Hannan-Quinn	-82,72302	
Rho	0,226378	Durbin-Watson	1,536626	

Excluding the constant, p-value was highest for variable 9 index of poverty

Fig. 2. Research on the impact of efficiency determinants on ROE

For these reasons, the model was simplified by eliminating determinants with no connections. Figure 3 shows the results of further tests.

After simplifying the model and leaving the variables significantly affecting the dependent variable, the analysis of the model to the data was analyzed - the results in this respect are presented in Table 1.

The standard error of the random indicator of the SE regression equation is 6%, which means that the model describes the reality well, while the matching factor is 75%, which indicates a good match of the explanatory variables to the explained variable, which allows for further analysis of verifications statistical econometric model, meaning significance.

For the purpose of verifying explanatory variables in terms of their existence, the level of significance was adopted $\alpha = 5\%$. In order to verify the model in terms of the significance of the regression coefficient system, we put two hypotheses.

H0: The factors included in the model are irrelevant

H1: The factors included in the model are significant

GRETL Model 4; OLS, using observations 1:1-18:1 (T=35)
Dependent Variable : ROE

	coefficient	Std. error	t-ratio	P-value	
Const	-8,84528	4,83124	-1,831	0,0778	*
GBP	0,0135786	0,00624305	2,175	0,0382	**
PL in EU	0,132131	0,0466066	2,853	0,0084	***
Money supply	-0,00058337	0,000133191	-4,380	0,0002	***
discount rate	-0,0120924	0,00384130	-3,148	0,0039	***
population	0,000223078	0,000127031	1,756	0,0900	*
Birthrate	0,509696	0,201995	2,523	0,0176	**

Mean dependent	0,080799	S.D. dependent	0,104817
Sum Squared resid	0,092478	S.E. of regression	0,057470
R-squared	0,752433	Adjusted R-squared	0,699383
F(13,21)	14,18348	P-value(F)	2,31e-07
Log-likehood	54,21948	Akaike criterion	-94,43895
Schwarz criterion	-83,55152	Hannan-Quinn	-90,68061
Rho	0,358337	Durbin-Watson	1,201644

Fig. 3. Study of the impact of adjusted determinants on ROE

Table 1. Matching the model to data – ROE

	Value
Matching error (S.E.)	0.057470
R-squared matching ratio	0.752433 = 75%

The empirical value of the statistics is $F = 2.31*e^{-7} \approx 0.0022$ and is smaller than the assumed level of significance α, allows to reject the hypothesis H_0, and accept H_1. Thus, the variables included in the model are significant, i.e. specified determinants of the economic efficiency of enterprises, have an impact on the selected measure of the economic efficiency of the enterprise.

However, the significance of individual coefficients should be examined, as presented in the Table 2.

Based on of the subsequent verification, it was considered that the explanatory variables PL in the EU, money supply, rediscount rate, population and population growth are important for the model, because their p value is lower than the assumed level of significance $\alpha = 5\%$ (Gładysz 3), whereas the variable natural increase did not show this significance any more.

The verification of the model and, hence, the significance of the specified determinants has not been completed on these tests, but also autocorrelation, normality and heteroscedasticity have been examined.

Research on Autocorrelation

An investigation of the autocorrelation of a random component based on Durbin-Watson statistics is carried out to determine to what extent a given word of the series depends on the previous words. The hypotheses were considered:

H_0: lack of autocorrelation of a random component,
H_1: there is an autocorrelation of a random component.

Table 2. Significance of individual regression coefficients -ROE

	Factor	P value	Interpretation
X_1 (const)	−8.84528	0.0778	Not important
X_2(PL in UE)	0.0135786	0.0382	Important
X_3(money supply)	0.132431	0.0084	Important
X_4 (reference rate)	−0.000883379	0.0002	Important
X_5 (population)	−0.0120924	0.0039	Important
X_6 (birthrate)	0.000223078	0.09	Not important
X_7 (const)	0.509696	0.0176	Important

The value of Durbin-Watson statistics in the model is 2.20, in the model there are 35 observations and 2 explanatory variables. From Durbin-Watson tables, for $\alpha = 0.05$, 35 observation and 7 explanatory variables, we read the values $d_L = 1.03424$ oraz $d_u = 1.76743$, thanks to which equality was determined:

$$1.76 < 2.20,$$

is saying that there is no reason to reject the hypothesis H_0, which indicates the lack of autocorrelation of the random component.

Normality
In order to verify the model for the normality of random components, the chi square test was carried out - χ^2 and hypotheses (Mercik 4):

H_0: The distribution of residues is not consistent with the normal distribution, $\chi^2 < \alpha$
H_1: The distribution of residues is in accordance with the normal distribution, $\chi^2 > \alpha$

The value of the coefficient is $\chi^2 = 6.750$. It is higher than the assumed significance level α, which means that the hypothesis should be rejected H_0, in favor of the H_1 hypothesis. Therefore, the random component in the model is in agreement with the normal distribution. The explanatory variable significantly affects the explained variable (Fig. 4).

Heteroscedasticity
In order to verify the model in terms of heteroscedasticity, White's test was carried out, based on hypotheses:

H_0: The random component is heteroskedastic
H_1: The random component is not heteroskedastic

The calculated value of the White's test statistic is 22.227047, and the level of significance level p = 0.676212 is higher than the value of assumed significance factor α, which leads to rejecting the H_0 hypothesis, in favor of the H_1 hypothesis. Therefore, the random component of the model is homoscedastic, the explanatory variables are uncorrelated with a random disorder, the functional form of the model is right (Fig. 5).

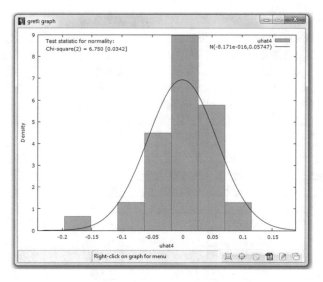

Fig. 4. Normality test – ROE

```
  Unadjusted R-squared = 0.635058

Test statistic: TR^2 = 22.227047,
with p-value = P(Chi-square(26) > 22.227047) = 0.676212
```

Fig. 5. Test of White – ROE

Conclusions

The conducted research, developed econometric models and their statistical verification allow to state that significant determinants of economic efficiency of enterprises affecting the value of ROE can be considered: PL in the EU (the polish membership to the European Union), money supply, rediscount rate, population.

4.2   The Impact of Efficiency Determinants on the Ratio Determining the Value of Profit Per One Share

The study of the relationship between the measure of the economic efficiency of enterprises in the form of the value of profit per share and external determinants of effectiveness was carried out according to the same methodology as in the analysis of the impact on the ROE indicator. The results of research in this area along with statistical verification are presented below.

Processed data using the GRETL program, indicate that some of the determinants have no impact on the variable explaining the profit per share, therefore for further testing they should be removed (Fig. 6).

Statistical tests were performed based on such explanatory variables as: inflation, rediscount rate, population number and tax base (Fig. 7).

GRETL Model 1; OLS, using observations 1:1-18:1 (T=35)
Dependent Variable : profit on shares

	coefficient	Std. error	t-ratio	P-value	
Const	72,7849	19,8927	3,659	0,0015	***
GBP	0,0351542	0,0223814	1,571	0,1312	
Reference rate	0,0263005	0,0241505	1,089	0,2885	
PL in EU	-0,0885797	0,0156615	-0,8387	0,4111	
Unemployment	-0,0332044	0,0237365	-1,399	0,1764	
Inflation	0,0359806	0,0203287	1,765	0,0920	*
Money supply	-0,00029796	0,000417417	-0,7138	0,4832	
Emplyment	3,00347e-05	9,83393e-05	0,3054	0,7631	
index of poverty	0,0237642	0,0134263	1,770	0,0913	*
discount rate	-0,0321292	0,0166802	-1,926	0,0677	*
population	-0,00187955	0,000516288	-3,641	0,0015	***
Birthrate	-1,13814	1,03009	-1,105	0,2817	
the tax base	1,73915e-05	4,63036e-06	3,756	0,0012	***
gross profit	7,03020e-09	9,92480e-09	0,07083	0,4865	

Mean dependent	0,373905	S.D. dependent	0,198646
Sum Squared resid	0,322847	S.E. of regression	0,123991
R-squared	0,759366	Adjusted R-squared	0,610401
F(13,21)	5,097639	P-value(F)	0,000502
Log-likehood	32,34083	Akaike criterion	-36,68166
Schwarz criterion	-14,90678	Hannan-Quinn	-29,16497
Rho	-0,440854	Durbin-Watson	2,880227

Excluding the constant, p-value was highest for variable 9 index of poverty

Fig. 6. Research on the impact of efficiency determinants on the profit/loss ratio

After the verification of explanatory variables it is possible to perform further tests that lead to verification whether the explanatory variable can have an effect on the explained variable.

Matching the model to the data

The standard error of the random indicator of the S.E. regression equation is 13%, which means that the model describes the reality well, while the matching index is 62%, which indicates a fairly good match of the explanatory variables to the explained variable (Gładysz 3) (Table 3).

Statistical verification of the econometric model - Significance

For the purpose of verifying explanatory variables in terms of their existence, the significance level $\alpha = 5\%$ has been adopted. In order to verify the model in terms of the significance of the regression coefficient system, we put two hypotheses.

H0: The factors included in the model are irrelevant
H1: The factors included in the model are significant

GRETL Model 3; OLS, using observations 1:1-18:1 (T=35)
Dependent Variable : Profit On shares

	coefficient	Std. error	t-ratio	P-value	
Const	52,4156	11,7209	4,472	0,0001	***
Inflation	0,0611130	0,0107338	5,694	3,29e-06	***
Rediscount rate	-0,0184299	0,00723236	-2,548	0,0162	**
Population	-0,00137076	0,000309811	-4,425	0,0001	***
Birthrate	6,75726e-06	2,44637e-06	2,762	0,0097	***

Mean dependent	0,373905	S.D. dependent	0,198646	
Sum Squared resid	0,511999	S.E. of regression	0,130639	
R-squared	0,618381	Adjusted R-squared	0,567499	
F(13,21)	12,15312	P-value(F)	5,45e-06	
Log-likehood	24,27082	Akaike criterion	-38,54164	
Schwarz criterion	-30,76490	Hannan-Quinn	-35,85711	
Rho	-0,219116	Durbin-Watson	2,421877	

Fig. 7. Research on the impact of adjusted determinants on the profit/loss ratio

Table 3. Adjusting the model to the data - earnings per share

	Value
Matching error (S.E.)	0.13
R-squared matching ratio	0.618381 = 62%

The empirical value of the statistics is $F = 5.45*e^{-6} \approx 0.014$ and is lower than the adopted level of significance α, which allows to reject the hypothesis H_0, in favor of the H_1 hypothesis. Thus, the variables included in the model are significant

Significance of individual regression coefficients (Table 4).

Variables explaining inflation, the rediscount rate, the number of population, the tax base and the constant (const) are important for the model, because their p value is lower than the assumed significance level $\alpha = 5\%$.

Research on autocorrelation

An investigation of the autocorrelation of a random component based on Durbin-Watson statistics is carried out in order to determine to what extent a given word of the series depends on the previous words. The hypotheses were considered:

H0: no autocorrelation of the random component,
H1: there is an autocorrelation of a random component.

The Durbin-Watson statistics value in the model is 2.41, in the model there are 35 observations and 5 explanatory variables. From Durban-Watson tables, for $\alpha = 0.05$, 35 observations and 5 explanatory variables, we read values $d_L = 1.16007$ and $d_u = 1.80292$ thanks to which equality was determined:

Table 4. Significance of individual regression coefficients - earnings per share

	Factor	P value	Interpretation
X_1 (const)	52.4156	0.0001	Important
X_2 (inflation)	0.0611130	0.0083	Important
X_3 (rediscount rate)	−0.0184299	0.0162	Important
X_4 (population)	−0.00137076	0.0001	Important
X_5 (tax base)	6.75726e−06	0.0097	Important

$$1.8 < 2.41$$

Indicating that there is no autocorrelation of the random component.[1]

Normality

in order to verify the model in terms of the normality of random components, the chi square test was carried out - χ^2 and hypotheses [4]:

H_0: The distribution of residues is not consistent with the normal distribution
$\chi^2 < \alpha$
H_1: The distribution of residues is in consistent with the normal distribution,
$\chi^2 > \alpha$

The value of the coefficient is $\chi^2 = 2.329$. This value is higher than the assumed significance level α, which means that the hypothesis H_0 should be rejected in favor of the H_1 hypothesis. Therefore, the random component in the model is in agreement with the normal distribution. The explanatory variable significantly affects the explained variable (Fig. 8).

Heteroscedasticity

In order to verify the model in terms of heteroscedasticity, White's test was carried out, based on hypotheses:

H_0: The random component is heteroskedastic, $\alpha > p$
H_1: The random component is not heteroscedastic, $\alpha < p$.

The calculated value of the White's test statistic is 27.19164, and the level of significance level $p = 0.18522$ is higher than the value of assumed significance factor α, which leads to the rejection of the H_0 hypothesis, in favor of the H_1 hypothesis. Therefore, the random component of the model is homoscedastic, the explanatory variables are uncorrelated with a random disorder, the functional form of the model is right. This proves that the value of the earnings per share ratio is determined by variables such as: inflation, rediscount rate, number of population, basis for taxation (Fig. 9).

After carrying out research and an attempt to construct econometric models, it can be concluded that only some of the selected efficiency factors affect two of the five performance indicators identified in the work.

---

[1] http://www.ekonometria.4me.pl/durbina.htm.

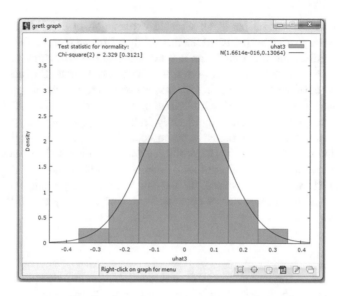

Fig. 8. Normality test - earnings per share

```
Unadjusted R-squared = 0.775119

Test statistic: TR^2 = 27.129164,
with p-value = P(Chi-square(14) > 27.129164) = 0.18522
```

Fig. 9. White's test - earnings per share

From an econometric model based on a variable explanatory ROE and a number of explanatory variables, it can be concluded that only the following indicators may affect the ROE: PL in the EU (Polish membership in the European Union), money supply, redistribution rate, population and natural increase, and adjustment to variables is at 75%. Further verification indicated that the impact of natural increase on ROE is not significant. Using the Durbin-Watson econometric test, it was found that there is no autocorrelation in the model, indicating the dependence of subsequent variables on each other. Conducting the normal distribution test indicated that the model is consistent with the normal distribution and the explanatory variables significantly influence the explained variable. White's test allowed the model to be verified for heteroscedasticity. The results of this study indicated that the model does not have heteroskadasticity, therefore the random component of the model is homoscedastic, the explanatory variables are not correlated with the random disorder, and the functional formula of the model is correct. This means that such factors as: Poland's accession to the European Union, money supply, rediscount rate and population number have a significant impact on the ROE ratio.

An econometric model based on variable explanation of earnings per share and many explanatory variables indicated the possibility of impact on such factors per share as inflation, rediscount rate, population and tax base, and the matching of variables to the model is shaped by 62%. After performing econometric tests: Durbin-Watson, the normal distribution test and White's test, it was concluded that the component of the model is consistent with the normal distribution, the model does not have autocorrelation, the explanatory variables significantly influence the explanatory variable, and the random component of the model is homoscedastic. This means that the functional formula of the model is correct. The profit per share is influenced by such efficiency determinants as: inflation, rediscount rate, population number and tax base. The conducted research were also focused on the study of the dependence of efficiency determinants and other measures of economic efficiency, namely ROA, ROS and the liquidity ratio.

4.3 Summary

The complexity of the research subject to this article should be emphasized. Despite attempts to comprehensively address the issue, not all of issues were presented. According to the authors: research on the determinants of effectiveness should be continued, because in particular the impact assessment still remains an unexplained, although it is a very important aspect of business management. The analysis of the impact of external factors conducted on the effectiveness of enterprises is only a preliminary outline of how to investigate the existence of causal relationships and is the result of research conducted for years.

References

Ćwiąkała-Małys, A., Nowak, W.: Wybrane metody pomiaru efektywności podmiotu gospodarczego. Wrocław, WUW, pp. 168–191 (2009)
Gładysz, B., Mercik, J.: Modelowanie ekonometryczne. Studium przypadku. Wydanie II, Oficyna Wydawnicza PWR, Wrocław (2007)
Mercik, J., Szmigiel, C.: Ekonometria, Oficyna Wydawnicza Politechniki Wrocławskiej, Wrocław, pp. 184–190 (2007)
Pilecki, B.: Ekonomika i zarządzanie małą firmą, Wydawnictwo Naukowe PWN, Warszawa-Łódź (1999)
Rummler, G.A., Brache, A.P.: Podnoszenie efektywności organizacji, Warszawa (2000)
Mitek, A., Micula, I.: Wspólczesne determinanty rozwoju przedsiebiorstw prywatnych, [w:]. In: Kryk, B. (red.) Wspólczesne wyzwania gospodarowania i zarzadzania. Studia i prace wydzialu nauk ekonomicznych i zarzadzania Nr 28., Wydawnictwo Naukowe Uniwersytetu Szczecinskiego, Szczecin (2012)

Intellectual Resources in Polish Enterprises in the SME Sector - Analysis of the Obtained Research Results

Aldona Dereń[1(✉)], Danuta Seretna-Sałamaj[2], and Jan Skonieczny[1]

[1] Wydział Informatyki i Zarządzania, Politechnika Wrocławska,
Wrocław, Poland
{aldona.deren, jan.skonieczny}@pwr.edu.pl
[2] Instytut Finansów, Państwowa Wyższa Szkoła Zawodowa w Nysie,
Nysa, Poland
danuta.seretna-salamaj@pwsz.nysa.pl

Abstract. In the modern market economy, intellectual resources are the main driving force behind the economic and technological development of enterprises in the SME sector. This issue is a key factor for the development of this type of enterprises in Poland. The article presents the results of research including published statistical data (Polish and European reports) and own surveys carried out in a group of 32 enterprises belonging to the sector of small and medium-sized enterprises. Questionnaire interview (oral communication, standardization of the message, individual respondent, presence of an interviewer) was directed to the managerial staff. The aim of the research was to determine the level of knowledge about intellectual resources and their role and importance for the functioning and development of the company. Statistical surveys and results obtained from interviews were illustrated with drawings. The analysis of the obtained results indicates low interest in practical issues of intellectual property management in the surveyed enterprises, which are again reflected in the limited use of these resources in order to build innovative potential and competitive advantage. In this area, our own research coincides with statistical data, including the European Innovation Scoreboard (2017).

Keywords: Intellectual resources · Intellectual property · Protection
SME sector

1 Introduction

Resources are what the organisation has (available resources) or may have (potential resources) as a result of the conducted economic operations. Examples of tangible resources of the organisation include buildings, locations, machines, tools, and real estate. The organisation can purchase these resources or sell them on the market. They have a price, and information about them is relatively well-known and available. Some resources, especially intangible ones, have specific properties – it is difficult to purchase them and sell them on the market, imitate or replace them. Such resources include know-how of employees, trademarks, copyrights, trade secrets, contracts and

© Springer Nature Switzerland AG 2019
Z. Wilimowska et al. (Eds.): ISAT 2018, AISC 854, pp. 49–59, 2019.
https://doi.org/10.1007/978-3-319-99993-7_5

licenses, software, databases, personal and organisational networks, reputation of the organisation and its products, as well as organisational culture, cumulated knowledge (Hall 1993).

Apart from resources, the organisation also possesses capabilities with the nature of processes. They are linked to activities, thanks to which organisations can gather, use and renew their resources. The higher the capabilities of the organisation, the more actively and cleverly it can accumulate, exploit and renew all its resources. It can acquire new streams of resources in response to the situation prevailing on the market, e.g. emergence of new market opportunities, convergence or division of the existing market segments, growth or disappearance of market demand (Eisenhardt and Martin 2000).

The resource approach in managing the organization is a starting point to present in this article the issues concerning the practical understanding and use of intellectual resources in Polish enterprises of the SME sector. The aim of the work is to present the results of analyzes of both statistical data and own research carried out in a group of 32 domestic enterprises belonging to the small and medium-sized sector. The presented results constitute a fragment of research for several years carried out by the authors, concerning entrepreneurship, innovation, creativity and protection of intellectual resources in Polish enterprises of the SME sector.

2 Concept of Intellectual Resources

In the praxeological perspective, resources are items used to achieve an objective: people, materials, tools (equipment), electric energy, etc. In the extended definition, resources also include time and space. Before commencing action, the resources allocated on achievement of a given objective are usually calculated, leaving some provisions (reserve) to be used in the future.

In the economic perspective resources can be divided into: capital and labour - to their contemporary perception as a complex and diverse, due to its numerous features, set of tangible and intangible assets under the control of the organisation. According to another classification, the economic perception makes a distinction into tangible, human and intangible resources. Tangible resources are natural resources, constituting a gift of nature, as well as capital resources in the form of physical and financial resources. Human resources are the characteristics and competences of employees. Intangible resources, on the one hand, are realised by people – their competences and, on the other hand, by the company itself – in the form of e.g. licenses, patents, know-how. Due to the fuzzy adopted classification, some authors combine the notion of human resources with the notion of creations of the human mind, such as inventions, works, industrial designs, etc., and define them as intellectual and human resources (Dollinger 2008). It may seem that such a point of view was based on the belief that the human intellect (mind) is a man's immanent feature that allows him for perform any work.

These are the resources we have adopted as the fundamental resources in the classification we propose, which covers distinction into: organic (primary) intellectual resources and acquisitive (secondary) intellectual resources (Dereń and Skonieczny 2016a, b).

Organic (primary) intellectual resources include: knowledge and experience of the founders, market contacts, talent and cognitive and behavioural skills of employees, trademark (logo, name), commercial mark, patents, trade secret, and organisational culture.

The aforementioned resources are the primary and main assets, which enable establishment of an organisation and commencement of its operations. These are the resources that may be used simultaneously in many places. They do not get devalued during their use, quite the contrary - usually they are enriched and strengthened in the process of the organisation's development. The discussed resources are the foundation for organisation and coordination of any processes in the company, in accordance with the vision or mission adopted by the founders.

On the other hand, acquisitive (secondary) intellectual resources are the resources created as a result of the organisation's operations in the process of transformation of primary resources into specific results, assuming the form of, e.g.: works (copyrights), performances (related rights), inventions (patents), utility models, industrial designs, trademarks, geographic indications, new varieties of plants, mask works, databases, and non-disclosed information (trade secrets, know-how, recipes, processes, technologies, organisational techniques, etc.).

These forms, as products of the human mind of intangible nature, constitute components of intellectual resources, determining the organisation's potential and its competitive power.

It should be emphasised that inventions and trademarks take up a special position in the set of intellectual resources. These resources may appear both at the stage of creating the organisation, as well as in the course of its operation. An invention patent may be the direct decisive factor for establishment of business operations, or it may be a result of these operations. Therefore, innovativeness can be considered an intellectual resource of dual nature, i.e. as a both organic and acquisitive resource. Table 1 presents and defines organic and acquisitive intellectual resources in an organisation.

Table 1. Organic and acquired intellectual resources in the company

Organic (primary) intellectual resources	Acquired (secondary) intellectual resources	Primary and acquired (dual) resources
- founders' knowledge and experience - market contacts - talent and behavioral skills of employees - brand names (logo, name)* - trademark - website - patents* - corporate culture	- new knowledge - copyright - ancillary rights - inventions (patents)* - utility models - industrial models - brand names (logo, names)* - geographical indications - rights to new plant varieties - mask works - databases - non-disclosed information	- patents* - brand names (logo, name)*

(*continued*)

Table 1. (*continued*)

Organic (primary) intellectual resources	Acquired (secondary) intellectual resources	Primary and acquired (dual) resources
	(trade secrets, know-how, recipes, processes, technologies, organizational techniques, etc., trade secrets) - license agreements - cooperation networks	

Depending on the business development stage, patents and brand names are either organic (primary) resource or acquired (secondary) resource.
Source: own study based on A. M. Dereń., J. Skonieczny, Strategies for Protecting Intellectual resources in a company, Problemy Eksploatacji, Maintenance Problems, Quarterly 2/2016 (101) p. 22.

In our opinion, the classification of intellectual resources in an organisation presented in such a way is of practical importance, since it allows for comprehensive identification and analysis of resources owned by the organisation. Identification of intellectual resources requires familiarity with the characteristics of intellectual resources such as: non-appropriability, non-divisibility, originality, commercialization ability, and ability to protect. All the above features affect the selection of appropriate actions in the process of transforming the intellectual resource into commercial intellectual products of the organization.

3   Intellectual Resources in Polish Enterprises of the SME Sector in the Light of European and Polish Statistical Surveys

The issues of creating and using intellectual resources treated as the basis for the development of innovative enterprises is a central point of interest for the European Commission. Research innovation is the key to research and analysis. The basic document in this respect is the European innovation scoreboard. The latest European Innovation Scoreboard report from 2017 presents the results of innovation systems in European countries. In order to create a European Innovation Scoreboard, four main types of indicators are distinguished (framework conditions, investments, innovation activities, impacts) and ten dimensions of innovation that together translate into 27 different indicators. Framework conditions are the main factors for innovation that are beyond the control of companies and cover three dimensions of innovation such as: Human resources (New doctorate graduates; Population aged 25–34 with tertiary education; Lifelong learning); Attractive research systems (International scientific co-publications; Top 10% most cited publications; Foreign doctorate students); Innovation-friendly environment (Broadband penetration; Opportunity-driven entrepreneurship). Investments mean public and private investment in research and innovation, and covers the two dimensions such like: Finance and support (R&D expenditure in the public sector; Venture capital expenditures); Firm investments (R&D expenditure in the business sector; Non-R&D innovation expenditures; Enterprises providing training to develop or upgrade ICT skills of their personnel).

Innovation activities are illustrated by innovation efforts at the enterprise level, included in the three dimensions of innovation such like: Innovators (SMEs with product or process innovations; SMEs with marketing or organizational innovations; SMEs innovating in-house), Linkages (Innovative SMEs collaborating with others; Public-private co-publications; Private co-funding of public R&D expenditures) and Intellectual assets (PCT patent applications; Trademark applications; Design applications).

The impacts include the impact of innovation activities in enterprises in two dimensions of innovation such like: Employment impacts (Employment in knowledge-intensive activities; Employment fast-growing enterprises of innovative sectors) and Sales impacts (Medium and high tech product exports; Knowledge-intensive services exports; Sales of new-to-market and new-to-firm product innovators).

Member States are classified on the basis of the average of results and classified into one of four groups. Based on the average of results calculated on the basis of the aggregate indicator - the total innovation indicator - Member States were divided into four groups.

The first group includes countries identified as Innovation Leaders whose innovation results are well above the EU average (Denmark, Finland, Germany, the Netherlands, Sweden and the United Kingdom).

The second group is called Strong Innovators, i.e. countries with results above or near the EU average (Austria, Belgium, France, Ireland, Luxembourg and Slovenia).

The third group includes the following countries: Croatia, Cyprus, Czech Republic, Estonia, Greece, Hungary, Italy, Lithuania, Latvia, Malta, Poland, Portugal, Slovakia and Spain. These are countries in which the level of innovation is below the EU average. These countries were therefore included in the group of Moderate Innovators.

The fourth group includes Bulgaria and Romania, which are referred to as the so-called Modest Innovators with results far below the EU average.

According to the data included in the European Innovation Scoreboard, Poland is in the so-called moderate innovators and taking the 25th place (Fig. 1). However, taking into account the index "Intellectual assets" our country occupies 18th place among EU countries (Fig. 2).

Coloured columns show Member States' performance in 2016, using the most recent data for the indicators in this dimension, relative to that of the EU in 2010. The horizontal hyphens show performance in 2015, using the next most recent data for the indicators in this dimension, relative to that of the EU in 2010. Grey columns show performance in 2010 relative to that of the EU in 2010.

Coloured columns show Member States' performance in 2016, using the most recent data for the indicators in this dimension, relative to that of the EU in 2010. The horizontal hyphens show performance in 2015, using the next most recent data for the indicators in this dimension, relative to that of the EU in 2010. Grey columns show performance in 2010 relative to that of the EU in 2010.

The analysis of Polish reports on the innovation of national startups was also focused on the issue of intellectual resources as the basis for the functioning of these

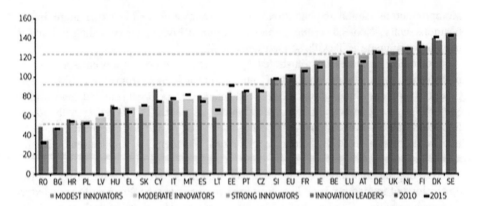

Fig. 1. Performance of EU Member States' innovation systems. Source: European Innovation Scoreboard (2017, p. 6).

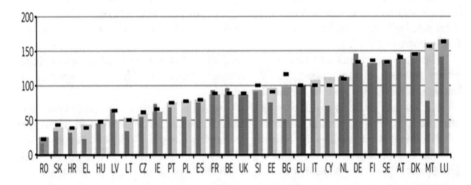

Fig. 2. Intellectual assets. Source: European Innovation Scoreboard (2017, p. 24).

organizations. Of key importance in this analysis was patent activity, which is an indicator of the level of innovation.

As far as patenting is concerned, the level of patent activity in total among Polish enterprises is very low compared to European and world leaders. Around the promise of Polish enterprises, he asks for reservations about his intellectual property or obtains them. In the group of 692 startups surveyed in 2016, only 14%, i.e. every seventh, have a patent or are in the process of patenting in the country (every third patenter) or abroad (2/3 patenting).

Patented startups are characterized by a shorter than the other surveyed internships on the market - more than half of them were established no earlier than in 2015. Every second defines its stage of development as a "Solution-Product Fit", or rather an early level of development. Only every fifth earns regularly - but where these revenues are already present, their increase in the last six months in 70% of cases exceeded 50%. Half of the patents are exported, mainly to the EU and the USA.

The discussed research indicates that the product is the main carrier of innovation for the respondents. Over half of respondents claim that they are creating a completely new product, and 43% indicated that their innovation consists in improving existing solutions. Only 4% of startups declared directly copying other products or services. The respondents' answers indicate the lack of in-depth knowledge about the essence and importance of intellectual resources in their business operations. What's more, the respondents did not stress the need to secure these resources in the market space.

The same test scenario was repeated in 2017. The research covered a group of 621 startups. In this group, 61% of respondents do not patent their developed solutions, and every fifth respondent does not see value in patenting. It should be noted that relatively few solutions related to the digital industry in general are patented by Polish entrepreneurs. Entrepreneurs say that they build their competitive advantage differently than by reserving intellectual property. At the same time, as in 2016, half of the respondents claim that they "create a completely new product" and a comparable percentage that "improves existing solutions". The discussed report, although it shows the upward trend in the number of patenting startups, also confirms the constant lack of knowledge and awareness in the scope of the need to protect intellectual property.

4 Intellectual Resources in Polish SMEs in the Light of the Research[1]

As part of the classes conducted in March 2017 at the Polish-American School of Business (34th Edition) – the Executive MBA Programme at the Wroclaw University of Science and Technology, the legal and organisational module covered the issues of intellectual resources management, in particular their conversion into intellectual products, protection of these resources and their trade. A group of 32 enterprises belonging to the small and medium-sized sector, participants in the study programme representing Polish SMEs, were subjected to surveys. The applied research method consisted in the managers answering 27 multiple choice questions. The vast majority of the questions concerned the respondents' understanding of the notion of intellectual resources, the methods of their protection, the procedures used to protect them, as well as the persons responsible for their management in the organisation. The questionnaire was supplemented by questions concerning the type of organisation, its size, as well as the scope and area of business activity. The survey was a pilot study, and the obtained results, in spite of their diversity, allow for formulating conclusions concerning the way Polish SMEs managers perceive intellectual resources and their management.

The majority of the surveyed managers (75%) represented limited liability companies. The business operations mainly focused on: production (28.12%), services (31.25%), trade (15.63%), multi-sector operations (25%). The majority of respondents represented organisations employing over 10 employees.

The survey was supposed to diagnose the managers' knowledge of the types of intellectual resources, as well as knowledge of their specific characteristics, strategies of

---

[1] A.M. Dereń, J. Skonieczny, *The Polish managers' perception of Intellectual resources in management of SMEs*, Raporty Wydziału Informatyki i Zarządzania Politechniki Wrocławskiej, Wrocław 2017, Seria PRE, Nr 12.

intellectual property management in organisations, and the forms of protection and security of intellectual resources used in market operations that are applied by these organisations.

The obtained results are also illustrated in the form of a map constituting Fig. 3. The respondents' answers are grouped according to the preferences adopted thereby – beginning with the values most representative for them. The respondents could choose more than one answer.

In the opinion of the survey's authors, managers of Polish SMEs have low and incomplete knowledge concerning intellectual resources and their management. Usually, the surveyed managers generally define intellectual resources as general

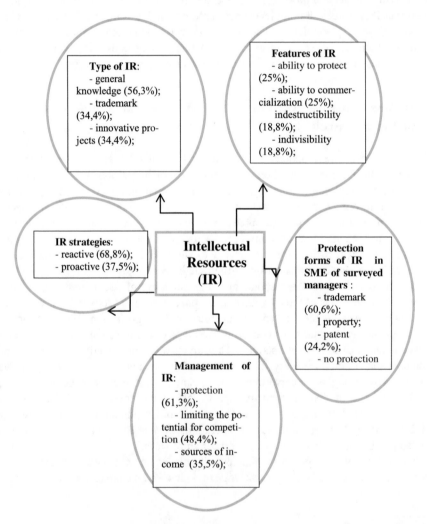

Fig. 3. Map of results. Source: A.M. Dereń, J. Skonieczny, *The Polish managers' perception of Intellectual resources in management of SMEs*, Raporty Wydziału Informatyki i Zarządzania Politechniki Wrocławskiej, Wrocław 2017, Seria PRE, Nr 12.

knowledge, without a division into diverse categories, trademark and innovative projects. According to the survey's authors, such limited understanding of intellectual resources as a basis for organisational development results from lack of education in this respect. Polish management and engineering study programmes have only for several years been teaching subjects covering intellectual property, its management and its protection. The authors of the article conduct such classes for students at the Wroclaw University of Technology and often encounter the environment's opinions that these issues should be limited only to legal protection of intellectual resources. In our opinion, such a narrow approach should be rejected, since these issues, apart from significant legal themes, should include the broadly understood management of these resources, as factors enabling building of competitive advantage, preparation of an innovative strategy and further development of the organisation.

The narrow understanding of the notion of intellectual resources by the surveyed managers is reflected in their answers to the other survey questions. The authors present these conclusions below.

When identifying types of intellectual resources, the respondents, next to technology, indicated trademarks and inventions as intellectual resources dominant in their organisations. The preference for trademark does not raise any doubts and means that Polish managers understand the significance of this category for the process of building the brand and the market image of the organisation. However, the lack of interest in inventions as an intellectual resource that may be used in the company is alarming.

The respondents emphasise lack of activity regarding protection and security of intellectual resources. In our opinion, the reasons for such a state of affairs may be found, among others, in the lack of patenting tradition in the group of SMEs and the low level of awareness among managers with regard to the benefits of application of industrial property protection instruments.

The answers to the question concerning intellectual resources should be regarded as fairly accidental and based on association. Managers, from many proposals of the intellectual resources, have chosen the following features: the ability to protect, the ability to commercialize, indestructibility and non-divisibility. An indication the feature the ability to protect as the first, we can interpret as a result of their own experiences in the filed of trade mark protection, as well as the result of the participation in training meetings on which the widely discussed issues the protection of intellectual property. Similarly, we can interpret the indication by managers the next feature like characteristics of commercialization. However, the indication the characteristics like invulnerability as the next, due to the simple statement of the concept of tangible and intangible resources. Out of the many proposed characteristics of intellectual resources, managers selected only three characteristics: indestructibility, non-divisibility and protection capability. Indication of indestructibility as the leading feature probably resulted directly from a straightforward juxtaposition of the notions of tangible and intangible resources. The former are commonly perceived as destructible resources, used up during their utilisation.

As for protection or security of intellectual resources, the respondents indicated protection rights for trademarks and patents as a form of this protection.

With regard to management of intellectual resources, the surveyed managers present a standpoint that can be identified as focused on actions inside the organisation (protection of resources; source of income), rather than as a set of external actions focused on obtaining competitive advantage or an element of market competition of the organisation.

Considering such an understanding of management of intellectual resources in an organisation, it is not surprising that practically the only strategy indicated by the respondents was the reactive strategy, namely the strategy of protecting intellectual resources while sustaining low costs.

5 Conclusion

Intellectual resources are nowadays a decisive factor in gaining an advantage competitive by enterprises, as they account for as much as 80% of global business.

The intellectual property system is very extensive - it includes industrial property (inventions, industrial designs, trademarks, etc.), copyright and related rights, but also personal rights, the right to the company, the right to the database, the right to new plant varieties and the company's secret (know-how). Very often, the accumulation of these rights occurs. In one product, technical solutions in the form of patented inventions can be found, whose market power is strengthened by a well-known and valued trademark among consumers, as well as the usability and aesthetics of the product achieved thanks to the use of modern industrial design.

In this situation, proper protection of intellectual property and effective management of exclusive rights simply seems a necessity. It is worth remembering that even if an enterprise does not use intangible assets in its activity and does not invest in it, it does not mean that it can operate freely on the market. There is a high risk that when introducing new products, it infringes the intellectual property rights of another company.

References

Beauchamp, M., Kowalczyk, A., Skala, A., Ociepka, T.: Polskie startupy, Raport 2017. Fundacja Startup Poland, Warszawa (2017). http://www.citibank.pl/poland/kronenberg/polish/files/Startup_Poland_raport_2017.pdf. dostęp 11 May 2018

Dereń, A.M., Skonieczny, J.: Zarządzanie twórczością organizacyjną. Podejście procesowe, Wyd. DIFIN, Warszawa (2016a)

Dereń, A.M., Skonieczny, J.: The Polish managers' perception of Intellectual resources in management of SMEs. Raporty Wydziału Informatyki i Zarządzania Politechniki Wrocławskiej, Wrocław, Seria PRE, Nr 12 (2017)

Dereń, A.M., Skonieczny, J.: Strategies for protecting intellectual resources in a company. Maintenance Problems, No. 2 (2016b)

Dereń, A.M.: Kultura ochrony własności intelektualnej MŚP, w: Przegląd Organizacji, Nr 1 (2017)

Dollinger, M.J.: Entrepreneurship: Strategies and Resources, pp. 32–62. Marsh Publications, Lombard (2008)

European Innovation Scoreboard. https://www.rvo.nl/sites/default/files/2017/06/European_Innovation_Scoreboard_2017.pdf. dostęp 11 May 2018

Kala, A., Kruczkowska, E.: Polskie Startaupy 2016, Fundacja Startup Poland, Warszawa, s. 56–61 (2016). http://www.citibank.pl/poland/kronenberg/polish/files/Startup_Poland_Raport_2016_16.pdf. dostęp 11 May 2018

Urząd Patentowy Rzeczypospolitej Polskiej, Raport Roczny 2015. Dostępny. http://www.uprp.pl/uprp/_gAllery/77/24/77242/raport_roczny_2015.pdf. dostęp 11 May 2018

IFI CLAIMS Announces Top Recipients of U.S. Patents in 2016. https://www.ificlaims.com/news/view/ifi-claims/ifi-claims-announces-2.htm?select=10. 14 May 2018

Quality of Investment Recommendation – Evidence from Polish Capital Market, Multiples Approach

Michał J. Kowalski[✉]

Wroclaw University of Science and Technology, Wrocław, Poland
michal.kowalski@pwr.edu.pl

Abstract. The aim of the paper is to assess the quality of practices used in the multiples approach in investment recommendations. A hundred investment recommendations were examined, coming from seven different brokerage houses, issued in year 2016. The results suggest that the cognitive value of the multiples valuation in the recommendations is very limited. In a statistically significant way, the results of multiples valuation do not differ significantly from the results obtained using the income method. In the majority of analyzed reports, multiples based valuation method has a direct impact on the final value of the shares resulting from the recommendation. Most reports use a multi-stage procedure, in which the valuations resulting from multiples were subject to subsequent transformations that can lead to the possibility of manipulating valuation results. The range of the applied multiples is very poor, in almost all cases P/E and EV/EBITDA measures are used, other multiples are used incidentally. On average, the valuations are based on about ten comparable firms, however the multiples have broad dispersion. The scope of disclosures in the area of multiples valuation is marginal and certainly insufficient. There is no information on the method of selecting both multiples, comparable companies and often very different transformations made on the input data. There is no justification that would ensure the objectivity of the proceedings.

Keywords: Business valuation · Investment recommendation
Multiples approach

1 Introduction

During the 38th International Conference on Information Systems Architecture and Technology, the author presented a paper concerning evaluation of quality of investment recommendations, where the income approach and DCF model valuation were analyzed. The research revealed low quality of valuations presented in the Polish investment recommendations. Especially, the level of disclosures and assumptions used in the financial models underlying financial forecasting and valuations were not sufficient. The aim of this paper is to present the results of further investigation into investment recommendations, extended to multiples approach. Investment recommendations usually make use of both income and multiples methods of valuation and final recommendations should take into account the comparison of both of them.

© Springer Nature Switzerland AG 2019
Z. Wilimowska et al. (Eds.): ISAT 2018, AISC 854, pp. 60–70, 2019.
https://doi.org/10.1007/978-3-319-99993-7_6

The aim of the author is to assess the quality of practices used in the analysts' estimates of investment recommendations based on multiples approach. The article is organized as follows: (i) short review of prior research focused on investment recommendations and procedures of valuations based on multiples, (ii) materials and methods, (iii) results and (iv) conclusions. In Materials and methods, a database of investment recommendations issued by Polish brokerage houses was presented and the research sample was characterized. Results were presented according to the analyzed valuation areas: types of multiples used, influence of the multiples valuation on finally recommended value and share price, selection of comparable firms, comparison of results of valuation conducted by different approaches. At the end of the article, short conclusions are presented.

2 Prior Research

Research into investment recommendations focuses on three main areas: (i) the impact of recommendations on market reactions, (ii) optimism of recommendations and (iii) accuracy of recommendations.

Investment recommendations influence market valuation, especially buying recommendations that have a positive impact on stock prices. Groth [6] showed that their implementation can, in the short term, bring abnormal returns for investors. Buying recommendations have a positive impact on stock prices and allow higher average market returns both in the short term and up to 90 days after their publication [1]. However, Jegadeesh et al. [8, 9] also indicated that stock recommendations are mainly prepared for companies with superior return, the so called "glamour stocks". Analysts make favorable reports for such firms and they select superior stocks for which recommendations will be positive.

Investigations of optimism of investment recommendations have a long history. Several studies have showed that most of published recommendations are "buy" rather than "sell" recommendations (Ertimur et al. [3]). Mainly the optimism was measured by comparing the analyst' forecast for a particular stock to the average forecast for the company made by other analysts in comparable forecast horizon (Morgan and Stocken [12]). Kowalski et al. [10] suggested that analysts are more likely to announce higher valuation than the historical value drivers and fixed growth model suggest. Other study conducted by Zaremba and Konieczko [15] documented over optimism among analysts. The authors examined returns after investing in securities, based on suggestions from stock recommendations. Trading strategy of purchasing (buying long) stocks with the most favorable recommendations by analysts achieved abnormal negative returns.

Kubik et al. [7] analyzed optimism forecast from analyst's carrier perspective, and they tried to explain differences between experienced and beginner analysts. The authors showed that the accuracy of the analyses for a particular analyst had a significant impact on their carrier. According to Kubik et al. [7], analysts were rewarded for optimistic forecasts that support profitable trading strategy of purchasing stocks with the most favorable recommendations. Similar research was performed by Ertimur et al. [3] on the sample of 97 000 recommendations.

Multiples approach procedures were the subject of a number of theoretical considerations, both on the ground of foreign and national literature. Feltham and Ohlson demonstrated that there is a strong theoretical basis for using multiples [4]. Empirical studies have shown that multiple-based valuation can give similar accuracy to discounted income (DCF) valuation. Kaplan and Ruback [14] showed that valuation using EBITDA multiples is as accurate as DCF. However, it should be emphasized that beside advantages of multiples methods, usually concentrated on its simplicity and rapidity, a numerous impairments and difficulties in their usage are mentioned in literature. Most of them are focused on choosing appropriate the multiples and the basis of comparison. Fernandez in empirical studies involving 1200 multiples from 175 firms concluded that multiples nearly always have broad dispersion, which may cause that valuation performed using multiples is highly debatable. The author proved that EBITDA and profit after tax are the most commonly used parameters as multiples are more volatile than equity value [5].

Patena [13], among the basic difficulties in defining the foundations for comparisons indicates: (i) market efficiency, (ii) compliance of accounting measures with economic reality, (iii) selection of multipliers depending on the financing structure and (iv) degree of similarity of companies in the group. The multiples method is based on the assumption about the efficiency of markets, and that the results assume a reliable form as long as the market price of the shares is close to the fundamental value. Accounting systems, or actually their dissimilarity, can cause significant differences in the values of multiples and lead to distortions in valuations.

The selection of companies for the valuation according to the multiples method has been the subject of numerous studies. Noticeable is the contrast between the approach followed by practitioners, who typically use a small number of closely comparable firms, and the academic literature which often uses all firms in an industry. Cooper and Cordeiro [2] found in their empirical research that using ten closely comparable firms is as accurate on average as using the entire cross-section of firms in an industry, and using five comparable firms is slightly less accurate.

Miedziński [11] points out that difficulties in applying comparative methods make them more attractive in manipulating the valuation of enterprises. This is due to the relatively large inaccuracy of the method and the possibility of manipulating valuation result, e.g. by tendentious selection of firms for comparisons, depending on the intentions of the valuer. It is especially important on the Polish capital market where the scope of disclosure of the content of the recommendations is not regulated at all. There are no formal standards or procedures indicating which methods or models of valuation should be used. The analysts have a free hand in taking decisions about choosing methods, making assumptions, sources of information, disclosures and presenting results.

In this context, research into the procedures used in investment recommendations on the Polish market, apart from having cognitive aspects is also significant from the point of view of the practice of economic life.

3   Methods and Materials

3.1   Research Design

A database dedicated to the project was set up from the investor recommendations issued by Polish brokerage houses. Each recommendation was analyzed individually to obtain information categorized in three areas: general information, results of valuations, and influence of multiples valuation on the final recommended value. Detailed data for each multiples was gathered. The summary of 50 information pieces was collected for each recommendation.

Each valuation was analyzed in terms of: (i) types of multiples, (ii) influence of the multiples valuation on the final recommended value, (iii) selection of comparable firms, (iv) comparison of results of valuation.

The popular view in literature indicates that different industries have different "best" multiples. Fernandez indicates examples of sectors and characterizes best multipliers [5]. He emphasizes that the selection of multipliers should be aimed at looking for measures with the smallest possible variability, which guarantees that the method accurately reflects the fundamental value. The aim of the research in the first stage was to assess which multipliers are used and procedures to select them, as well as if disclosures in this field are implicit in recommendations.

The second step of the research procedure focused on the analysis of the impact of the multiples valuation on the final price of the shares indicated in the recommendations. Fernandez indicated that multiples are useful in the second step of valuation, after performing valuation using another method. Multiples approach let to identify differences between the firm valued and the firms it is compared with. In the course of the research, a schema was analyzed according to which the multiples valuation influences the final price that results from the recommendations and the assumptions that were made at this stage.

The third step of the research procedure included the analysis of a number of comparable firms, which were used to determine the multiples value. The dispersion measures for multiples used in the recommendations were also analyzed.

In the last step, the results of the multiples valuation were compared with the income valuation, the current price of shares and the recommended valuation. The differences were analyzed statistically, using t-statistic.

Statistical tools were used to analyze the collected material. Analyzes were mainly carried out with the use of contiguous tables, multiple response and dichotomies tables, and analysis of variance. The analysis of the relationship between variables was examined by Pearson's chi-squared test. Additionally, the strength of the relationship between variables was identified by the Yule's coefficient, the Pearson's convergence coefficient and the Spearman's rank correlation.

The research was performed on the stock recommendations sample issued by the Polish brokerage houses. The summary of 100 investment recommendations were examined, coming from seven different brokerage houses, issued in years 2016.

We investigated only those recommendations that used comparative valuation. The research material initially included 120 positions, so about 17% of the valuations analyzed did not use a comparative approach. The information for the recommendations analyzed was obtained from websites:

- www.bankier.pl
- www.bgzbnpparibas.pl
- www.dm.pkobp.pl
- www.bdm.com.pl

The entities subject to the valuation were divided into 22 branches, grouped for the needs of the study into 10 sectors of the economy.

4   Results

4.1   Types of Multiples

The analyzed recommendations used four types of multipliers: P/E (96% cases), EV/EBITDA (84%), EV/EBIT (21%), and P/BV (17%). The results obtained with the use of distribution tables indicate that P/E and EV/EBITDA are always used, and the other two multipliers are used depending on the industry. EV/EBIT is more often used in the valuation of companies belonging to the light industry, while P/BV (at low level of significance) for companies in the real-estate and finance sector. The analyzes confirm then that in recommendations at least some of the multipliers are used selectively. However, further research, including brokerage houses, indicates that this principle is applied only by some of them.

The descriptive statistics on the number of multipliers used are presented in Table 1.

Table 1. Numbers of used multiples

Statistics	No.	AVG	MIN	Q1	MED	Q3	MAX
value	100	2,2	1,0	2,0	2,0	2,0	3,0

No. – number of observations, AGV – average, MIN – minimum value, Q1, Q3, quartile 1st and 3rd, MED – median, MAX – maximum value

The valuation used, most often, two multiples (in 73% of cases). The results of statistical analyzes did not show any relation between the number of multiples and the sector, while the strong relationship between the number of multipliers and the brokerage house was shown (χ^2 Pearson 54.127, p = 0.000). Four of the analyzed brokerage houses always used the same multiples, regardless of the company.

Despite the fact that there is a reason behind choosing proper multiples in a given sector, the valuations did not reveal why these and not other multipliers were used. The results of the multiples' volatility analyzes or other arguments in favor for their use were not well explained.

4.2   The Influence of the Comparative Method on the Final Valuation

In the majority of analyzed reports (74%), the valuation carried out using the multiples method has a direct impact on the final value of the shares resulting from the recommendations. However, the impact varies, and the relationship between the value of multiples for compared companies and the final recommended valuation is complex. Most reports use a multi-stage procedure, in which the valuations resulting from multiples were subject to subsequent transformations in order for the final result to emerge. Usually a four-step procedure is used, and such a procedure is presented schematically in Fig. 1.

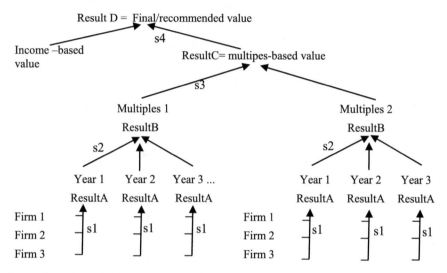

Fig. 1. How the multiples for comparable firms and influence the final valuation, steps in procedures used by brokerage houses.

The step 1 usually concerns multiples value on the basis of collected data for comparable companies (s1). To extract one value from the collected sample in the vast majority of cases the median was used in 54% of cases, in 28% the arithmetic mean was used, in other cases the data presented in the recommendation did not allow to determine the applied procedure.

The analysis of relationships carried out with the use of contingency tables did not reveal any significant links between the procedures used in step 1 and the sector in which the company is being valued. However, the relationship between the applied procedure and the brokerage houses was recorded at any low level of confidence (χ^2 Pearson 74.704, p = .000). This confirms once again that brokerage houses use the same procedures automatically regardless of the entity being valued.

Typically, data for companies with a similar business profile come from 3 periods (67% of cases), one year with historical results and two additional forecasts; in other cases, historical data is most often rejected and the valuation is based on two years of forecast. In one case four periods were included - one historical and three forecasted. In the second step of the procedure (s2), one multiples value is determined, which is a result taking into account the data from all analyzed periods. Disclosures regarding the calculation procedure used were presented in 45% of cases. The remaining recommendations present the results of the valuation for several periods, but it is not clear how these data are later processed.

The most frequently used procedure assumed equal weights of each period (nearly 77.8% of cases). In general, it should be stated that the weights used for particular years are very different. The list of observed combinations of weights and their numbers are presented in Table 2. A detailed analysis of the recommendation documents did not allow to specify the reasons for which individual periods are assigned different weights, or why certain periods have been eliminated.

Table 2. Shares of periods pattern according companies sectors

Sector	Shares for periods							
	50–50	40–60	10–40–50	20–40–40	25–25–50	33–33–33	50–30–20	25–30–30–15
% of cases	28,9	2,2	2,2	6,7	6,7	48,9	2,2	2,2
Construction	3					6		
Developers	2		1			1		
Finance		1			1			1
Retail						3		
Other services	4			1	1	1		
Chemical	2		1			3		
Heavy industry					1	5	1	
Light industry				1		1		
Petrol	2					2		

In the title of columns a schema of proceed in step 2 was presented. For instance 40–60 mean that two periods are used with w weights 40 and 60% respectively

Interesting conclusions are provided by the analysis of the applied methods in the sector perspective, presented in the same table. They indicates that there is no regularity in the procedures used. There are very different approaches for companies belonging to a given sector, which further emphasizes that the expert method is dominant. Unfortunately, the collected empirical material, after the rejection of companies that did not make disclosures, is too small and does not support this observation statistically. The obtained results indicate that the used procedures are entirely dependent on the analysts.

In addition, the analysis according to brokerage houses showed that 4 out of 7 surveyed brokerage houses always used the same procedures regardless of the company being investigated. The Chi-squared test indicated that in the case of 4 brokerage houses, there is no reason to reject the hypothesis that the same procedure is applied regardless of the sector of the company belonged to.

The purpose of the third stage of the procedures used in investment recommendations is to determine the value of the company or shares according to the multiples approach. This step requires a comparison of the results obtained with various multiples used. Unfortunately, in 46% cases, the valuation methods were not revealed. Most often (44%) the results were averaged over the arithmetic mean, but other approaches were also used. In Table 3 combinations of weights used for various multiples are presented. As in the other analyzed aspects, the recommendations do not explain or justify the solution applied.

Table 3. Shares of multiples pattern

Shares of multiples	ND	33-33-33	50-50	70-30	75-25	90-10
% of cases	46	14	30	4	4	2

In the fourth stage, the results of the valuation using the multiples method are compared with those obtained by other methods. In 60% of cases, the multiples valuation has a direct impact on the final recommended value. That is, the recommended value is the average - most often weighted - of multiples valuations and valuations made using other methods, usually income valuation. The share of valuation results according to the multiples method in the final value is different. The most frequently used share is 30%, but the range observed in the sample is between 10 and 95%. Descriptive statistics for this factor are presented in Table 4. The highest 95% share was recorded in the case of a venture capital company although the application of a 5% impact of income valuation based on dividend may raise objections.

Table 4. Shave multiples approach in recommended value, descriptive statistics

Statistics	No.	AVG	MIN	Q1	MED	Q3	MAX
Value [%]	60	34,6	10,0	30,0	30,0	50,0	95,0

In 83% of cases in which the multiples valuation had a direct impact on the recommended value, no justification for the applied share of this valuation was indicated. In other cases, the justification was to indicate that the income valuation due to the fact that it takes into account more specific value factors will have a greater impact on the final recommended value or that due to the fact that business cycles run differently on different markets, the results obtained by the multiples method will have smaller importance. In none of the cases was the justification of the applied multiples contribution used, which would be based on research or empirical evidence. It can therefore be concluded that the selection of weights was discretionary and based on the expert judgment.

4.3   Selection of Comparable Firms

The recommendations used most often 8 benchmarks, with companies predominating from foreign markets and commonly only one comparable company was chosen from the Polish market. Descriptive statistics on the number of recommendations used by comparable companies for all analyzed multiples are presented in Table 5.

Table 5. Numbers of comparable firms

	No.	AVG	MIN	Q1	MED	Q3	MAX
Sum	205	10,9	2,0	6,0	8,0	13,0	33,0
Polish	205	1,8	0,0	0,0	1,0	3,0	8,0
Foreign	205	9,1	0,0	4,0	8,0	13,0	30,0

There were slight differences between the number of companies compared for various multiples, but the research did not confirm their statistical significance. In terms of the number of benchmarks used, the procedures used in recommendations realize the conclusions resulting from empirical research, indicating that a selection from 5 to 10 comparable companies is sufficient.

Assessing the selection of benchmarks is not easy. The selection is subjective and expert. However, you can evaluate the dispersion of results. The variability of multiples was analyzed using relative standard deviation (RSD) and the extreme values of multiples were additionally compared too. The results are presented in Table 6.

Table 6. Volatility of multiples

	No.	AVG	MIN	Q1	MED	Q3	MAX
RSD	631	37,1	0	22,9	31,6	46,3	122,1
(Max/min)-1	579	1,99	0	0,88	1,42	2,52	8,06

The results indicate that the variability of multiples should be considered moderate, towards a significant one. On average, the highest value differs from the lowest in the sample by two times. The extreme values were eliminated during the analysis. In the raw data, the differences between the largest and the smallest values were even 40 times.

4.4   Comparison of Results of Valuation

The last stage of the research was to compare the results of the valuation obtained by various methods. The aim was to verify to what extent the multiples method enriches the valuation process. The relative difference between multiples valuations (M) income measurement (I), recommended valuation resulting from the valuation (R) and the current market valuation of the analyzed Company (C) were analyzed. The relative difference was determined by means of a dependence: for example, for the comparison

Table 7. Comparison of valuation results

	AVG [%]	SDv [%]	No	t-statistic	df	p-value
M/I	1,2	26,0	88	0,419051	87	0,676211
M/R	1,5	23,9	88	0,585904	87	0,559457
I/R	0,3	8,5	97	0,371942	96	0,710756
M/C	20,6	46,9	91	4,180454	90	0,000067
I/C	14,5	24,6	98	5,825912	97	0,000000

of M/C-1 results of valuation based on multiples and income approach. In Table 7, the results were presented.

The results indicate that in investment recommendations, the multiples valuation differs from the income estimate by an average of only 1.2%. The value of t-statistics indicates that there is no reason to reject the hypothesis that these valuations are equal in the analyzed sample. The procedure used to determine the recommended value means that with statistical significance, the values recommended in the analyzed sample do not differ from the results of multiples and income valuation. At the same time, both valuations indicate results significantly higher than the current market valuation by 21 and 14%, respectively for the multiples and income approach. In this respect, the results are supported by numerous previous studies indicating that the stock price increases.

5   Conclusions

In the light of the obtained results, it should be recognized that the basic functions of the multiples approach, which is to constitute a reference point for results obtained by other methods, are not implemented in the investment recommendations. In most cases, multiples valuation directly affects the ultimately recommended share price. Data regarding comparable companies are subject to multistage and complex transformations. The manner of these transformations is subjectively based on the expert knowledge.

The scope of disclosures in the area of multiples valuation is marginal and certainly insufficient. There is no information on the method of selecting both multiples, comparable companies and often very different transformations made on the input data. There is no justification that would ensure the objectivity of the proceedings. The range of applied multiples is very poor, in almost all cases P/E and EV/EBITDA are used, other multiples are used incidentally though there are examples of sectors for which analysts reach for other measures, which was shown with statistical significance.

The analysis of the relationships shows that some brokerage houses use the same method, regardless of the sector the company belong to. Only comparable companies are changed. On average, the valuations are based on about ten multiples, which should be considered in accordance with the standards confirmed in empirical studies. The values of the multiples are moderate at the border of strong variability.

As a result, the results of multiples valuation do not differ statistically from the results obtained using the income method. This suggests that the cognitive value of the multiples valuation in the recommendations is very limited.

References

1. Bjerring, J.H., Josef, L., Theo, V.: Stock prices and financial analysts' recommendations. J. Finance **38**(1), 187–204 (1983)
2. Cooper, I.A., Cordeiro, L.: Optimal equity valuation using multiples: the number of comparable firms (2008)
3. Ertimur, Y., Zhang, F., Muslu, V.: Why are recommendations optimistic? Evidence from analysts' coverage initiations (2010)
4. Feltham, G., Ohlson, J.A.: Valuation and clean surplus accounting for operating and financial activities. Contemp. Account. Res. **11**, 689–731 (1995)
5. Fernandez, P.: Valuation using multiples: how do analysts reach their conclusions?. Working Paper, IESE, University of Navarra, 29 January 2003
6. Groth, J.C., et al.: An analysis of brokerage house securities recommendations. Financ. Anal. J. **31**, 32–40 (1979)
7. Hong, H., Kubik, J.D.: Analyzing the Analysts: Career Concerns and Biased Earnings Forecast. Stanford Business School, Stanford (2001)
8. Jegadeesh, N., Kim, J., Krische, S.D., Lee, C.M.C.: Analyzing the analysts: When do recommendations add value? J. Finance **59**, 1083–1124 (2004)
9. Jegadeesh, Narasimhan, Kim, Woojin: Value of analyst recommendations: international evidence. J. Financ. Mark. **9**(3), 274–309 (2006)
10. Kowalski, M.J., Praźników, J.: Investment recommendation optimism – results of empirical research on polish capital market. In: Information Systems Architecture and Technology: Proceedings of 37th International Conference on Information Systems Architecture and Technology – ISAT 2016 - Part IV. Springer, Berlin (2017)
11. Miedziński, B.: Nadużycia w wycenie przedsiębiorstw, Warszawa (2012)
12. Morgan, J., Stocken, P.: An Analysis of Stock Recommendations. Princeton University, University of Pennsylvania, Princeton (2001)
13. Patena, W.: W poszukiwaniu wartości przedsiębiorstwa. Metody wyceny w praktyce. Oficyna a Wolters Kluwer Business, Warszawa (2011)
14. Kaplan, S.N., Ruback, R.S.: The valuation of cash flow forecasts: an empirical analysis. J. Finance **50**, 1059–1093 (1995)
15. Zaremba, A., Konieczka, P.: Skuteczność rekomendacji maklerskich na polskim rynku akcji. Zeszyty Naukowe Uniwersytetu Szczecińskiego nr 803 Finanse, Rynki Finansowe, Ubezpieczenia nr 66, Szczecin (2014)

Time-Dependent Biases in Analysts' Forecasts in Polish Recommendations Published in 2005–2012

Katarzyna Włodarczyk[✉]

Wroclaw University of Science and Technology, Wrocław, Poland
katarzyna.wodarczyk@pwr.edu.pl

Abstract. The main goal of this article is an attempt to indicate the analysts' bias in forecasting financial parameters published in Polish over the calendar years 2005 through 2012, in divided into two sub-periods – bull and bear market and years with lower and higher economic growth. The first part of the article assesses the accuracy of analysts' forecasts. The second part examines the analysts' tendency of underestimation or overestimation of forecasts. This paper provides evidence that there are statistically significant differences in analysts' biases between analysts' forecasts published during bull and bear market - for net profit. And also between analysts' forecasts from period with lower economic growth and greater economic growth for the following parameters: revenue, EBIT, net profit and net working capital.

Keywords: Stock recommendations · Forecasts · Bias · Accuracy
Tendency

1 Introduction

For many years, plenty of researches confirm that analysts have a clear tendency to publish more positive recommendations than pessimistic recommendations [1, 8, 13, 20 and many others]. What is more, analysts' recommendations are usually characterised by high level of optimism [e.g.: 8, 14, 22].

Probably, both above-mentioned trends are the most common tendencies which analysts have characterised. However, the sell-side analysts, similarly as investors, succumb some behavioural aspects, and their products so as forecasts, can be burdened with some biases.

These biases are caused by e.g.: the type of company for which they work [4, 6, 7], the country from which analysts come [2, 19], the type of industry which they analyse [12], their internal beliefs [11, 16] or due to analysts' conflict of interest [8]. What is more, Schipper [20] noted that analysts' underreaction to good news could result in forecast pessimism and many surveys focus on underreaction-related optimism in the face of bad news. Easterwood and Nutt [9] demonstrated that analysts underreact to negative information, but overreact to positive information, in both cases suggested optimism. Despite that, thousands of forecasts are created each year, similarly as recommendations.

© Springer Nature Switzerland AG 2019
Z. Wilimowska et al. (Eds.): ISAT 2018, AISC 854, pp. 71–80, 2019.
https://doi.org/10.1007/978-3-319-99993-7_7

Some researchers state that sophisticated investors do not take into account recommendations, because they are aware of the analysts' conflicts of interest who published them, therefore, investors attach more attention to forecasts [17, 18]. What is more, it has been empirically documented that analysts' opinions are reflected in stock prices [19, 22 and many others]. This suggests that, in an efficient capital market, analysts' forecasts and also recommendations play a significant role. Given the importance of these forecasts, errors can reduce informational efficiency within stock markets. Gu and Wu [10] argue that accuracy is one of the most important aspects of analysts' forecast performance, and forecast bias can be a result of analysts' efforts to improve forecast accuracy and part of forecast bias can be caused by excessive analysts' efforts to improve forecasts accuracy. Loh and Mian [15] proved that analysts who issue more accurate earnings forecasts also issue more profitable stock recommendations. This study was extended by Ertimur et al. [8], who also confirmed this result. Call et al. [3] evidenced that analysts' earnings forecasts are more accurate when they also issue cash flow forecasts. Earnings forecasts and cash flow forecasts are published together in Polish stock recommendations, in which there is a detailed valuation model – mostly discounted cash flow model (DCF).

None of the above-mentioned studies link analysts' bias in forecasting with time-dependency. Looking at the number of published recommendations on Polish stock market, it is possible to observe interesting regularity.

Over the years from 2005 to 2012, on the Polish stock market, has been published over three thousand free recommendations, it presents Table 1.

Table 1. Number of published recommendation and listed companies on Polish stock exchange in 2005–2012

Recommendations/Year	2005	2006	2007	2008	2009	2010	2011	2012	Total
Positive	127	154	171	318	306	197	231	146	1 650
With DCF model	62	92	125	153	188	102	159	83	62
Neutral	81	123	120	168	183	147	95	94	1 011
With DCF model	23	49	71	67	80	54	55	37	436
Negative	32	59	61	47	130	87	49	50	515
With DCF model	12	25	51	27	64	51	22	19	271
IPO recommendations	17	13	11	9	1	2	0	1	54
Total	257	349	363	542	620	433	375	291	3 230
With DCF model	97	166	247	247	332	207	236	139	1 671
No of companies on Polish Stock Exchange	255	284	351	374	379	400	426	438	

More than half of the reports recommended buying or accumulating assets. Negative recommendations – recommended reducing or selling - accounted for only 16%. It is consistent with finding observed from years in on other markets. Moreover, during the bear market[1] on Polish stock exchange, the number of published stock

---

[1] Years: 2005, 2008, 2009 and 2011.

recommendations was higher than during the bull market[2]. Furthermore, the number of published recommendations has fallen since 2009, one year after the largest fall in the stock market in the analysed period.

It is really surprising that the number of published recommendations has fallen, and at the same time the number of listed companies on the Polish stock exchange has increased over the years. It indicates trends in releasing of free recommendations, as well as detailed forecasts. Base on this, it can be expected that decrease of published recommendations may contribute to increasing the forecasts accuracy include therein.

That is why, this study examines the analysts' bias in forecasting financial parameters in the first year of detailed forecast, published in Polish over the calendar years 2005 through 2012, in two sub-periods – bull and bear market and years with lower[3] and higher economic growth[4].

For the purposes of this analysis, the term *forecasts* refers to the estimates of financial parameters included in brokerage recommendation such as: revenue, net profit, earnings before interest and taxes, depreciation, capital expenditure, net working capital and free cash flow provided by sell-side analysts. To explore the factors that affect an analysts' biases in forecasting, this paper examines two different properties of analyst forecasts in two time divisions. Forecast *accuracy*, is defined as an absolute value of the percentage difference between forecasted parameter and realised value of parameter. *Analysts' tendency* refers to the difference between forecast and realized value of parameters, and measures overestimation and underestimation of forecasts by analysts.

The paper is organized as follows. In section Research Sample is presented source of the date base and sample selection. In the Methodology section is defined the measure of forecast accuracy and analysts' tendency in forecasting taking into account analysed sub-periods. The section Results comprises main findings from researches and possible explanations of particular behaviour. Results are presented accordingly to the analysed parameters. Summary provides concluding remarks.

2 Research Sample

Taking into consideration all published Polish recommendations only recommendations with detailed DCF forecast are in the scope of this research; initial sample consists of 1671 recommendations. In order to provide a clear picture, from this sample were rejected those reports which: included DCF model with missing forecast parameters, were issued for companies for which financial statements were not available, issued for companies which financial statements with the fiscal year end were different than the forecasted year presented in the analysts' recommendations, issued for companies which have financial statements in the currency other than the currency used in the analysts' recommendations, recommendation released before public listing on the

---

[2] Years: 2006, 2007, 20010 and 2012.

[3] Years: 2005–2008.

[4] Years: 2009–2012.

Polish stock exchange. The final sample therefore is composed of 1593 recommendations issued from 2005 to 2012 by the following fourteen institutions: BM BGŻ, BM BPH, DI BRE, DM AmerBrokers, DM IDMSA, DM BDM, DM BOŚ, DM BPS, DM BZ WBK, DM Mercurius, DM Millennium, DM Noble Secure, DM PEKAO and DM PKO for 194 companies. All recommendations have been down-loaded from broker houses websites or financial portal www.bankier.pl.

Extreme observations were rejected in next step of the study, after calculated forecast errors, in order to minimise the potentially detrimental effects.

3   Methodology

Analysts' bias in forecasting parameters included in the stock recommendations, has been analysed in two divisions of time for the following parameters: revenue, net profit, earnings before interest and taxes (EBIT), depreciation, capital expenditure (CAPEX), net working capital (NWC) and free cash flow (FCF). First period is related with Polish stock index – WIG (Warsaw Stock Exchange Index). Research period has been divided into two subsamples: years in which WIG Index was lower than the median calculated based on the data pulled out of all analysed years – bear market (years: 2005, 2008, 2009, 2011) and years in which WIG Index was greater than the median calculated based on the data pulled out of all analysed years – bull market (years: 2006, 2007, 2010, 2012).

Second period is related with Polish GDP (gross domestic product). Research period has been divided into two subsamples: for years in which GDP was lower than the median of all analysed years – lower economic growth period (years: 2005–2008) and for years in which GDP was greater than the median of all analysed years – greater economic growth period (years: 2009–2012).

Table 2 presents values of WIG index and GDP for each analysed year and divisions for sub-periods.

Table 2. Division into research periods

Year	2005	2006	2007	2008	2009	2010	2011	2012	Median
WIG Index	35 601	50 412	55 649	27 229	39 986	47 490	37 595	47 461	43 723
Bull/bear - period	Bear	Bull	Bull	Bear	Bear	Bull	Bear	Bull	
GDP bn. PLN	983	1 060	1 177	1 276	1 345	1 417	1 528	1 596	1 310
Greater/lower economic growth	Lower	Lower	Lower	Lower	Greater	Greater	Greater	Greater	

Source: The Warsaw Stock Exchange, https://www.gpw.pl, Central Statistical Office, https://stat.gov.pl

To assess analysts' forecast accuracy between two different periods defined by WIG index and GDP absolute forecast errors were calculated for each parameters: revenue, net profit, EBIT, depreciation, CAPEX, NWC and FCF, according with formula:

$$AFE_{it} = |F_{it} - A_{it}| / |A_{it}|$$ (1)

where F_{it} is a parameter forecast for firm i in year t, A_{it} – is actual value of parameter for firm i in year t.

Forecast parameters were taken from the detailed forecast in stock recommendations and except revenues and net profit all of them are parameters from the DCF model. Although revenues and net profit are not directly included in the valuation model, they are closely related to the other parameters. The actual values of all parameters were taken from the financial statements of the companies for appropriate year of forecast. For each parameter for each firm, the forecast error was calculated only for the first year of detailed forecast included in the stock recommendation.

This measures allowed to check during what period of time the analysts' forecasts were less accurate.

In the next step in verifying accuracy between sub-periods, a parametric ANOVA test and non-parametric Mann-Whitney U test were used in order to indicate statistically signification in the average forecast errors between sub-periods.

To assess analysts' tendency between two different sub-periods defined by WIG index and GDP, test whether analysts had tendency to overestimate, underestimate or publish completely accurate forecasts (for each analysed parameter) during those sub-periods. As per below formula signed forecast errors have been calculated:

$$SFE_{it} = F_{it} - A_{it}$$ (2)

Where F_{it} and A_{it} as above. SF_{it} - positive value means overestimated forecast, negative - underestimated forecast, equal zero – accurate forecast.

Counting up the number overestimated, underestimated and equal zero forecasts allowed to calculate the percentage of overestimated, underestimated and equal zero forecasts in each sub-periods. In the next step, a contingency table approach were used and has been calculated the Chi-square test for independence for comparison of magnitude overestimated, underestimated between the groups.

4   Results

4.1   Forecast Accuracy

Table 3 below shows number of absolute forecast errors for each of the analysed parameters for bull as well as bear markets and for lower and greater economic growth periods. Panel A of Table 3 presents averages (mean and median) absolute forecast errors, standard deviations and statistics for two researched sub-periods – bull and bear market – for each analysed parameters. On average, absolute forecast errors indicate that analysts' accuracy in forecasting during bear market was worst that during bull market, in case almost each parameters except CAPEX and FCF. But statistically significant difference between analysts' accuracy in those sub-periods was only in case net profit and net working capital – both type of test confirm significance. In case FCF, during bull market, on average, analysts' accuracy in forecasting was higher than during bear market and, this results is statistically significant, at 10% level of significance only by the parametric test. Analysing sub-periods defined by the level of WIG Index, analyst's accuracy in forecasting is not strongly noticeable other than in relation to net profit and net working capital.

Panel B of Table 3 shows averages (mean and median) of the absolute forecast errors, standard deviations and statistics for two researched sub-periods – bull and bear market – for each analysed parameters. Both mean and median confirm that except deprecation and FCF, analysts had the tendency to make higher forecasts error during second half on the researched period i.e. in years with greater economic growth. This tendency has not been confirmed by parametric test for EBIT and net profit, but Mann-Whitney U test confirmed the statistical significance, which means that this analysts' bias in these two cases is not so strong. Interesting finding is that deprecation during sub-period with greater economic growth had better forecasts' accuracy and it is statistically significant at 1% level. Similar tendency concerns FCF, but the result is not statistically significant. Recommendations published in the later years are characterized by greater inaccuracy of analysts, although the number of published recommendations in this period was smaller than in the first half of the research period.

Additional finding, that links both time divisions is noteworthy, those parameters which should be forecasted firstly as revenue, EBIT have lower forecast error than more complex parameters as net working capital or free cash flow.

Table 3. Accuracy forecasts for analysed sub-periods for two division of time

Panel A: Absolute forecast errors for periods defined by WIG Index

$|(F - A)|/|A|$	N		Mean		Median		Std. dev		ANOVA	Test U M-W
Parameter	Bear	Bull	Bear	Bull	Bear	Bull	Bear	Bull	p-value	p-value
REVENUE	840	687	0,076	0,073	0,051	0,046	0,078	0,079	0,562 (0,4534)	1,334 (0,1823)
NET PROFIT	797	659	0,371	0,313	0,197	0,182	0,439	0,367	7,246 (0,0072***)	1,851 (0,0641*)
EBIT	808	666	0,298	0,284	0,164	0,170	0,354	0,322	0,621 (0,4307)	−0,304 (0,7612)
DEPRAC.	865	696	0,132	0,128	0,082	0,077	0,140	0,136	0,251 (0,6164)	0,144 (0,8851)
CAPEX	822	656	0,557	0,580	0,429	0,433	0,517	0,531	0,690 (0,4063)	−0,744 (0,4569)
NWC	784	649	1,181	1,059	0,997	0,919	0,848	0,807	7,690 (0,0056***)	3,267 (0,0011***)
FCF	716	598	1,013	1,098	0,824	0,875	0,861	0,938	2,879 (0,0900*)	−1,477 (0,1398)

Panel B: Absolute forecast errors for periods defined by GDP

$|(F - A)|/|A|$	N		Mean		Median		Std. dev		ANOVA	Test M-W U
Parameter	L	G	L	G	L	G	L	G	p-value	p-value
REVENUE	682	845	0,071	0,078	0,044	0,051	0,076	0,080	2,934 (0,0869*)	2,089 (0,0367**)
NET PROFIT	627	829	0,324	0,360	0,175	0,197	0,394	0,419	2,670 (0,1025)	1,868 (0,0618*)
EBIT	653	821	0,284	0,298	0,153	0,182	0,349	0,333	0,662 (0,416)	1,984 (0,0473**)
DEPRAC.	688	873	0,145	0,119	0,090	0,071	0,146	0,131	13,365 (0,0003***)	−3,572 (0,0003***)
CAPEX	659	819	0,539	0,590	0,418	0,452	0,490	0,547	3,403 (0,0653*)	1,139 (0,2545)
NWC	645	788	1,071	1,171	0,916	0,984	0,820	0,839	5,073 (0,0245**)	2,926 (0,0034***)
FCF	582	732	1,079	1,030	0,875	0,818	0,945	0,858	0,935 (0,3338)	−0,378 (0,7051)

*, **, *** indicate that the differences are significant respectively at 10, 5 and 1 percent levels of significance. Shortcut M-W U test means Mann-Whitney U test. Shortcut of sub-periods: L – lower economic growth, G – greater economic growth.

4.2 Analysts' Tendency

Panel A Table 4 presents magnitude of overestimated, underestimated and equal zero forecasts in each sub-period divided by the level WIG Index – bull and bear market - for each analysed parameters. Analysts had the tendency to underestimate forecasts during bear market and overestimate during bull market for most parameters. This tendency is statistically significant only for two parameters: revenues and net profits, in case other parameters this bias is not statistically significant. This finding suggests that during bear market analysts were not as optimistic as during bull market.

Panel B Table 4 presents magnitude of overestimated, underestimated and equal zero forecasts in each subperiod divided by the level of GDP – lower and greater economic growth - for each analysed parameters. In this case the time division into these sub-periods, analysts' bias is strongly noticeable. For revenue, net profit, EBIT, NWC and FCF analyst underestimated forecasts during greater economic growth and overestimated forecast during lover economic growth period. This evidence is

Table 4. Analysts' tendency – overestimated and underestimated forecasts

Parameter	Type of bias SFE	Panel A: Analysts' tendency in periods defined by WIG Index							Panel B: Analysts' tendency in periods defined by GDP						
		n			Percentage [%]			χ² test p-value	n			Percentage [%]			χ2 test p-value
		Bear	Bull	Total	Bear	Bull	Total	SFE < 0 vs SFE > 0	L	G	Total	L	G	Total	SFE < 0 vs SFE > 0
Revenue	SFE < 0	378	388	766	56	47	51	0,00036***	314	452	766	46	54	51	0,00308***
	SFE > 0	297	442	739	44	53	49		359	380	739	53	46	49	
	SFE = 0	3	2	5	0	0	0		3	2	5	0	0	0	
	Total	678	832	1510					676	834	1510				
Net profit	SFE < 0	344	378	722	53	48	50	0,07897*	273	49	722	44	55	50	0,00004***
	SFE > 0	309	409	718	47	52	50		349	369	718	56	45	50	
	SFE = 0	2	6	8	0	1	1		3	5	8	0	1	1	
	Total	655	793	1448					625	823	1448				
EBIT	SFE < 0	325	367	692	49	46	47	0,1734	288	404	692	44	50	47	0,05793*
	SFE > 0	329	429	758	50	53	52		353	405	758	54	50	52	
	SFE = 0	7	7	14	1	1	1		7	7	14	1	1	1	
	Total	661	803	1464					648	816	1464				
Depreciation	SFE < 0	359	441	800	52	52	52	0,97755	364	436	800	54	51	52	0,36342
	SFE > 0	293	361	654	43	43	43		282	372	654	42	43	43	
	SFE = 0	33	47	80	5	6	5		31	49	80	5	6	5	
	Total	685	849	1534					677	857	1534				
CAPEX	SFE < 0	240	284	524	37	35	35	0,5137	220	304	524	33	37	35	0,13401
	SFE > 0	417	530	947	63	64	64		436	511	947	66	62	64	
	SFE = 0	0	8	8	0	1	1		3	5	8	0	1	1	
	Total	657	822	1479					659	820	1479				
NWC	SFE < 0	311	393	704	46	49	48	0,4329	298	406	704	45	50	48	0,0646*
	SFE > 0	354	412	766	53	51	52		361	405	766	55	50	52	
	SFE = 0	4	1	5	1	0	0		3	2	5	0	0	0	
	Total	669	806	1475					662	813	1475				
FCF	SFE < 0	250	305	555	41	41	41	0,96521	216	339	555	36	45	41	0,00164***
	SFE > 0	355	431	786	58	58	58		374	412	786	63	55	58	
	SFE = 0	4	2	6	1	0	0		5	1	6	1	0	0	
	Total	609	738	1347					595	752	1347				

*, *** indicate that the differences are significant at 10 and 1 percent levels of significance, respectively. Shortcut of sub-periods: L – lower economic growth, G – greater economic growth.

significant at 1% level of significance, except EBIT and NWC wherein there is 10% level of significance. CAPEX also has the same result, but not statistically significant. In case depreciation, this tendency is reversed but not statistically significant. In division into publication in 2005–2008 and 2009–2012, there are more visible analyst's tendency to overestimation in the first period and underestimation in the second period.

Comparing the forecasts accuracy and analysts' tendency in the division defined by the WIG Index level, during bear market analysts published forecasts underestimated and thus their forecast were more inaccurate. During bear market analysts were more accurate in their forecasts and more optimistic, because they had the tendency to overestimate forested parameters.

Comparing the forecasts accuracy and analysts' tendency in the division defined by the GDP level, analysts' forecasts published in 2005–2008 were more accurate and included more analysts' optimism by overestimated forecasts. Forecasts published within 2009–2012 were more inaccurate but analysts had the tendency to underestimate forecasts. The above results are statistically significant only for individual parameters, mainly net profit.

Analysts' bias is more related to economic growth than stock market cycles. Nevertheless, it is to be noted that optimism in analysts' forecasts was related to the stock market cycle.

5 Summary

Examining parameters included in Polish brokerage recommendations over the calendar years 2005 through 2012, this paper provides the evidence that analysts in their forecasts were biased for some parameters in sub-periods.

Those biases were related to greater inaccuracy during the bear market than during the bull market – statistically significant results have only been observed for net profit and NWC. The bear market was characterised by underestimated forecasts and the bull market by overestimated parameters in forecasts – statistical significance difference is only for revenue and net profit.

While analysts' forecasts published during lower economic growth were characterised by more accuracy for almost each forecasted parameters. It is to be noted that these forecasts were mostly overestimated, in contrast to the worst accuracy and lower optimism in published forecasts during the higher level of economic growth period.

Analysts' biases in forecasting for periods defined by GDP are more visible and they are related with more parameters.

What is more, the parameters that are not required in the valuation model of company (DCF) as revenue and net profit, which serve as an information role in this case, they are the most dependent to analysts' bias than other parameters required in the valuation model. It proves that the parameters of the DCF model are quite stable forecasted, which is in contrast to the main categories of company earnings – revenues and net profit.

Additionally, it has been shown that the size of the analyst forecast error increases along with the complexity of the estimated parameter. It seems to be logical, if analysts keep the order of forecasting the next parameters.

How write Gu and Wu [10] analysts are biased even when they are rational forecasters, which truthfully and unselectively prepare their forecasts. This bias can be considered in many aspects, as well as an impact of this analysts' bias on markets.

References

1. Barber, B., Lehavy, R., Trueman, B.: Comparing the stock recommendation performance of investment banks and independent research firms. J. Financ. Econ. **85**, 490–517 (2007)
2. Bradshaw, M., Huang, A., Tan, H.: Analyst target price optimism around the world. Midwest Finance Association 2013 Annual Meeting Paper (2013)
3. Call, A., Chen, S., Tong, Y.: Are analysts' earnings forecasts more accurate when accompanied by cash flow forecasts? Rev. Acc. Stud. **14**(2), 358–391 (2009)
4. Carleton, W., Chen, C., Steiner, T.: Optimism biases among brokerage and non-brokerage firms' equity recommendation agency cost in the investment industry. Financ. Manag. Assoc. Int. **27**(1), 17–30 (1998)
5. Central Statistical Office. https://stat.gov.pl. Accessed 15 Mar 2018
6. Cliff, M.: Do independent analysts provide superior stock recommendations? Working Paper, Virginia Tech (2004)
7. Das, S., Saudagaran, S.: Accuracy, bias, and dispersion in analysts' earnings forecasts: the case of cross-listed foreign firms. J. Int. Financ. Manag. Account. **9**(1), 16–33 (1998)
8. Ertimur, Y., Sunder, J., Sunder, S.: Measure for measure: the relation between forecast accuracy and recommendation profitability of analysts. J. Account. Res. **45**(3), 567–606 (2007)
9. Esterwood, J., Nutt, S.: Inefficiency in analysts' earnings forecasts: systematic misreaction or systematic optimism? J. Financ. **54**(5), 1777–1979 (1999)
10. Gu, Z., Wu, J.: Earnings skewness and analyst forecast bias. J. Account. Econ. **35**, 5–29 (2003)
11. Hugon, A., Musulu, V.: Market demand for conservative analysts. J. Account. Econ. **50**, 42–57 (2010)
12. Hutira, S.: Determinants of analyst forecasting accuracy. Joseph Wharton Research Scholars. http://repository.upenn.edu/joseph_wharton_scholars/15. Accessed 26 Apr 2018
13. Jegadeesh, N., Kim, J., Krische, S., Lee, C.: Analyzing the analysts: when do recommendations add value? J. Financ. **59**, 1083–1124 (2004)
14. Kowalski, M., Praznikow, J.: Investment recommendation optimism - results of empirical research on Polish capital market. In Proceedings of 37th International Conference on Information Systems Architecture and Technology, ISAT 2016, Part IV. Springer, Cham (2017)
15. Loh, R., Mian, G.: Do accurate earnings forecasts facilitate superior investment recommendations? J. Financ. Econ. **80**, 455–483 (2006)
16. Luo, T., Xie, W.: Individual differences and analyst forecast accuracy. Rev. Account. Financ. **11**(3), 257–278 (2012)
17. Malmendier, U., Shanthikumar, D.: Do security analysts speak in two tongues? Rev. Financ. Stud. **27**(5), 1287–1322 (2014)
18. Mikhail, M., Walther, B., Willis, R.: When security analysts talk who listens? Working paper, Northwestern University (2006)

19. Moshirian, F., Ng, D., Wu, E.: The value of stock analysts' recommendations: evidence from emerging markets. Int. Rev. Financ. Anal. **18**, 74–83 (2009)
20. Schipper, K.: Commentary on analysts' forecasts. Account. Horiz. **3**, 105–121 (1991)
21. The Warsaw Stock Exchange. https://www.gpw.pl. Accessed 4 Feb 2018
22. Womack, K.: Do brokerage analysts' recommendations have investment value? J. Financ. **51**, 137–167 (1996)

Generalized Model of Regularity of Cash Flow Generation by Innovative Products Generations in Digital Economy

Vladimir F. Minakov[ID] and Oleg S. Lobanov[✉][ID]

St. Petersburg State University of Economics,
21, Sadovaya Street, 191023 St. Petersburg, Russian Federation
thelobanoff@gmail.com

Abstract. This article verifies a mathematical model that generalizes the processes of development and commercialization of innovations, representing a modernized product, with such essential distinctive features that its entry to the market leads to a devaluation of consumer properties and the value of the previous one. Such innovations are essentially a new generation of a product that is consistently in demand on the market. In this paper, we solve the problem of developing an economic and mathematical model for the dissemination of innovations in high-tech sectors of the economy, characterized by the periodicity of bringing products to the market that represent a new generation with significantly higher consumer qualities, and generating cash flow as recyclable products and again commercialized. The model of this innovation process is typical for high-tech sectors of the economy: engineering, automotive, computer technology, software and a number of others that are key to the digital economy. Therefore, the generalized mathematical model makes it possible to increase the validity of management decisions in the management systems of the most dynamically developing innovations of high-tech industries in the derivation and commercialization of new generations of products.

Keywords: Innovation process · Generalized mathematical model
Replacement of innovations · Generation of innovative products

1 Introduction

The gross domestic product of the Russian economy (GDP), the state budget, the exchange rate of the national currency, the competitiveness of export products now depend significantly on the conjuncture of prices for hydrocarbons in world markets. In fact, the standard of living of most of the country's population is correlated with the prices of oil and gas on world and national exchanges. Given the cyclical nature of energy prices, it is easy to predict the inevitability of their cyclical decline, and, consequently, a synchronous decline in both the country's GDP and budget, as well as the standard of living. It is natural that in such a situation it is necessary to increase the degree of processing of the produced products, and accordingly, the added value, as a factor of economic development.

© Springer Nature Switzerland AG 2019
Z. Wilimowska et al. (Eds.): ISAT 2018, AISC 854, pp. 81–90, 2019.
https://doi.org/10.1007/978-3-319-99993-7_8

It is obvious that the cyclical appearance of super profits from the export of oil and gas at high prices allows investing in the development of processing and high-tech industries, innovative industries [6, 7]. The presumption of innovative development is a concept that has been realized to date, which is the basis of the strategy for the development of the information society and the digital economy. The possibility of stimulating advanced innovative development in the Russian economy is already finding practical implementation. In the last decade, a rather large number of venture and innovation funds, state corporations, supporting innovative projects have been created [9]. As a result of such institutional dynamics, there is an unprecedented increase in investment in innovation. Levels of annual spending only on fundamental and applied research in Russia have reached 7 billion euros. The budgets of the funds that support innovative activity have grown hundreds of times over the past 13 years [1].

However, to date, the economic effect of innovation has remained low [10]. With annual financing of innovative projects in Russia in 1.3 trillion. rub. (of which 950 billion rubles are financing from the state budget), the volume of innovative products is calculated in units of percent [4]. At the same time, the elasticity of innovative products at costs does not exceed 0.001. Comparison of this indicator with the results of investment in innovation by the best campaigns of the high-tech sector shows that the potential for investing in innovation, on the contrary, is much higher than investment in traditional sectors of the economy and the elasticity of innovative products reaches up to 9 [11].

An actual problem, therefore, is the improvement of methods for modeling the distribution of high-tech innovative products and its commercialization. Considering the decisive importance of these processes for the digital economy, solving the problem of modeling, it is important to take into account the peculiarities of innovations in the field of information and communication technologies, consisting in developing and launching new generations of products that supplant the foregoing.

In this paper, we solve the problem of developing an economic and mathematical model for the dissemination of innovations in high-tech sectors of the economy, characterized by the periodicity of bringing products to the market that represent a new generation with significantly higher consumer qualities, and generating cash flow as recyclable products and again commercialized.

2   Theory, Materials and Methods

It is important to note that the life cycle of innovation processes include the following consistently implemented stages [2]: fundamental research; research and development; primary development; widespread adoption (distribution of innovation); operation (use); modernization; leaving the market (or disposal). However, only these stages are connected with the generation of cash flows: entering the stage of widespread adoption (distribution of innovation), operation (use), modernization of innovations (in innovative products, where it is possible, for example, software of computer systems), and leaving the market. These stages are responsible for success of innovative projects commercialization. Hence follows the importance of clarifying the mathematical description of the listed stages of innovation processes.

The dynamics of innovation spreading into the consumer environment is inherently diffusional in nature (diffusion of innovation) [3]. The essence of diffusion of innovation consists in spreading into the environment of innovative products consumption because of unsaturation of the product in this area, up to a saturated state. Innovation spreading in an unsaturated environment is caused by loss of leading or competitive positions by entities that do not have innovative products in comparison with innovative companies, and, accordingly, a natural desire to gain such positions. The volume of diffusing products during the time dt (diffusion processes) is proportional to the current number of innovations already used in the market (their users gain advantages over other entities) and the share of free market unsatisfied product:

$$dV = r \cdot V(1 - V/V_m) \cdot dt \tag{1}$$

where V is the current volume of innovations used in the environment;

V_m is the level of maximum volume of innovation spread (for example, sales) in the environment (when it is saturated with an innovative product);

r is the rate of diffusion of innovation (the diffusion coefficient of innovation, is determined by the method of least squares based on empirical data: for high-tech innovations related to information technology $r = 0.08$ 1/day).

The solution of Eq. (1) is the logistic function (sigmoid):

$$V = V_m \cdot \left(1 + e^{p-r \cdot t}\right)^{-1}, \tag{2}$$

where p is the indicator of delay in the beginning of innovation diffusion relative to the beginning of the reference frame (the lag of the innovation's exit to the market in the calculated coordinate system);

e is the base of the natural logarithm.

It should be noted that the differential Eq. (1) and its solution – sigmoid (2) – characterize the spread of innovations precisely because of the advantages that innovative products possess. This can be improved performance, higher cost-effectiveness, enhanced functionality, improved quality, new properties, etc.

Additional factors of promoting innovative products like marketing costs, creating tax preferences, government regulation and incentives can be taken into account in Eq. (1) by additional components. They provide additional advancement of innovative products in the markets in larger volumes, often with greater speed. In these cases, the addition of the right-hand side of (1) by the parameters that quantitatively express the factors stimulating the innovation processes is quite justified. In these cases, it is advisable to use models of innovations spread of a more general type: Mansfield, Floyd, and others. However, both the quantitative indicators of such factors and the results of their impact can always be singled out in separate equations, which makes it possible to assess their effectiveness. At the same time, the processes of innovation spread into the consumption environment described by Eqs. (1) and (2) remain the basic factor of innovation spreading, due to the needs of this environment in innovations (their useful properties and effects).

In addition, in some cases, due to insufficient information, for example, on the capacity of the market for innovative product implementation (V_m), there is no need for

clarification (1) and (2). On the contrary, there is a need for simplifications of mathematical description of innovations diffusion, as it was done by the authors of a number
of models (Perla). Moreover, simple approximating curves (for example, the Gompertz
curve) are sometimes used, and descriptions of the factors influencing the processes of
innovation spread are omitted.

In the phases of innovation spread among consumers in the market, the solution (2),
presented in Fig. 1 of curve 1, agrees well with the results obtained by Rogers [12, 14] –
curve 2.

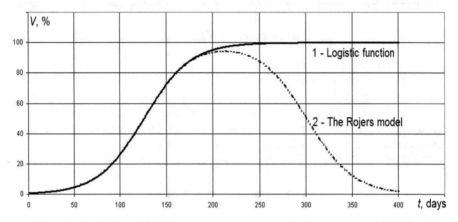

Fig. 1. Models of the phase of commercialization of innovations

However, sigmoid does not describe the phase of innovative products exit from the
market. So we have to add these factors in developed model in order to achieve higher
correlation with real market conditions.

3   Model and Results of Its Applying

The author's approach to modeling the phases of decline in volume of innovation (for
example, sales of an innovative product) consists of a mathematical description of
leaving a product from the market by a cascade of sigmoids that reflect the processes of
innovation spread noted by Sahal [13], with replacement of innovations by its next
generation, which has higher consumer qualities than the previous ones.

Indeed, despite the widespread use of video recorders and recording players of
analog formats for recording and reproducing video images (innovations of the late
twentieth century – sigmoid curve 1 in Fig. 2), the emergence of such an innovative
product as digital video recording media and formats, CD-ROM drives, that was followed by a DVD ROM (curve 2 in Fig. 2), led to the replacement of the first product by
the second one. Obviously, the sale of each DVD player replaces and displaces analogue media. Consequently, the sigmoid of distribution of digital video recording and
playback systems should be subtracted from the distribution function of analog ones.

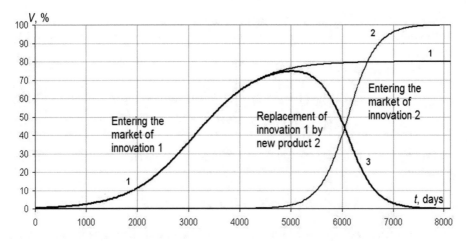

Fig. 2. Replacement the previous innovation by subsequent innovation

So the proposed model of diffusion of innovation and replacement of its new innovation has the form:

$$V_1 = V_{m1} \cdot \left(1 + e^{p_1 - r_1 t}\right)^{-1} - k_2 \cdot V_{m2} \cdot \left(1 + e^{p_2 - r_2 t}\right)^{-1}, \tag{3}$$

where V_{m1}, V_{m2} are the volumes of market saturation with previous and subsequent innovations (for example, the maximum sales volume);

k_2 is the share of subsequent innovative product in replacing the previous one (displaced).

If the innovation is replaced by several subsequent products with saturation volumes V_{m2}, V_{m3}, ..., V_{mn}, then the innovation diffusion model acquires a generalized form:

$$
\begin{aligned}
V_1 =\ & V_{m1} \cdot \left(1 + e^{p_1 - r_1 t}\right)^{-1} - k_2 \cdot V_{m2} \cdot \left(1 + e^{p_2 - r_2 t}\right)^{-1} \\
& - k_3 \cdot V_{m3} \cdot \left(1 + e^{p_3 - r_3 t}\right)^{-1} - \cdots - k_n \cdot V_{mn} \cdot \left(1 + e^{p_n - r_n t}\right)^{-1}.
\end{aligned}
\tag{4}
$$

If innovation is an improvement, then the previous product is not replaced by the next one, but, on the contrary, it is supplemented with products with saturation volumes V_{m2}, V_{m3}, ..., V_{mn}. The diffusion model of the innovation cascade acquires the following form:

$$
\begin{aligned}
V_n =\ & V_{m1} \cdot \left(1 + e^{p_1 - r_1 t}\right)^{-1} + k_2 \cdot V_{m2} \cdot \left(1 + e^{p_2 - r_2 t}\right)^{-1} + k_3 \cdot V_{m3} \\
& \cdot \left(1 + e^{p_3 - r_3 t}\right)^{-1} + \cdots + k_n \cdot V_{mn} \cdot \left(1 + e^{p_n - r_n t}\right)^{-1}.
\end{aligned}
\tag{5}
$$

In order to calculate cash flows generated by an innovative product when it enters the market, and to take into account the discounting of non-recurrent costs and incomes, it is required to know the current volumes of innovation that generate income, for example, daily, weekly, etc. To determine them, we differentiate the sigmoid (2):

$$
\begin{aligned}
dV/dt &= r \cdot V_m \cdot (1 + e^{p-r \cdot t})^{-1} \left[1 - V_m \cdot (1 + e^{p-r \cdot t})^{-1} \cdot V^{-1} \right] \\
&= r \cdot V_m \cdot (1 + e^{p-r \cdot t})^{-1} \left[(1 + e^{p-r \cdot t} - 1) \cdot (1 + e^{p-r \cdot t})^{-1} \right] \\
&= r \cdot V_m \cdot e^{p-r \cdot t} \cdot (1 + e^{p-r \cdot t})^{-2}.
\end{aligned}
\tag{6}
$$

During the accounting period T the volume of innovation spread will be the following:

$$
V_T = T \cdot dV/dt = T \cdot r \cdot V_m \cdot e^{p-r \cdot t} \cdot (1 + e^{p-r \cdot t})^{-2}.
\tag{7}
$$

In accordance with (7), sales cause the generation of cash flows. At the price of an innovative product P, the income from the sale of an innovative product for the accounting period T will be the following:

$$
C_T = P \cdot T \cdot r \cdot V_m \cdot e^{p-r \cdot t} \cdot (1 + e^{p-r \cdot t})^{-2}.
\tag{8}
$$

Then the discounted cash flow generated by the innovation, for N time intervals T (at a discount rate of d) will be:

$$
C_\Sigma = \sum_{j=1}^{N} \frac{P \cdot T \cdot r \cdot V_m \cdot e^{p-r \cdot t_j}}{(1 + e^{p-r \cdot t_j})^2 (1 + d)^j},
\tag{9}
$$

where j is the number of accounting time interval.

4   Discussion

It was said that in video recording systems, analogue devices are being squeezed from the market by digital ones. The same is also true in production of sound recording and sound reproduction devices (replacement of tape and cassette tape recorders by digital recorders). Even faster was replacement of analog mobile phones and mobile communication systems with digital gadgets, which quickly captured the market through convergence with digital sound recording and reproducing devices, cameras, video recording and playback devices, and others. Graphic interpretation of the proposed model of innovation diffusion and its displacement in subsequent by new innovative solution is the classical curve of innovation spread and its exit from the market (curve 3 in Fig. 2, where, in comparison, 1 shows the spread of innovation 1, 2 shows the spread of innovation 2 that displaced the previous one).

The result obtained from (3) (Fig. 2) is in good agreement with the actual data. The second important advantage of the economic-mathematical model of innovation cascade diffusion (succession of substitutions the previous innovative products with new ones) is the possibility of presenting long-run innovative products. These include an innovative product of the beginning of the twentieth century, such as an incandescent lamp, whose sales period at the saturation level lasted about 100 years. And only after the emergence of several innovative solutions, namely low-pressure fluorescent lamps, then high-pressure fluorescent lamps, supplementing them by start-up systems with socles compatible with commonly used lamps, LED lamps, which have unquestionable energy-saving properties, the replacement of incandescent lamps took place. This process was further stimulated by government regulatory actions (the law on energy saving). For the commercial phase of the life cycle of incandescent lamps, model (4) allows us to obtain an adequate result that is shown on Fig. 3.

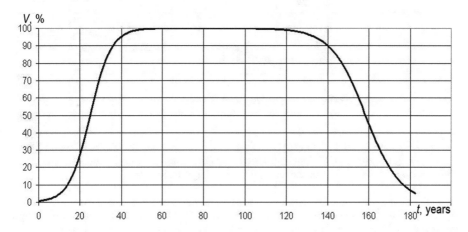

Fig. 3. Replacement the previous innovation by subsequent innovation

A feature of innovations of this type is the excess of the duration of the phase of successful operation over the phase of innovation spread.

To illustrate the adequacy of innovation cascade model Fig. 4 presents the actual data [15] of innovative product spread, such as a landline phone, and simulation results. This figure illustrates well the adequacy of proposed model, especially at the time intervals of bends in the phone's spread curve.

The results shown in Fig. 4 in the form of an economic-mathematical model that is a series of sigmoid $V_T = 530 \cdot e^{12.7 - 0.77t} + 177 \cdot e^{4.1 - 0.635t}$, validated to the economic processes of distribution of innovative products, namely, communication equipment: the first sigmoid – in the analog data transfer format, the second sigmoid – through digital channels and voice communication protocols. This allows telecommunication companies to determine the moment of market saturation with available innovation and to make a timely decision on need to bring a new product to the market and, accordingly, to invest in innovative processes of creating such a product to create an advantage over competing companies. So did the companies that mastered the technology of providing digital content and, accordingly, data transmission services in digital format. Thus, verification the validity of developed economic-mathematical model during the analysis

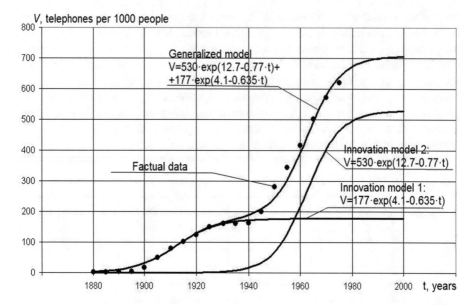

Fig. 4. Dynamics of fixed telephones spreading: actual data and simulation results (innovation 1 is initial design, innovation 2 is modernization of telephones and telephone stations in connection with the replacement of electromechanical devices with electronic ones)

of real market data related to information and telecommunication spheres showed high accuracy, sufficient for the justification and adoption of management decisions by companies in this industry.

5   Conclusion

In this research a generalized model (9) of cash flow generation is obtained, taking into account its discounting in the phases of market use of the cascade of innovations generations. Such a model allows us to adequately assess the effect of the commercialization of innovations, which is important, for example, for the selection of the fund for promoting the development of small forms of enterprises in the scientific and technical sphere. It is essential that the generalized model takes into account the key innovations of the digital economy, in the field of information and communication technologies, which consist of the development and launch of new generations of products that displace the previous ones.

Thus, the economic-mathematical models (3–5, 7 and 9), in comparison with the known ones, describe not only the growth phase in innovations spread, but also the phase of leaving the market. In addition, these models are more adequate in the asymmetric phases of the growth of innovations spread and their displacement from commercial use by representing them in a sequence of innovations with their own time parameters and rates of substitution [5]. Models describe the most important effect of innovation activity which is the cash flow generated by them. Discounting the cash flow in the model and bringing it to the point of time when an innovative project is selected for investment makes it possible to take into account the timing of the implementation of the projects being compared. The use of this model, firstly, allows us to determine the limits of growth of each innovative product, and, consequently, the choice of the time period when a change in strategy is needed in the form of a transition from the "cream-skimming" phase of commercial innovation product to the phase of investment in the innovative product modernization [8]. In this way, the developed model is expedient for use in decision-making process concerning strategic issues of development of companies in high-tech spheres of economy.

Acknowledgements. The study has been developed within the framework of research projects implementation funded by the St. Petersburg State Economic University.

References

1. Dynkin, A.V., Ivanova, N.I.: Innovative Economy. Science, Moscow (2004)
2. Dyatlov, S.A., Lobanov, O.S.: NBIC convergence as a stage of transition of Saint-Petersburg's e-government information space to the sixth techno-economic paradigm. Commun. Comput. Inf. Sci. **745**, 347–361 (2017). https://doi.org/10.1007/978-3-319-69784-0_30
3. Dyatlov, S.A., Lobanov, O.S., Selischeva, T.A.: Information space convergence as a new stage of e-governance development in Eurasian economic space. In: ACM International Conference Proceeding Series, Part F130282, pp. 99–106 (2017). https://doi.org/10.1145/3129757.3129775
4. Economy of Russia: In: Wikipedia. https://en.wikipedia.org/wiki/Economy_of_Russia. Accessed 29 Dec 2017
5. Martino, J.: Technological prediction. Progress, Moscow (1977)
6. Minakov, V.F., Barabanova, M.I., Lobanov, O.S., Schugoreva, V.A.: The rate of geometric progression of evolutionary innovations diffusion. In: Modern Economics: Problems and Solutions, vol. 3, pp. 20–28 (2016). http://dx.doi.org/10.17308/meps.2016.3/1406
7. Minakov, V.F., Lobanov, O.S., Makarchuk, T.A., Minakova, T.E., Leonova, N.M.: Dynamic management model of innovations generations. In: Proceedings of 2017 XX IEEE International Conference on Soft Computing and Measurements (SCM), pp. 849–852 (2017). https://doi.org/10.1109/scm.2017.7970743
8. Minakov, V.F., Lobanov, O.S., Minakova, T.E., Makarchuk, T.A., Kostin, V.N.: The law of diminishing marginal productivity in the model of pure discounted income of innovations. Int. J. Econ. Res. **14**(14), 435–441 (2017)
9. Minakov, V.F., Minakova, T.E.: Remote monitoring systems for quality management metal pouring. In: Evgrafov, A. (ed.) Advances in Mechanical Engineering. Lecture Notes in Mechanical Engineering, vol. 22, pp. 63–71. Springer, Cham (2015). https://doi.org/10.1007/978-3-319-15684-2_9
10. Minakov, V.F., Sotavov, A.K., Artemyev, A.V.: Model of integration of analog and discrete indicators of innovative projects. In: Scientific and Technical Sheets StPSPU, Series "Economic Sciences", vol. 6, no. 112, pp. 177–186 (2010)
11. Minakov, V.F., Galstyan, A.Sh., Piterskaya, L., Radchenko, M., Shiyanova, A.A.: Innovative investment trends in the Northern Caucasus. Central Asia Caucasus **17**(1), 61–70 (2016). https://doi.org/10.5901/mjss.2015.v6n3s6p307

12. Rogers, E.: Diffusion of Innovation, 5th edn. Free Press, New York (2003)
13. Sahal, D.: Technical progress: concepts, models, assessments (translation). In: Finance and Statystics, Moscow (1985)
14. Skiba, A.N.: Demand on radical grocery innovations: structurally functional features and dynamics. Econ. Anal. Theory Pract. **26**(191), 43–51 (2010)
15. Twiss, B.: Forecasting for technologists and engineers. In: Practical Guidance for Making Better Decisions (translation), Parsec-NN, N. Novgorod (2000)

Operational and Information Risk Management System Based on Assessment Model of Risk Culture Level

Vlada A. Schugoreva(iD), Oleg S. Lobanov$^{(\boxtimes)}$ (iD),
and Vladimir F. Minakov(iD)

St. Petersburg State University of Economics,
21, Sadovaya Street, 191023 St. Petersburg, Russian Federation
thelobanoff@gmail.com

Abstract. The article describes research in the field of approaches to calculation of risk culture in banking organizations in terms of operational and information risk management system. It propose a decision-making system based on various tactics, which based on strengthening zones of influence on the risk culture of organizations. Proposed classifications on the model used in one of the biggest commercial bank in Russian Federation with a separate structure of risk compliance and information risk. In framework of this study, ways and approaches to optimize the operational risk management process were defined, a model for parametrizing and assessing the level of risk culture was developed, and the relationship between risk culture and bank damage as the main driver of operational risk management system was explored. The main goal of introducing a risk culture is to develop a company's behavior that determines collective ability to identify, analyze, openly discuss and respond to organization's existing and future risks to ensure long-term and sustainable reduction in risk-related losses. The result of implementation of the developed model is the optimization of the bank's policy in the field of standardization, HR and diversification of banking risks.

Keywords: Operational risk · Information risk · Banks · Risk-management
Assessment model

1 Introduction

Currently, both in Russia and around the world, taking into account strong economic shocks and crises, risks and risk management are in the constant focus of central banks, banking regulators and supervisors of virtually all countries in the world. Legislation is changing, regulators are raising requirements both for the level of capitalization of financial institutions and for the quality of risk management systems. Banks must quickly adapt to new business conditions in the face of stringent regulatory constraints [8].

In modern conditions of digital economy for the banking industry, there is a situation when a stable system of risk management is not only a reputational bonus and strict compliance with requirements of the Central Bank of Russia in order to avoid revocation

© Springer Nature Switzerland AG 2019
Z. Wilimowska et al. (Eds.): ISAT 2018, AISC 854, pp. 91–100, 2019.
https://doi.org/10.1007/978-3-319-99993-7_9

of the license, but also the necessary system of a multilevel protection line against various threats, arising not only in credit but also in operating activities [9].

One of the main tasks for banks is the effective risk management inherent in all types of activities that must be continuously implemented at all levels of the business model and is one of the key factors for the success of the organization [11, 12].

In the framework of the study, ways and approaches to optimize the operational risk management process were defined, a model for parametrizing and assessing the level of risk culture was developed, and the relationship between risk culture and bank damage as the main driver of the operational risk management system was explored [13].

The main goal of introducing a risk culture is to develop a company's behavior that determines the collective ability to identify, analyze, openly discuss and respond to the organization's existing and future risks to ensure long-term and sustained reduction in losses from risk realization [14].

2 Theory, Materials and Methods

Within the framework of the research, it was assumed that before understanding at what level the risk management in the organization is built and what can be done to make the system more optimal and effective, it is necessary to assess, preferably parametrically, at what specific level the so-called risk culture of the organization. After all, the word "culture" was chosen not accidentally - it is the culture of the bank employees, starting from the highest to the lowest levels, it is possible to provide natural protection to prevent risks, when not only the systems or specific manager participate in the defense, but every employee in his workplace provides a risk protection within his competence and functionality [14].

In order to effectively manage the risk system, it is proposed to use a method from general to specific. I.e., if we consider the issue fundamentally, then banking risk needs to be clearly classified and divided into more specific grouping species by different sources of classification in order to determine various theoretical and practical approaches to managing different risks.

Evolution of the classification of types of banking risks can be observed both in the works of Russian and international researchers, and at the state and regulatory levels. The main requirements of regulators in the 80–90's of XX century on the accounting for risk in the capital of the bank only related to credit risk, while in the late 90's. already the main risks were recognized not only credit, but also market and operational [1].

At the modern level, taking into account all the experience of risk management research in recent decades, the Bank of Russia gives quite a complete classification [2]. Here the risks are primarily allocated to groups consisting of financial and non-financial risks.

In this context, banking risks can be classified according to different criteria as it is shown on Fig. 1, depending on the purposes of risk grouping and further actions for their management.

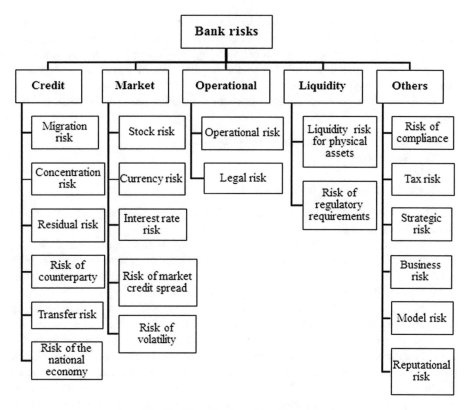

Fig. 1. Classification of banking risks

Banking risks can be classified according to the following criteria [8]:

- degree of uncertainty;
- sources of origin;
- level of materiality (potential consequences, damage);
- probability of occurrence.

From the point of view of the uncertainty degree, risks can be identified with *partial* and *total* uncertainty. Risks with partial uncertainty the bank consciously accepts, clearly evaluates (as it is able to determine the main characteristics of such risks) and pawns in the risk-premium. Risks with total uncertainty can not be standardized and effectively assessed from a quantitative point of view.

By sources of occurrence, risks can be divided into external and internal.

External risks are risks not depending on the bank. The emergence of external risk can be caused by local and global socio-economic trends, changes in the political regime, technological and technological progress, etc. Bank losses can also arise due to force majeure – military conflicts, natural disasters (earthquakes, floods and fires), nationalization, etc. An example of external risk for a bank is the credit risk, determined in large part by the dynamics of the counterparty's activity.

Internal risk is a risk that the bank largely controls and is able to effectively limit. Internal risk can be associated with unqualified financial management, excessive adherence to risky (aggressive) operations with high profit margins, deterioration of the bank's business reputation, violation of the requirements of legislation and internal standards of activity, and so on.

By the level of potential consequences for the bank's profit and capital, the risk may be:

- moderate (risk is below average, acceptable risk);
- increased (average risk);
- high;
- critical (very high risk, unacceptable risk).

Also, risks depend on the characteristic of *probability* (frequency) of realization - small, medium or high.

This classification of risks is offered directly by banks, developing their own internal regulatory documents and standards [10]. Considering primarily large banks in which there is a high probability of realizing a large number of different risks, let us detail the classification of one of the largest commercial banks in Russia.

It can be concluded that the main feature of this classification is the materiality regarding the occurrence of possible damage, as well as the requirements for accounting in the bank capital adequacy ratio [6], which, taking into account the current requirements of the banks, is the most important criterion for the formation of an effective banking management system risks. At the same time, it should be noted that this classification was developed in 2014 and was the most complete (exhaustive) of the proposed and studied classifications, covering the full range of possible risks at the time of development.

If we consider in more detail the system of proposed risks, then the identified risks can be reduced to the five following enlarged categories.

Credit risk is the risk of loss arising from failure to fulfill, untimely or incomplete performance of financial obligations by the debtor in accordance with the terms of the contract.

When lending foreign counterparts, there may also be a *risk of the national economy* and *transfer risk*.

Market risk is the risk of loss due to adverse changes in the market value of financial instruments, as well as foreign exchange rates and (or) precious metals.

Operational risk is the risk of loss arising from the unreliability of internal management procedures, employee bad faith, failure of information systems or the impact of external events.

Liquidity risks are the risks manifested in the inability of banks to finance their activities, that is, to ensure the growth of assets and fulfill obligations without incurring losses in amounts that are unacceptable for financial sustainability.

Other risks are the risks of losses that do not fall into the categories of credit, market, operational and liquidity risk. Other risks include such types of risks as strategic risk, business risk, compliance risk, tax risk, regulatory risk, and others.

Studies based primarily on profit and loss data [5] showed that classification used in fact covers all the significant risks that arise for the global banking sector. At the same time, such a classification today requires certain clarifications and restructuring in terms of enlargement and the importance of a number of risks [7].

Thus, the sectoral and blocking sanctions established by the OFAC[1] and EU countries in 2014 for Russian enterprises and financial organizations [4], and at the same time development of new types and methods of illegal banking transactions, require to release of compliance risk into a separate group of risks that is divided by the sanction risk, information risk, risk of non-compliance with the requirement to counteract legalization (laundering) of proceeds from crime and financing of terrorism (AML/CFT) and other compliance risks, which also include risks associated with corruption, with disclosure of insider information and business ethics (Fig. 2).

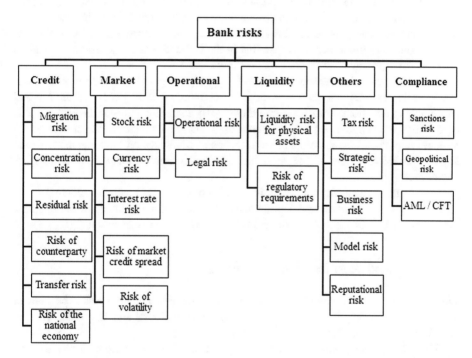

Fig. 2. Classification of banking risks with new elements

Sanction risk can be defined as the risk associated with activities of enterprises and financial organizations, as well as citizens for whom the US, EU and several other countries have imposed economic sanctions [8]. It is worth noting that the sanction risk is currently a risk with significant substantial damage, for violation of sanctions, there are significant fines that have been experienced by many European and Russian banks.

---

[1] OFAC – The Office of Foreign Assets Control (OFAC) of the US Department of Treasury administers and enforces economic and trade sanctions based on US foreign policy and national security goals against targeted foreign countries and regimes, terrorists, international narcotics traffickers, those engaged in activities related to the proliferation of weapons of mass destruction, and other threats to the national security, foreign policy or economy of the United States.

According to Internet sources [9] in 2009, the French bank Credit Agricole paid the US authorities $ 787 million in the settlement of the case of violation of sanctions against Iran, Cuba and other states, in 2012 HSBC paid $ 1.9 billion for settling claims from the side US Department of Justice, which found that HSBC violated the US sanctions against Iran, in 2014 the French bank BNP Paribas was found guilty of violating the US sanctions regime against Sudan, Cuba and Iran and paid a $ 9 billion penalty to the US", also in 2014 The Office of Financial Assets Control of the US Treasury Department (OFAC) fined the Bank of Moscow for $ 9.4 million for violation of sanctions against Iran. Since 2009, more than ten banks, mostly European, have paid the US authorities a total of $ 14 billion for violations of sanctions requirements. Investigations about violations of the sanctions regime were also conducted against the German Deutsche Bank, the French Societe Generale and the Italian UniCredit.

The above examples of AML/CFT risks and sanctions risks, suggest that commercial banks are implementing a risk that in the last decade causes serious damage to banks up to withdrawal of licenses or bankruptcies.

Thus, now there is a reason to believe that, given the growing information tensions in the world, a significant expansion of list of countries with respect to which sectoral and blocking sanctions have been introduced [10], including Russian Federation; this type of risk as compliance takes an increasingly significant share in the structure of a possible major bank loss, comparable to the generally accepted banking risks, which requires the development of a separate management system for this type of risk and its prevention.

One of the main tasks for banks is the effective risk management inherent in all types of activities that must be continuously implemented at all levels of the business model and is one of the key factors for the success of the organization [3].

3   Model and Results of Its Applying

The main task was to calculate the ability to anticipate potential risks, be able to understand the level of information about the risks realized, and the availability of built-in programs that are configured to minimize risks.

Thus, the primary model for calculating the level of risk culture is the derived normalized function of the above parameters – *Rcult*.

As a result of the conducted research and the proposed model for calculating an integrated assessment for identifying risks of banking processes, results have been obtained that enable the new and principally key parameter (Z) – $Rcult = F$ (SA, Inf, Act, Z), which gave a more accurate look at the calculations and confirmed the assumption of a high correlation of this indicator with the damage to the bank with this parameter.

An additional parameter to refine the model is the estimation of the maturity level of the business process (M), based on the approaches of the CMMI model.

The calculation model takes the following form:
   Period is 1 month, i is business block, n is the number of blocks;
   SA_i is block self-estimation, dimension $(0...1)$;

Inf$_i$ is level of awareness (%);

Act$_i$ is sufficiency of actions (zone: green (1), yellow (0.5), red (0.3) – expert);

Z_i is business process risk (normalized, 0...1), 0 – process is high-risk, 1 – process without risks;

M_i is level of maturity of the process (1...5);

$M_i = g\ (M_j)$, where M_j is the maturity level of the *j*-th process in the *i*-th block;

α_i is the fraction of damage to the *i*-th block in the last period (0...1).

β_k is the coefficient of risk culture element. max $\beta_k = 1$, by default $\beta_k = 0.2$, can vary depending on the parameterization of the model applicable to different banks, also configurable for stability of the model.

$$\text{Rcult} = \sum\nolimits_{i=1}^{n} (\text{SAi} * \beta_1 + \text{Infi} * \beta_2 + \text{Acti} * \beta_3 + \text{Zi} * \beta_4 + \text{g(Mi)} * \beta_5/5) * \alpha_i \quad (1)$$

The end result is a normalized (from 0 to 1) rational number (rounding to two decimal places) and is a relative indicator of risk culture level that can be used in comparison to the same indicator of the previous period in order to determine the dynamics of the value change, also in comparison with similar indicators of other elements in organization structure.

Relativity of this indicator taking into account direct correlation with damage to the bank shows the dependence of damage on individual elements in the model, which allows determining the subdivisions of the zone of impact on bank damage through the adjustment of the model elements.

4   Discussion

In 2014–2015, studies were conducted and measurements of risk culture level were made on the basis of the established described parameters of valuation (*SA, Inf, Act, Z$_i$, g*) in different areas of banking activities of one large bank in Russia. As a result, for each activity, the level of risk culture was assessed on a periodic basis and suggestions were made for a specific improvement of this indicator. Further research was aimed at determining how far the possible activities to increase the level of risk culture can really affect the damage (Fig. 3).

The obtained results led to the conclusion that systemic work on prevention of risk is a key task in the field of countering emerging risks for financial companies, while the proposed mechanisms and methods for assessing risk culture allow to talk about the impact on the assessment of the level in bank, and with a significant increase in the level of risk culture, we can see a decrease in risk losses, despite the presence of independent external factors.

Despite the high correlation between the level of risk culture and the damage of bank, these parameters for calculating the level of risk culture first of all assess the internal readiness of the bank's employees in the real need to prevent emerging threats without direct and sole involvement of the top management of the bank.

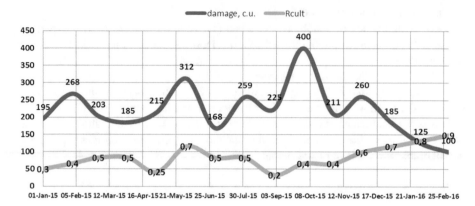

Fig. 3. Correlation of damage and risk culture level

5 Conclusion

In the framework of this study, ways and approaches to optimize the operational risk management process were defined, a model for parametrizing and assessing the level of risk culture was developed, and the relationship between risk culture and bank damage as the main driver of the operational risk management system was explored.

The main goal of introducing a risk culture is to develop a company's behavior that determines the collective ability to identify, analyze, openly discuss and respond to organization's existing and future risks to ensure long-term and sustained reduction in losses from risk realization [15].

The updated author's model of risk management allows to formulate and outline the main priorities of the risk management process, which in Russian banks should be transformed into principles, policies, standards and other normative documents that meet the following characteristics:

- allocation of resources taking into account all significant risks, which implies compliance with regulatory standards for the adequacy of equity (capital) and development of a system of both reactive and proactive measures to avoid risk in the future;
- ensuring efficient use of resources, which implies an improvement in the ratio of resources invested in the risk management system and the profitability of the bank;
- measurement (evaluation) of all significant types of risk and, based on measurement results, initiate actions for hedging, diversification or transfer of risk to maintain it within the limits established by management.

It should also be noted that improving the operational risk management system in the bank helps to stabilize its financial position, increase profitability, reduce capital reserves for operational risk and increase competitiveness. In addition to direct impact on the bank's profits/costs, operational risk affects reputation of the bank, development of human capital as the company's core intangible assets, allows mobilization of hidden reserves to increase employee productivity. In addition, the high-quality management of operational risk can positively influence the prospects of the bank's activity, including optimization of the client base, increase of investors' interest in the bank's activity [5].

References

1. Consulting services in the field of compliance with sanctions regimes. Review of Russian practice. http://www.pwc.ru/en/forensic-services/assets/pwc-global-sanctions-revisited.pdf. Accessed 28 Feb 2018
2. Disclosure of information by credit institutions - Central Bank of the Russian Federation. http://www.cbr.ru/credit/transparent.asp. Accessed 28 Feb 2018
3. Dyatlov, S.A., Lobanov, O.S.: NBIC convergence as a stage of transition of Saint-Petersburg's e-government information space to the sixth techno-economic paradigm. Commun. Comput. Inf. Sci. **745**, 347–361 (2017). https://doi.org/10.1007/978-3-319-69784-0_30
4. Dyatlov, S.A., Lobanov, O.S., Selischeva, T.A.: Information space convergence as a new stage of e-governance development in Eurasian economic space. In: ACM International Conference Proceeding Series Part F130282, pp. 99–106 (2017). https://doi.org/10.1145/3129757.3129775
5. Hovostik, E.: Bank Agricole will pay $ 800 million for Iran and Sudan. http://www.kommersant.ru/doc/2836471. Accessed 28 Feb 2018
6. International convergence of capital measurement and capital standards: refined framework approaches - Central Bank of the Russian Federation. http://www.cbr.ru/today/ms/bn/basel.pdf. Accessed 28 Feb 2018
7. Istomin, E.P., Abramov, V.M., Sokolov, A.G., Burlov, V.G., Slesareva, L.S.: Knowledge database in geoinformation management of the territory development. In: International Multidisciplinary Scientific GeoConference Surveying Geology and Mining Ecology Management, SGEM, vol. 17, no. 21, pp. 951–959 (2017). https://doi.org/10.5593/sgem2017/21/s07.120
8. Kulik, V.V., Vedyakhin, A.A.: Fundamentals of Risk Management, 2nd edn. Tutorial, Moscow (2016)
9. Lobanov, O., Minakov, V., Minakova, T., Schugoreva, V.: NBIC convergence of geoinformation systems in Saint-Petersburg's information space. In: International Multidisciplinary Scientific GeoConference Surveying Geology and Mining Ecology Management, SGEM, vol. 17, no. 21, pp. 471–478 (2017). https://doi.org/10.5593/sgem2017/21/s07.060
10. Martoch, T, Kodym, O.: Production information systems based on the internet of things. In: 13th SGEM GeoConference on Informatics, Geoinformatics and Remote Sensing, pp. 207–212 (2013). https://doi.org/10.5593/sgem2013
11. Minakov, V.F., Lobanov, O.S., Minakova, T.E., Makarchuk, T.A., Kostin, V.N.: The law of diminishing marginal productivity in the model of pure discounted income of innovations. Int. J. Econ. Res. **14**(14), 435–441 (2017)
12. Minakov, V.F., Lobanov, O.S., Makarchuk, T.A., Minakova, T.E., Leonova, N.M.: Dynamic management model of innovations generations. In: Proceedings of 2017 XX IEEE International Conference on Soft Computing and Measurements (SCM), pp. 849–852 (2017). https://doi.org/10.1109/scm.2017.7970743

13. Nelson, R.R.: Economic development from the perspective of evolutionary economic theory. Oxford Dev. Stud. **36**(1), 9–21 (2008)
14. Politics of risk management of the Bank of Russia. http://www.cbr.ru/today/risk/policy.pdf. Accessed 28 Feb 2018
15. Strelnikov, E.V.: Peculiarities of the application of the Basel II and Basel III standards in Russian banks. Supervisor **1**(41), 7–11 (2013)

Impact of Volatility Risk on the TAIEX Option Return

Jui-Chan Huang[1], Wen-I Hsiao[2], Jung-Fang Chen[2], Ming-Hung Shu[3],
Thanh-Lam Nguyen[4(✉)], and Bi-Min Hsu[5]

[1] Yango University, Fuzhou 350015, China
[2] Department of Business Administration,
National Kaohsiung University of Science and Technology,
Kaohsiung 807, Taiwan
[3] Department of Industrial Engineering and Management,
National Kaohsiung University of Science and Technology,
Kaohsiung 807, Taiwan
[4] Office of International Affairs, Lac Hong University, Biên Hòa, Vietnam
green4rest.vn@gmail.com
[5] Department of Industrial Engineering and Management,
Cheng Shiu University, Kaohsiung, Taiwan

Abstract. The purpose of this study is to gauge the impact of volatility risk on TAIEX Option return using regression models for analyzing 12-month cross-sectional option data. Our empirical results indicate that disparate volatility-risk factors have considerable effects on abnormal returns, and expect to create risk premium; specifically, the market risk premium, policy rewards, and fear index can lessen the return of 7.806%, 6.336%, and 1.294% per year, respectively.

Keywords: TAIEX · Option return · Stock exchange · Volatility risk

1 Introduction

In the last few years, we have observed the fast development of financial markets where traditional financial tools such as stock, bonds, and real estate are replaced with financial derivatives such as option, forward contract, futures, credit default swap, and collateralized debt obligations which are quite risky, highly leveraged and complicated. Specifically, in December 2001, Taiwan Futures Exchange (TFE) launched Taiwan weighted index options (TXO) which is the first option on the open market in Taiwan and launched stock options in January 2003. TXO is currently the most actively traded options market in Taiwan, but virtually no stock options trading volume due to the release of warrants market. Warrants market and individual stock options have high homogeneous and better mobility to affect the stock options market. Moreover, TFE launched the MSCI index of dollar-denominated options in March 2006. And, cabinet choices and the non-payment of electricity options were listed in October 2007. Later, in January 2009, TFE launched the NT dollar-denominated gold options which are the first product options in Taiwan. Especially, the trading volume of Taiwan index option is

Z. Wilimowska et al. (Eds.): ISAT 2018, AISC 854, pp. 101–110, 2019.
https://doi.org/10.1007/978-3-319-99993-7_10

ranked the sixth in the global market. Though Taiwan options market started late, it has developed quite fast and becomes one of the interesting research objects in the field.

However, the low-interest environment in recent years has made the investment of general financial portfolios fail to provide attractive return expected by the investors. This paper aims at investigating the influence of volatility risk on option return of Taiwan Stock Exchange Capitalization Weighted Stock Index (TAIEX) which is capitalization-weighted index of all listed common shares traded on the Taiwan Stock Exchange.

2   Literature Review

Several scholars have proposed various approaches in modeling the relationships among stock return and its affecting factors. For example, Shilling [1] argued that the *ex-ante* expected risk premiums on stocks appear to be too large for their risk to be explained by standard econometric models, relative to the returns on common stocks or fixed income securities, which was called "risk premium puzzle in stock". The apparent discrepancy between investors' extreme unwillingness to accept variations in stock returns without an average earning on a high-risk premium [2] and less averseness when it comes to investing in real estate [1] is a fundamental demonstration of the stock premium puzzle. Unless we can advance our knowledge about this basic risk-return tradeoff in stock market, it is not easy to understand why they demand a higher risk premium. Merton [3] first used risk ratio to estimate the expected returns of common stocks. The study suggested that market returns can be predicted to some degree by using "reward-to-risk ratio" as a measure of certain function of the variance of market returns. It was suggested that more accurate variance estimation models could be developed to leverage on the relationship between risk and return [3].

Since the seminal study, the relationship between mean and volatility of returns has become a central research issue in mainstream finance literature. Many asset pricing models have been developed based on the relationship between expected returns and risk. Several researchers such as French et al. [4], Campbell and Hentschel [5], Glosten et al. [6] and Conover et al. [7] used the GARCH model to investigate the long-run co-movement of risk and returns over time. However, unlike the theoretical framework, the empirical intertemporal relationship between the mean and volatility of common stock returns is less conclusive as the results vary according to models, risk measure techniques and exogenous predictors used to draw inferences [8]. The mixed results are also evident in the stock literature. For instance, Liow et al. [9] concluded that there is no significant relationship between expected returns and conditional volatility for Asian property stocks; however, by examining EREIT in relation with the market portfolio, Conover et al. [7] found that its systematic risk β is significant in explaining returns during bull market months but there is no significant relationship between β and returns during bear markets.

Moreover, several researchers have also proposed different approaches to identify critical factors affecting the stock price as well as its trend for the next periods. Like, Whaley [10] proposed Volatility Index (VIX) as investor panic pointer (Investor fear gauge) to reflect investor expectations of 30-day volatility of future stock price. Xing et al. [11] found that the smile curve of United States stock option (Volatility Smirk - VS) can be used to predict stock prices; particularly, by using multiple regression for a wide range of control variables, they found that the future stock prices of high VS will fall whereas low Volatility Smirk will rise. Cremers and Weinbaum [12] found that the deviations of Put-call parity can be used to predict future stock price movements and Volatility Spread has strong prediction ability even when the liquidity of target stock is poor, and the liquidity of its options is better, which was further approved by Xing et al. [11].

In an empirical study, Ang et al. [13] found that high risk stocks have lower rates of remuneration. Because this behavior violates the expectation of high risk and high reward, is defined as "low volatility anomaly". Furthermore, they investigated the effects of investor sentiment for the low volatility anomaly, resulting in three important findings, including: (1) The low volatility effect still exists even after the risk factors are adjusted; (2) By adding investor sentiment as a risk factor, investor sentiment has positive effects on low volatility forecasts; (3) Only under the period of economic expansion, investor sentiment can positively predict the effects of low volatility; while in the recession period, insufficient evidence of investor emotions affecting the low volatility is found. Especially, to test the robustness of these findings when the volatility clustering samples are in very small sample sizes, changing the way of constructing portfolios shows that similar results are obtained.

Ang et al. [13] used assets pricing model to investigate the relationship between investment behaviors and affecting factors of stock return, and they found that investors cannot accept the loss risk more than profit because individual stock normally requires risk premium to control pricing factors such as company's market value, scale factor, liquidity risk and so on. Fama and MacBeth [14] pointed that there is a positive relationship between stock return and β which obviously further supports the performance of the traditional theory of CAPM.

Moreover, Fama and French [8] discussed the relationship of average paid rate of the stocks of non-financial companies in terms of cross section β value, and company scale, book market, benefits ratio and lever degree (A/ME, A/BE). By considering that company scale and the book market can effective find out the variation of individual company stock paid rate, they found that there is a negative relationship between company scale and paid rate whereas there is a positive relationship between book market and paid rate and the paid rate of value type stock (High BE/ME) is higher than growth type stock (Low BE/ME). In additional, book market ratio can be used to sufficiently explain the impact of benefit ratio and leverage degree on the average rate of return; therefore, value stocks take a greater financial risk, and investors require higher return ratio for this kind of stock.

By investigating the historical data of 30 years from 1963 to 1990 of stocks of non-financial company including NYSE, AMEX and the NASDAQ, Fama and French [8] found that there is no obviously relationship between β value and stock average paid ratio because their approach resulted in insignificant statistical figures. However, they found that company scale (SMB) and the book market (HML) have significantly impact on stock average paid ratio, and market factor (market excess return) as well as can be used to appropriately explain the underlying relationship existing in the examined dataset. Their findings of the key risk factors, also called common risk factors, including market factor, company scale and book market are the cores in the so-called Fama-French three factor model.

Financial researchers have traditionally utilized the Capital Asset Pricing Model (CAPM) to assess the risk of investment projects. The CAPM theory [15–17] proposed that the expected return on a risky asset is composed of the risk-free rate plus the risk premium, where the risk premium, in a portfolio sense, is the excess market return over the risk-free rate multiplied by the level of systematic risk for the specific investment. Or, it can be mathematically written as:

$$E(R_j) = R_f + \beta_j[E(R_M) - R_f]$$ (1)

where: $E(R_j)$ and $E(R_M)$ respectively denotes expected returns on the jth portfolio and on the capitalization weighted portfolio of all assets; R_f denotes risk-free return; and

$$\beta_j = Cov(R_j, R_M)/Var(R_M).$$ (2)

It is important to note that the model is a normative equilibrium model which assumes that all investors are rational and markets are efficient. Under these circumstances, only the non-diversifiable risk of an asset would be rewarded. The only thing one would ever need to know about an asset would be its beta relative to the market portfolio, since the beta fully determines its expected return.

Despite its significant advancement in financial theory, Roll [18] claimed that CAPM is very hard to validate empirically, in terms of expected returns which are not directly observable, and the "market portfolio", which contains all assets (not restricted to those traded in financial markets), is impossible to quantify. Additionally, practical implementation of this model could be flawed by its simplifying assumption that (1) beta is assumed to be stationary (2) the correlation between the different assets arises from a common source, i.e. the index. The major value of the CAPM must therefore not be sought in its formalization of the return generating equation, but mainly in the fact that it helped to structure the risk quantification and decomposition. As a result, the main use of the single index CAPM model is typically concentrated in hedging away the benchmark-related risk in portfolios or books using index-based futures or derivatives. Very few organizations or academics would model asset risk based on one explanatory variable. However, it is now generally recognized that risk is multidimensional and needs to be explained using a multitude of characteristics of attributes. This has naturally led to multiple factor models.

Literally, volatility can be effectively measured with the rolling standard deviation [19], calculating the sum of squared deviations of the returns on the market portfolio from its mean return [3] or using downside risk measures [20–22]. The volatility of portfolio returns depends on the variances and covariances between the risk factors of the portfolio, and the sensitivities of individual assets to these risk factors. Thus, most researchers and market practitioners agree that the volatility of aggregate stock markets is not constant, but changes overtime. A number of stylized facts about the volatility of financial asset returns have emerged over the years and been confirmed in numerous studies.

1. Volatility exhibits persistence. According to Mandelbrot [23] and Fama [24], large changes in the price of an asset are often followed by other large changes and small changes are often followed by small changes, which was further confirmed by Baillie et al. [25], Chou [26] and Schwert [27].
2. Volatility is mean reverting, a normal level of volatility to which volatility will eventually return [28]; i.e. the long run forecasts of volatility should eventually converge to this same normal level of volatility, regardless of when they are made. Moreover, mean reversion in volatility implies that current information has no effect on the long run forecast.
3. Return volatility is asymmetric; i.e. positive and negative shocks may not have the same impact on the volatility. This asymmetry is sometimes ascribed to a leverage effect and sometimes to a risk premium effect. The "leverage effect" posits that a firm's stock price decline raises the firm's financial leverage, resulting in an increase in the volatility of equity while the "risk premium effect" suggests that this negative relationship between returns and return volatility stems from natural time- variation in the risk premium on stock returns [6, 29–32].
4. Exogenous variables may influence volatility. Obviously, financial asset prices fail to evolve independently of the market around them, so other variables may contain relevant information for the volatility of a series. Thus, it is hardly developed any kind of consensus model of how it changes over time and the fundamental economic forces guiding it, a model's ability to reproduce above stylized facts have become the guide to the specification [33].

3   Research Methods

This study investigated the volatility risk effect on abnormal returns; using regression to explore its effects and information content, as shown in the following equation.

$$AR_{i,t} = \beta_{i,0} + \beta_{i,MP_t}MP_t + \beta_{i,MP_{t-1}}MP_{t-1} + \beta_{i,AVR_t}AVR_t + \beta_{i,AVR_{t-1}}AVR_{t-1} + \varepsilon_{i,t} \quad (3)$$

where $AR_{i,t}$ refers to the ith period stock and exceed non-risk interest rate in t day rate of return; MP_t refers to over market rate; AVR_t refers to the total risk proxy variable.

Table 1. Taiwan weighted index option transaction data in June 2015

Due month	Strike price	Call delta	Put delta
2015-06	8950	0.963	−0.037
2015-06	9050	0.852	−0.148
2015-06	9150	0.635	−0.365
2015-06	9250	0.365	−0.635
2015-06	9350	0.148	−0.852
2015-06	9450	0.037	−0.963

Source: Taiwan Stock Exchange, 6/2015

This paper aims at building a buying across the site through the Taiwan weighted index options (TXO) for the transaction object. Buy due date within 28 days of the price level for investment in cross-site. For example, if there is an option on June 16, 2015 and the market matures on June 15, 2015, we use the June 16, 2015 closing price bought by Delta-neutral add up to a price level strategy. Using the same conditional sale and purchase, the right characteristics of Delta-phase is equal to one to get Delta-neutral buys right to buy weights for the right to sell the absolute value of the Delta, Delta to buy weights for the right to buy the right to sell (Table 1). Using Delta-neutral to avoid the forecast of policy on index change, to achieve the Zero-Beta effect [34].

This study investigated the fluctuation risk on the financial markets from January 2, 2002 to December 31, 2014, after critical events of SARS in 2002, high oil prices in 2007, United States subprime wind blast in 2008, and the European debt crisis in 2011. Specifically, this study explores the change of Taiwan index options prices and return ratio; also, verifies the target asset changes and economic changes at domestic and abroad, based on the following two types of data sources: (1) TFE website: monthly futures settlement price and (2) Taiwan Economic News Database: the option data, stock indexes, and trading volume of derivatives; such as income per capita, the interest rate and so on. Taiwan economic news only got the closing database of option, each trading transactions are assumed to be closed right after TFE is closed. As TFE is closed from 1:45 PM, so all trading assumptions are sold at 1:45 PM.

Table 2 shows that the right of option trading volume from 2005 to stabilize, because of simple cross-site only used near the price level to the nearest option, sample period refers to the right to choose by Taiwan officially in TFE from January 2002 to December 2014. All stocks and options closing data were downloaded from the Taiwan economic news database, and then rearranged the order by transaction date, strike price.

Table 2. Trading volume of TAIEX options

Annual	Trading volume	Days	Average volume/day
2001	5137	6	856
2002	1,566,446	248	6316
2003	21,720,083	249	87,229
2004	43,824,511	250	175,298
2005	80,096,506	247	324,277
2006	96,929,940	248	390,847
2007	92,585,637	247	374,841
2008	92,757,254	249	372,519
2009	72,082,548	251	287,181
2010	95,666,916	251	381,143
2011	125,767,624	247	509,181
2012	108,458,103	250	433,832
2013	109,311,515	246	444,356
2014	151,620,546	248	611,373

Sources: Taiwan Futures Exchange and Taiwan
Economic News Database

It is obvious that the distance of 28 days of the expiration date of the options is mainly due to the characteristics of Taiwan index option. TFE for options and futures settlement price is calculated by Taiwan stock exchange and the R.O.C. over-the-counter securities day trading time within 30 min with due date before the close of the target index provided by arithmetic average prices. Preceding target index of arithmetic average price is obtained by using simple arithmetic average calculation, and the closest futures contract quotes minimum lifting units integer times of numerical set of prices are then taken.

4 Empirical Results

By using multiple regression models of Fama and MacBeth [14] to measure the total volatility risk, this paper investigates the influence of the following variables: (1) Delta-neutral price levels across policy rewards (DN); (2) Strategy for Crash-Neutral and Delta- neutral under the price level for panic across policy rewards (CNDN), and fear index variation (ΔVIX). To fix in time series regression analysis may lead to arguments among collinearity, violating independence assumptions, and errors might be self-relative, we use estimation of Newey and West [35].

$$AR_{i,t} = \beta_{i,0} + \beta_{i,DN_t}DN_t + \beta_{i,CNDN_t}CNDN_t + \beta_{i,\Delta VIX_t}\Delta VIX_t + \varepsilon_{i,t} \qquad (4)$$

$$AR_{i,t} = \beta_{i,0} + \beta_{i,MP_t}MP_t + \beta_{i,DN_t}DN_t + \beta_{i,CNDN_t}CNDN_t + \beta_{i,\Delta VIX_t}\Delta VIX_t + \varepsilon_{i,t} \qquad (5)$$

With the multiple regression analysis, this paper considers the adjusted R^2 values to judge the explanatory power of independent variables to the model of the dependent variable. Table 3 shows that the explanatory power of the three investigated models are respectively 2.1%, 21.8% and the F values for the statistical models are all significant at the level of 5%, indicating that two regression models have significant explanatory power. Hence, the inclusion of market risk premium significantly improves the explanatory power of abnormal returns.

Table 3. Overall volatility risk premium

Variables	Model 1	Model 2
β_0	3.788 (3.463)[a]	0.464 (2.294)[b]
$DN_{i,t}$	−0.135 (−0.563)	−0.175 (−1.373)
$CNDN_{i,t}$	−0.487 (−0.871)	−0.414 (−2.647)[b]
$\Delta VIX_{i,t}$	−1.042 (−0.875)	−1.020 (−4.293)[a]
$MP_{i,t}$		0.130 (2.293)[b]
Ad. R^2 (%)	2.1	21.8
F-value	2.617[b]	19.731[a]

[a]1% significance level
[b]5% significance level
Standardized coefficients

In company's overall risk factor analysis, the three independent variables were found negatively correlated with abnormal returns, which is consistent with the research results of Ang et al. [13], Bakshi and Madan [36], Pan [37], and Eraker et al. [38]. It can be possibly explained by the fact that investors using option market and stock market as a tool can avoid risk, so abnormal returns decline when volatility risk increases. Overall volatility risk has less influence on abnormal returns before more control variables are added. However, overall volatility risk has significant affect when market risk premium is added in the three-factor model of Fama and French [8]. Among these factors, the fear index (VIX) is the most obvious and has the greatest impact on the return. It is also found that in the three-factor model of Fama and French [8], Delta-neutral price levels across policy payment, Crash-Neutral and Delta-neutral price levels across policy rewards influence on the abnormal returns and the change of fear index.

Table 4 shows that when the average Beta values and standard deviation of Delta-neutral- price level across the policy paid (DN) for the market risk premium (price volatility) are respectively −0.0057 and 0.0843, with the regression coefficient of −0.463; indicating that within two standard deviations, Delta-neutral paid price levels across a strategy expected to bring down the rate of 7.806% per year. Similarly, the average beta values and standard deviation of Crash-Neutral and Delta-neutral price levels across policy rewards (CNDN) for the risk premium (price volatility) are respectively −0.0103 and 0.1022 with the regression coefficient of −0.310, showing that within two standard deviations, Crash-Neutral and the Delta-neutral-price level across the policy rewards are expected to bring down the rate of 6.336% per year. And, the average beta values and standard deviation of fear index (ΔVIX) are respectively 0.0012 and 0.0061 with the regression coefficient of −1.061, showing that within two standard deviations, fear changes in the indices are expected to bring down the rate of 1.294% per year.

Table 4. Average value and Standard deviation of overall volatility risk premium

Variables	Average value	Standard deviation
$DN_{i,t}$	−0.0057	0.0843
$CNDN_{i,t}$	−0.0103	0.1022
$\Delta VIX_{i,t}$	0.0012	0.0061

5 Conclusion

This study aims at measuring the influence of volatility risk on TAIEX Option return by employing a multiple regression analysis for cross-sectional data on a 12-month study period. Our empirical results show that different volatility risk factors have significantly effects on abnormal returns and will cause risk premium. However, traditional financial theories consider that high risk should have high return which is contrast to the result of this study. The reason is investors trading behaviors are irrational and nonrandom as well as financial markets are inefficient, leading to high risk with less payoff.

For future research, we will investigate different groups of investors across preference industry categories (electronic, financial, biography produced, and health technical medical, etc.), investment type (value type or growth type), and standard of assets scale different (high unit or low-price unit). As a matter of fact, the current trading volume of each stock option domestically is low, and lack of research data; thus, larger trading volume can help to increase different kind of data connotations in practice.

References

1. Shilling, J.D.: Is there a risk premium puzzle in real estate. Real Estate Econ. **4**, 501–525 (2003)
2. Mehra, R., Prescott, E.: The equity premium: a puzzle. J. Monet. Econ. **2**, 145–162 (1985)
3. Merton, R.C.: On estimating the expected return on the market. J. Financ. Econ. **8**, 323–361 (1980)
4. French, K.R., Schwert, G.W., Stambaugh, R.F.: Expected stock returns and volatility. J. Financ. Econ. **19**, 3–29 (1987)
5. Campbell, J., Hentschel, L.: No news is good news: an asymmetric model of changing volatility in stock returns. J. Financ. Econ. **31**(3), 281–318 (1992)
6. Glosten, L., Jagannathan, R., Runkle, D.: On the relation between the expected value and the volatility of the nominal excess returns on stocks. J. Finance **5**, 1779–1802 (1993)
7. Conover, M.C., Friday, H.S., Howton, S.H.: An analysis of the cross section of returns for EREITs using a varying-risk beta model. Real Estate Econ. **1**, 141–163 (2000)
8. Fama, E.F., French, K.R.: Multifactor explanations of asset pricing anomalies. J. Finance **51**(1), 55–84 (1996)
9. Liow, K.H., Ooi, J.T.L., Gong, Y.: Cross market dynamics in property stock markets-some international evidence. In: European Real Estate Society Conference Paper, pp. 10–13 (2003)
10. Whaley, R.E.: Understanding VIX. J. Portf. Manag. **35**(3), 98–105 (2009). https://doi.org/10.3905/JPM.2009.35.3.098
11. Xing, Y., Zhang, X., Zhao, R.: What does the individual option volatility smirk tell us about future equity returns? J. Financ. Quant. Anal. **45**, 641–662 (2010)
12. Cremers, M., Weinbaum, D.: Deviations from put-call parity and stock return predictability. J. Financ. Quant. Anal. **45**(2), 335–367 (2010)
13. Ang, A., Hodrick, R.J., Xing, Y., Zhang, X.: The cross-section of volatility and expected returns. J. Finance **61**(1), 259–299 (2006)
14. Fama, E., Macbeth, J.: Risk, return and equilibrium: empirical tests. J. Polit. Econ. **81**(3), 607–636 (1973)
15. Sharp, W.F.: Capital asst prices: a theory of market equilibrium under conditions of risk. J. Finance **3**, 425–442 (1964)
16. Lintner, J.: The valuation of risk assets and the selection of risk investments in stock portfolios and capital budgets. Rev. Econ. Stat. **1**, 13–37 (1965)
17. Mossin, J.: Equilibrium in a capital asset market. Econometrica **34**, 768–783 (1966)
18. Roll, R.: A critique of the asset pricing theory's tests' part I: on past and potential testability of the theory. J. Financ. Econ. **4**(2), 129–176 (1977)
19. Officer, R.R.: The variability of the market factor of New York Stock Exchange. J. Bus. **46**, 434–453 (1973)

20. Bawa, V.S., Lindenberg, E.B.: Capital market equilibrium in a mean-lower partial moment framework. J. Financ. Econ. **5**, 189–200 (1977)
21. Sivitanides, P.: A downside risk approach to real estate portfolio structuring. J. Real Estate Portf. Manag. **4**, 159–168 (1998)
22. Stevenson, S.: Emerging markets, downside risk and the asset allocation decision. Emerg. Mark. Rev. **2**, 50–66 (2001)
23. Mandelbrot, B.B.: The variation of certain speculative prices. J. Bus. **36**, 392–417 (1963)
24. Fama, E.F.: The behavior of stock-market prices. J. Bus. **38**, 34–105 (1965)
25. Baillie, R.T., Chung, C.F., Tieslau, M.A.: Analyzing inflation by the fractionally integrated ARFIMA-GARCH model. J. Appl. Econom. **11**, 23–40 (1996)
26. Chou, R.Y.: Volatility persistence and stock valuations: some empirical evidence using GARCH. J. Appl. Econom. **3**, 279–294 (1988)
27. Schwert, G.W.: Why does stock market volatility changes over time? J. Finance **44**, 1115–1153 (1989)
28. Medova, E.A., Smith, R.G.: Does the firm-specific asset volatility process implied by the equity market revert to a constant value. Judge Institute of Management Working Papers (2004)
29. Karolyi, G.A.: Why stock return volatility really matters? In: Strategic Investor Relations, pp 1–19 (2001)
30. Christie, A.A.: The stochastic behavior of common stock variances: value, leverage, and interest rate effects. J. Financ. Econ. **10**, 407–432 (1982)
31. Nelson, D.: Conditional heteroskedasticity in asset returns: a new approach. Econometrica **59**, 323–370 (1991)
32. Engle, R.F., Ng, V.: Measuring and testing the impact of news on volatility. J. Finance **48**, 1749–1778 (1993)
33. Campbell, J.Y., Lettau, M., Malkiel, B.G., Xu, Y.: Have individual stocks become more volatile? An empirical exploration of idiosyncratic risk. J. Finance **1**, 41–43 (2001)
34. Coval, J.D., Shumway, T.: Expected options returns. J. Finance **56**, 983–1009 (2001)
35. Newey, W.K., West, K.D.: A simple, positive semi-definite, heteroskedasticity and autocorrelation consistent covariance matrix. Econometrica **55**(3), 703–708 (1987)
36. Bakshi, G., Madan, D.: Spanning and derivative security valuation. J. Financ. Econ. **55**, 205–238 (2000)
37. Pan, J.: The jump-risk premia implicit in options: evidence from an integrated time-series study. J. Financ. Econ. **63**, 3–50 (2002)
38. Eraker, B., Johannes, M., Polson, N.: The impact of jumps in volatility and returns. J. Finance **58**(3), 1269–1300 (2003)

Simulation-Based Analysis of Penalty Function for Insurance Portfolio with Embedded Catastrophe Bond in Crisp and Imprecise Setups

Maciej Romaniuk[(✉)]

Systems Research Institute, Polish Academy of Sciences,
ul. Newelska 6, 01-447 Warsaw, Poland
mroman@ibspan.waw.pl

Abstract. In this paper, important properties of an insurer's portfolio, which consists of a catastrophe bond and a reinsurance contract, are numerically analysed. Because of stochastic nature of considered processes, simulations and the Monte Carlo (MC) methods are applied. Special attention is paid to estimation and optimization of an average value of the portfolio and to probability of the insurer's ruin. Moreover, we assume that the ruin event is related to additional expenses, which are modelled by various penalty functions. In the considered numerical examples, apart from strictly crisp sets of parameters, also fuzzy numbers are used to model the possible penalties, which are related to the ruin event.

Keywords: Risk process · The Monte Carlo simulations · Insurance portfolio
Catastrophe bond · Fuzzy numbers

1 Introduction

Financial and economical consequence of natural catastrophes, like tsunamis, earthquakes, floods, hurricanes etc. are serious problems, not only for the whole countries, but also for financial and insurance institutions. Standard approaches, like reinsurance contracts, are not adequate in case of present-day challenges. In order to mitigate these risks, insurers developed many new instruments, which combine financial and insurance aspects. A catastrophe bond (usually abbreviated as a cat bond) is an example of such an instrument (see, e.g., [6, 7, 11, 12, 14]).

Of course, development of these new instruments does not mean, that more classical methods and instruments should be completely neglected. Therefore, a creation of an insurer's portfolio, which consists of a few different instruments, is an useful way to mitigate insurance risks caused by nowadays problems. Many authors point out effectiveness of the catastrophe bonds as a diversification tool in the case of financial, transport and infrastructure industries (see, e.g., [1, 8]). However, an analysis of behaviour of such a portfolio poses a serious challenge. If more complex financial or insurance instruments are taken into account, we have to generalize the classical risk process and then apply simulation methods to describe the respective portfolio (see, e.g., [9, 11, 12]).

© Springer Nature Switzerland AG 2019
Z. Wilimowska et al. (Eds.): ISAT 2018, AISC 854, pp. 111–121, 2019.
https://doi.org/10.1007/978-3-319-99993-7_11

In this paper, we further develop the ideas proposed in [9, 11, 12]. Instead of the classical risk process, its generalized form is used. Apart from flows related to insureds' premiums and a claim process, additional layers, like a catastrophe bond and a reinsurance contract, are taken into account. Moreover, we assume that there is dependency between time and money. This dependency is not always taken into account in the classical insurance models. Due to the mentioned assumption, computer-aided simulations of behaviour of the insurer's portfolio are necessary.

In this paper, our main aim is optimization of a construction of the portfolio, taking into account the insurer's point of view. We focus on minimization of probability of the insurer's ruin and maximization of the expected value of the whole portfolio, if number of the issued catastrophe bonds is treated as a parameter. To solve these problems, simulations and the Monte Carlo (MC) methods are applied. We assume that the ruin event is related to additional expenses, modelled by various penalty functions. But in real-life problems, some data are usually given in an imprecise way. As we further point out, the same applies to respective parameters of the penalty functions. Therefore, apart from a strictly "crisp" case (i.e. described by real numbers), we also analyse the second, imprecise instance, as the fuzzy numbers are applied. As far as we know, this imprecise setting is not considered in the literature (see e.g., [5]).

A contribution of this paper is fourfold. Firstly, we focus our attention on numerical problems, concerning minimization of the probability of the insurer's ruin and maximization of the expected value of the whole portfolio. In order to solve these problems, trajectories of the respective risk process are generated. Secondly, special penalty functions, related to the insurer's ruin, are introduced. Behaviour of the insurer's portfolio for both the crisp and the fuzzy cases for these penalty functions are numerically analysed, using the MC methods. Thirdly, apart from a cat bond, an additional layer in the insurer's portfolio, in the form of a reinsurance contract, is discussed. Fourthly, a new, more complex intensity function is used during the MC simulations of number of catastrophic events.

This papers is organized as follows. In Sect. 2, necessary assumptions and models concerning the insurer's portfolio are discussed. In Sect. 3, the penalty functions are introduced and the aims of the portfolio optimization are stated. Section 4 is devoted to simulation-aided analysis of some examples of the insurer's portfolios.

2 Generalization of the Risk Reserve Process

In the insurance industry, the risk reserve process R_t is defined as a model of financial reserves of an insurer depending on time t, i.e.

$$R_t = u + pt - C_t^*, \tag{1}$$

where u is an initial reserve of the insurer, p is a rate of premiums paid by the insureds per unit time and C_t^* is a claim process, which is given by

$$C_t^* = \sum_{i=1}^{N_t} C_i, \tag{2}$$

where $C_1, C_2 \ldots$ are *iid* random values of claims. These claims are traditionally identified with losses U_i, which are caused by natural catastrophes, so we have $C_i = U_i$ (see, e.g., [9, 11] for other approaches). The process of number of the claims $N_t \geq 0$ is usually driven by the homogeneous Poisson process (abbreviated as HPP), or the non-homogeneous Poisson process (NHPP). In this paper, we apply a non-homogeneous intensity function with linear, cyclic and exponential parts, in the form of

$$\lambda(t) = a + bt + c\sin(2\pi(t+d)) + m\exp\left(\cos\left(\frac{2\pi t}{\omega}\right)\right). \tag{3}$$

This function, proposed in [4], is a more refined form of the intensity function, which was introduced in [3] and applied in [9, 11, 12] to describe behaviour of the respective insurance portfolio. The parameters of (3) were fitted in [4] to real-life data, collected by PCS (Property Claim Services), which is a division of the Insurance Services Office, USA. This data is related to insurance losses caused by pre-specified natural catastrophes, so it can be used in this paper for our purposes. Therefore, we set $a = 24.93$, $b = 0.026$, $c = 5.6$, $d = 7.07$, $m = 10.3$, $\omega = 4.76$ and apply the thinning method to generate times of catastrophic events. Moreover, a respective random distribution, which models the single claim, is necessary. Based on the approach, considered in [4], we apply the lognormal distributions with parameters $\mu = 18.58$ and $\sigma = 1.49$.

In this paper, the premium in (1) is modelled as a constant function with a specified deterministic time T, i.e.

$$p(T) = (1 + v_p)EC_i \int_0^T \lambda_{NHPP}(s)ds, \tag{4}$$

where v_p is a safety loading (or security loading) of the insurer. In practical situations, this value is about 10%–20% (see also a respective discussion in [9]).

In the following, the classical risk reserve process is generalized. Firstly, we take into account additional financial and insurance instruments, like a catastrophe bond and a reinsurance contract. Secondly, we assume that value of money in time is modelled using some stochastic interest rate. In this paper, the one-factor Vasicek model with parameters, which are fitted in [2], is applied. These two generalizations of the process (1) are described in more detailed way in [9, 11, 12].

Let us briefly describe cash flows, which are related to a cat bond and a reinsurance contract. When the catastrophe bond is issued, the insurer pays an insurance premium p_{cb} to a special agent in exchange for a coverage, when a triggering point (usually some catastrophic event, like an earthquake) occurs. Then, the investors purchase the cat bonds. They hold these assets, whose future payments depend on the occurrence of the mentioned triggering point. If such a catastrophic event occurs during the specified period, then the special firm, which acts as agent, compensates the insurer and the cash flows for the investors are lowered, even to zero. However, if the triggering point does not occur, the investors usually receive the full payment (see, e.g., [6–9]).

In order to improve their financial stability, the insurers transfer some part of their risks to reinsurers. A standard reinsurance contract is an *excess-of-loss policy*. In this case, in exchange for some premium p_{reins}, the reinsurance company pays a part of value of claim to the insurer, if the total value of claims C_t^* till the moment T exceeds an attachment point A_{reins}. There is also a maximum limit of the possible flow, a cap level B_{reins}. Beyond this level, the reinsurer pays only the fixed value $B_{reins} - A_{reins}$ regardless of the reported claims. Then, the payment function for the insurer is equal to

$$f_{reins}^i(C_T^*) = \begin{cases} B_{reins} - A_{reins} & \text{if } C_T^* \geq B_{reins} \\ N_T^* - A_{reins} & \text{if } C_T^* \geq A_{reins} \end{cases}, \tag{5}$$

And the new, generalized risk process is given by

$$R_T = FV_T(u - p_{cb} - p_{reins}) + FV_T(p(T)) - FV_T(C_T^*) + n_{cb}f_{cb}^i(C_T^*) + f_{reins}^i(C_T^*), \tag{6}$$

where $f_{cb}^i(N_T^*)$ is a payment function of the single cat bond for the insurer, p_{cb} is an insurance premium, which is related to the issued catastrophe bonds, p_{reins} is a price of the reinsurance contract. We assume, that p_{cb} is proportional to both a part α_{cb} of a whole price of the single catastrophic bond I_{cb}, and to number of the issued bonds n_{cb}, so that $p_{cb} = \alpha_{cb}n_{cb}I_{cb}$. In this paper, we set $\alpha_{cb} = 0.1$, so 10% of the whole cat bond price is "refunded" by the insurer. Because of the introduced assumption about the value of money in time, the respective future values, denoted by $FV_T(.)$, are calculated using the considered process of the interest rate.

In the following, a piecewise function as the payment function $f_{cb}^i(C_T^*)$ is applied (see also [6, 7, 9] for further details). Then, we have

$$f(C_T^*) = Fv\left(1 - \sum_{i=1}^n \frac{\min(C_T^*, K_i) - \min(C_T^*, K_{i-1})}{K_i - K_{i-1}} w_i\right), \tag{7}$$

where Fv is a face value of the cat bond, $w_1, \ldots, w_n > 0$ are payoff decreases (satisfying the requirement $\sum_{i=1}^n w_i \leq 1$), and $0 \leq K_0 \leq K_1 \leq \cdots \leq K_n$ are the triggering points. In the considered setting, we set $Fv = 1$ (one monetary unit assumption), and

$$K_0 = Q_{NHPP-LN}^{loss}(0.9), \quad K_1 = Q_{NHPP-LN}^{loss}(0.95), \tag{8}$$

where $Q_{NHPP-LN}^{loss}(x)$ is x-th quantile of the cumulated value of the losses, if number of the losses is given by the NHPP and value of the single loss is modelled by the lognormal distribution (see also [6, 7, 9]). The payoff decrease is given by $w_1 = 1$ and we apply one year time horizon, so $T = 1$. It means that, if after one year the cumulated value of the losses surpasses K_1, then the bond holder receives nothing. In order to find the price of this cat bond, the simulation-aided approach, which was considered in [6, 7, 9], is applied. Then, we obtain $I_{cb} = 0.896552$.

We assume that the lower and upper limits for the reinsurance contract are also related to the quantiles of the cumulated value of the losses, namely

$$A_{reins} = Q^{loss}_{NHPP-LN}(0.8), \quad B_{reins} = Q^{loss}_{NHPP-LN}(0.9). \tag{9}$$

It means, that payments, related to the reinsurance contract, occur before the triggering point of the issued catastrophe bond. The respective price of the reinsurance contract p_{reins} can be also calculated using the MC method and the simulated trajectories of the process (2). Usually, the price p_{reins} is increased by a security loading for the reinsurer v_{reins}. In our case, we set $v_{reins} = 0.05$.

3 Bankruptcy Event and a Penalty Function

In insurance mathematics, we are interested in an evaluation of probability of an ultimate ruin (i.e. a ruin with an infinite time horizon) of the insurer, which is given as

$$\psi(u) = Pr(inf_{t \geq 0} R_t < 0) \tag{10}$$

and probability of a ruin before the time T (i.e., a ruin with a finite time horizon)

$$\psi(u, T) = Pr\left(inf_{t \in [0,T]} R_t < 0\right). \tag{11}$$

In the following, because of the complex form of the generalized version of the risk reserve process (6), we estimate the probabilities of the ruin using simulations of the trajectories of (6) and the MC approach.

In real-life situations, the insurer is interested in both minimization of the probability of his ruin and maximization of the expected value of his portfolio, i.e., the maximization of ER_T. Therefore, we also analyse these parameters of the insurer's portfolio. Moreover, a bankruptcy can be more dangerous for the insurer, than only simple lack of necessary funds. It could lead to additional financial penalties, bankruptcy of other enterprises directly connected with the insurer, further problems with law etc. Therefore, we focus on simulation-aided estimation of the expected value of the portfolio with an embedded penalty function, which is related to a bankruptcy event. In a simple case, the considered value is given by

$$ER_T - b_{pen}Pr(R_T < 0), \tag{12}$$

where b_{pen} is some penalty factor, which is related to the occurrence of the ruin. Because of the applied simulation-aided approach, it is possible to analyse a more sophisticated case, too. Then, instead of the simple penalty factor, a more complex function of the value of the ruin can be used, so

$$ER_T^* = ER_T - Ef_{pen}(\max(-R_T, 0)) = ER_T - Ef_{pen}(-R_T)_+, \tag{13}$$

where $f_{pen}(.)$ is a penalty function, which is related to the value of the insurer's portfolio, if the bankruptcy occurs. If this penalty function is equal to a constant parameter, then (13) reduces to (12). Otherwise, e.g., if

$$f_{pen}(R_T) = a_{pen}\max(-R_T, 0) + b_{pen}I(R_T < 0), \qquad (14)$$

where $I(.)$ is the characteristic function, then not only the probability of the ruin, but also the size of this ruin is taken into account in a linear way. In (14), $a_{pen} > 0$ is a respective linear parameter and $b_{pen} > 0$ is some constant parameter.

Moreover, the insurer may be interested in maximization of (13) for some parameters of his portfolio. Because of the stochastic nature of (6), a special approach is necessary then. There are many procedures to solve a maximization problem in a stochastic setting, like, e.g., the Kiefer–Wolfowitz method (see, e.g., [13]). In the following, we simulate bunch of the trajectories of (6), using the respective Euler scheme for the introduced interest rate process. Then, the same bunch of the trajectories is used to find the MC estimators of ER_T^* for different values of the considered parameter. Due to this approach, both overall variability of the obtained estimators and error of the MC procedure are lowered (see also [12]).

4   Simulation-Aided Analysis of Different Examples of Portfolios

After the introduction of the necessary notation and the applied simulation approach, we focus on an analysis of different examples of the insurer's portfolios. Three kinds of the penalty functions will be further used during simulation-based estimation of the expected value (13), i.e., no penalty, the constant and the linear penalty functions. Due to the applied MC method and the previously mentioned generation of bunch of the trajectories, a comparative analysis of these three functions is possible. Moreover, two different types of the insurer's portfolio will be considered. The first one is related to the process of the insureds' premiums and to the issued catastrophe bonds, so this portfolio has two layers. The second portfolio includes the reinsurance contract, too. Then, this portfolio has three layers. The respective parameters of the components of these exemplary portfolios were described in Sect. 2.

Let us start from the insurer's portfolio with the two layers. Using the MC approach, the expected value (13) can be estimated for various values of the number of the issued catastrophe bonds n_{cb}. And, in order to provide appropriate practical conclusions, a wide range of the possible values of n_{cb} is taken into account during our analysis. The simulation-based estimators of ER_T^*, denoted further by $\widehat{ER_T^*}$, can be seen in Fig. 1. Estimators of the probabilities of ruin before $T = 1$, denoted further by $\widehat{\psi}$, can be obtained in the same way, too (see Fig. 2).

If no penalty function is used, then the average value of the portfolio is a strictly decreasing, almost linear function of n_{cb} (see Fig. 1, the plot denoted by circles). In the case of the constant penalty function, we set $b_{pen} = 2000$. Therefore, this value of the penalty factor is a rather significant one, if it is compared to the average value of the

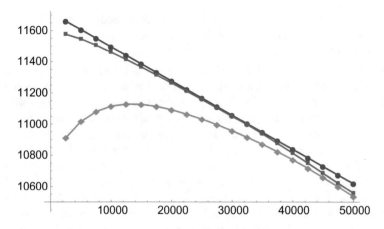

Fig. 1. Estimators of ER_T^* in Example I.

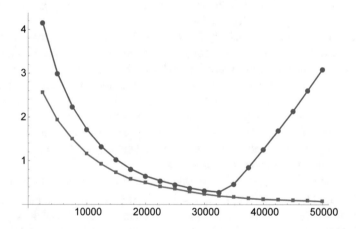

Fig. 2. Estimators of the ruin probabilities before $T = 1$ in Example I and Example II.

whole portfolio. As it is seen (Fig. 1, squares), $\widehat{ER_T^*}$ is also a decreasing function in this case, too. But now, due to the additional penalty bias, this function is below the plot of $\widehat{ER_T^*}$ obtained in the first case. It is worth noting, that both of these functions are very close to each other, when, approximately, $n_{cb} = 3500$. This phenomena will be explained later. In the case of the linear penalty function, we set $a_{pen} = 2$ and $b_{pen} = 50$. It means that, if a bankruptcy event occurs, then for each lacking unit of money, the insurer has to pay twice that value. Moreover, the constant penalty factor is far less influencing, if it is compared to the whole value of the portfolio now. And the obtained average value of the portfolio (see Fig. 1, diamonds) is a more complex function than in the two previous cases. For the smaller values of n_{cb}, it is an exponentially increasing function, then, for the bigger values of the number of the issued bonds, it is a strictly exponentially decreasing function. It means that the previously mentioned maximization problem for (13) can be directly solved using the MC approach in this case. And in the considered example, $n_{cb} = 13250$ seems to be the optimal value of the number of the issued cat bonds.

Estimates of the ruin probabilities $\widehat{\psi}$ are also an U-shaped function of n_{cb} (see Fig. 2, the plot denoted by circles). And shape of this function is the same, regardless of the considered penalty function. For the smaller values of n_{cb}, the probability of the ruin is an exponentially decreasing function. Then, it is fruitful for the insurer to issue more catastrophe bonds, but not too many of them. As it is seen, when the values of n_{cb} are greater than about 32500, the probability of the final ruin is almost a linearly increasing function. This complex behaviour can be directly explained. First, the ruin probability decreases, due to increased possible cash flows, which are related to the bigger number of the issued bonds. Then, the constant costs arising from the necessity of paying some part of price of the issued bonds, become more important than possible further cash flows. The shape of this function explains also the proximity of the plots of $\widehat{ER_T^*}$ (in the case of no penalty and constant penalty functions), which was visible in Fig. 1. Moreover, it is possible to point out the optimum value of n_{cb} for which minimum of the ruin probability is attained.

In practical situations, the penalty function can be given in an imprecise form. For example, due to regulations or law, some parameters of the penalty function can be stated as "about", not as "strictly precise" values. Then, it is straightforward to apply fuzzy numbers in this setting. We focus our considerations on triangular fuzzy numbers, which are commonly used to describe various real-life models, but other types of fuzzy numbers can be also applied together with the simulation-based approach, which is considered here (see, e.g., [6, 9–11] for notation, definitions and further details). Let $\widehat{a} = [a_L, a_C, a_R]$ denote a triangular fuzzy number, where a_L is its left end of a support, a_R – its right end of a support, and a_C – a core.

Let us assume, that $n_{cb} = 1000$ (so, one thousand of cat bonds is issued) and that $\widehat{a}_{pen} = [1, 2, 3]$, or $\widehat{a}_{pen} = [1, 2, 4]$, $\widehat{b}_{pen} = [45, 50, 60]$. In the first case, the linear parameter of the penalty function is "about 2 plus/minus 50%", and it is given by a symmetric triangular fuzzy number. In the second case, both the linear and the constant parameters are given by right-skewed triangular numbers (e.g., "about 2, plus 100%, minus 50%", if $\widehat{a}_{pen}$ is considered). Using the MC approach and simulations of the consecutive α-cuts (where $\alpha \in [0, 1]$, see [6, 10–13] for further details), it is possible to obtain respective measures for the insurer's portfolio.

The obtained fuzzy approximations of ER_T^* are then plotted in Fig. 3. In the first case (when $\widehat{a}_{pen} = [1, 2, 3]$, the plot denoted by circles), the fuzzy output is also a symmetric fuzzy number, almost a triangular one. In the second case (Fig. 3, squares), the fuzzy approximation of ER_T^* is a strictly left-skewed fuzzy number, which is also close to a triangular number. Other practically useful characteristics of the portfolio, like the ruin probability, can be obtained and analysed in the similar way, using the same MC approach.

In the second example, the reinsurance contract is added to the considered portfolio, apart from the cat bond. But, even in this complex case, it is possible to apply the MC approach to find estimators of the characteristics of the insurer's portfolio.

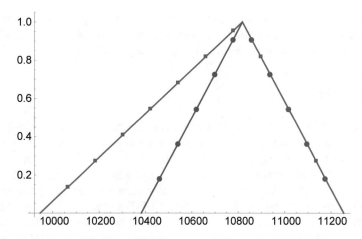

Fig. 3. Fuzzy approximations of ER_T^* in Example I

The plots of $\widehat{ER_T^*}$ as functions of n_{cb} for the three considered cases of the penalties functions (see Fig. 4) are very similar to the respective plots in Example I (see Fig. 1). The relative differences are about 0.2–0.3% of the whole value of the portfolio. Still, there are some minor divergences. In the case of no penalty function, values of $\widehat{ER_T^*}$ are smaller for the whole considered range of n_{cb} than in Example I. In the case of the constant penalty function, first the values of $\widehat{ER_T^*}$ are smaller, then greater. And in the case of the linear penalty function, the situation is quite opposite.

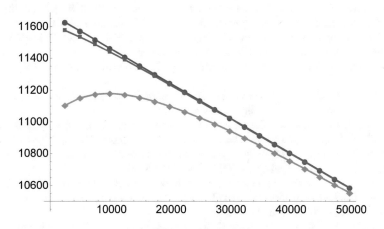

Fig. 4. Estimators of ER_T^* in Example II.

The plot of $\widehat{\psi}$ behaves in a more interesting way (see Fig. 2, the plot denoted by squares). As it is seen, the estimated probabilities are a strictly decreasing function of n_{cb}, in contrary to the U-shaped function from Example I. Then, due to the additional reinsurance contract, it is possible to lower the ruin probability even further, than if the catastrophe bond is used exclusively. However, for bigger values of n_{cb}, differences among the obtained ruin probabilities are almost negligible.

5 Conclusions

In this paper we have focused on simulation-based minimization of the probability of the insurer's ruin and maximization of the expected value of the whole portfolio, if the number of the issued catastrophe bond is treated as a parameter. We have also introduced three kinds of penalty functions (without penalty, the constant and the linear functions), which are related to the additional insurer's expenses, if the ruin event occurs. Both the crisp and the fuzzy examples of these penalty functions have been numerically analysed, using the MC methods. Then, based on these examples, some practically useful conclusions concerning influence of number of issued catastrophe bonds on behaviour of the portfolio have been drawn. In general, it seems that the insurer should issue "not too low and not too high" number of the catastrophe bonds, and its precise value can be found using the simulation approach considered in this paper.

References

1. Carayannopoulos, P., Perez, M.F.: Diversification through catastrophe bonds: lessons from the subprime financial crisis. Geneva Pap. Risk Insur. Issues Pract. **40**(1), 1–28 (2015)
2. Chan, K.C., Karolyi, G.A., Longstaff, F.A., Sanders, A.B.: An empirical comparison of alternative models of the short-term interest rate. J. Finance **47**, 1209–1227 (1992)
3. Chernobai, A., Burnecki, K., Rachev, S., Trueck, S., Weron, R.: Modeling catastrophe claims with left-truncated severity distributions. Comput. Stat. **21**, 537–555 (2006)
4. Giuricich, M., Burnecki, K.: Modelling of left-truncated heavy-tailed data with application to catastrophe bond pricing. https://ssrn.com. http://dx.doi.org/10.2139/ssrn.2973419. 10 Nov 2017
5. Goda, K.: Seismic risk management of insurance portfolio using catastrophe bonds. Comput. Aided Civ. Infrastruct. Eng. **30**(7), 570–582 (2015)
6. Nowak, P., Romaniuk, M.: Catastrophe bond pricing for the two-factor Vasicek interest rate model with automatized fuzzy decision making. Soft. Comput. **21**(10), 2575–2597 (2017)
7. Nowak, P., Romaniuk, M.: Valuing catastrophe bond involving correlation and CIR interest rate model. Comput. Appl. Math. **37**(1), 365–394 (2018)
8. Pizzutilo, F., Venezia, E.: Are catastrophe bonds effective financial instruments in the transport and infrastructure industries? Evidence and review from international markets. Bus. Econ. Horizons **14**(2), 256–267 (2018)
9. Romaniuk, M., Nowak, P.: Monte Carlo methods: theory, algorithms and applications to selected financial problems. ICS PAS, Warszawa (2015)
10. Romaniuk, M.: On simulation of maintenance costs for water distribution system with fuzzy parameter. Maint. Reliab. **18**(4), 514–527 (2016)

11. Romaniuk, M.: Analysis of the insurance portfolio with an embedded catastrophe bond in a case of uncertain parameter of the insurer's share. In: Wilimowska, Z., Borzemski, L., Grzech, A., Świątek, J. (eds.) Information Systems Architecture and Technology: Proceedings of 37th International Conference on Information Systems Architecture and Technology—ISAT 2016—Part IV, pp. 33–43. Springer, Berlin (2017)
12. Romaniuk, M.: Insurance portfolio containing a catastrophe bond and an external help with imprecise level—a numerical analysis. In: Kacprzyk, J., Szmidt, E., Zadrożny, S., Atanassov, K.T., Krawczyk, M. (eds.) Advances in Fuzzy Logic and Technology 2017: Proceedings of: EUSFLAT 2017, vol. 3, pp. 256–267. Springer, Berlin (2018)
13. Romaniuk, M.: Optimization of maintenance costs of a pipeline for a V-shaped hazard rate of malfunction intensities. Maint. Reliab. **20**(1), 46–56 (2018)
14. Zhang, X., Tsai, C.C.-L.: The optimal write-down coefficients in a percentage for a catastrophe bond. N. Am. Actuar. J. **22**(1), 1–21 (2018)

Selecting the Efficient Market Indicators in the Trading System on the Forex Market

Przemysław Juszczuk[1(✉)] and Lech Kruś[2]

[1] Faculty of Informatics and Communication, Department of Knowledge Engineering,
University of Economics, 1 Maja 50, 40-287 Katowice, Poland
przemyslaw.juszczuk@ue.ue.katowice.pl
[2] Systems Research Institute, Polish Academy of Sciences,
Newelska 6, 01-447 Warsaw, Poland
krus@ibspan.waw.pl

Abstract. A concept of the trading system is proposed in which only selected market indicators generating the most beneficial recommendations for traders are included. The proposal can be implemented as an additional module included in the existing trading systems. We present a procedure, in which for a given initially considered indicators, only the selected ones being the most beneficial for a trader for a given time window are derived. In the procedure, an efficiency of initially considered indicators is verified in each time window on the basis of their past performance. Using the verification results the ranking and selection of the best ones is made.

This procedure is implemented and included in the trading system on the Forex market. To evaluate our approach we have performed detailed experiments including real-world data covering the period of almost three years and ten different technical indicators.

The presented results show, that the proposed approach is promising. The set of indicators included in trading systems should be limited and changed sequentially using the presented procedure.

Keywords: Trading system · Forex · Selection of market indicators
Efficiency of market indicators

1 Introduction

Nowadays an intensive growth of trading on different markets is observed. The Forex market with daily turnover reaching $5.3 trillion in January 2014 and the cryptocurrency market belong to the most popular ones. The single instrument of trading on the Forex market is the currency pair, which is the ratio of one currency to another. Data present on this market can be observed as chaotic one without any visible point attractor [1], what effects difficulties in effective trading. One should not expect, that in a longer time window a single price trend direction could be identified.

© Springer Nature Switzerland AG 2019
Z. Wilimowska et al. (Eds.): ISAT 2018, AISC 854, pp. 122–133, 2019.
https://doi.org/10.1007/978-3-319-99993-7_12

With a large number of different instruments along with an easy access to the market, the Forex becomes a market, on which a wide range of different trading techniques and respective trading systems can be applied. The fully automatic systems with algorithmic trading and the manual trading systems including elements of decision support are the most popular.

A general description of the algorithmic trading is presented in [2]. In this type of trading the final order decision is made automatically by the system. Such systems are comfortable in use however their application on real markets encounter many limitations.

The second most popular approach to the trading on the Forex market consists in the application of the manual systems with decision support. Such systems only suggest possible decisions but the final decision is made by the decision maker – trader. This approach assures sovereignty of the decision maker.

Methods involved in the trading process can be divided into three main groups using a different type of analysis: fundamental, technical, and sentiment. The technical analysis deals with the prediction of future prices on the basis of historical market data. Such quantitative data are mostly used in the process of finding new trading rules [3]. The fundamental analysis primarily looks at the economic conditions and interest rates. In this analysis information related to geopolitics, financial environment, macroeconomics, financial reports are used. One of the newest works investigating this subject can be found in [4]. The sentiment analysis tries to find recommendations for trading using news from opinion leaders in media and social networks like Twitter, see [5]. In the trading system information about the market is presented in the form of market indicators. The market indicators are formulated in the technical (or fundamental) analysis together with some trading rules. When the rules are satisfied a trading signal is generated. By the signal, we understand a recommendation for the trader to make an order (for example the BUY order). In the automatic system, an order is generated automatically.

We purport, that the set of market indicators deriving the most perspective signals about the market may dynamically change and propose a sequential selection of such indicators in trading systems. We analyze a number of signals derived by each considered market indicator. The efficiency of the single indicator is calculated on the basis of the number of positive signals (profitable orders) to an overall number of generated signals. The proposed approach is implemented and included in the trading system developed by the authors. It can be used in both: automatic and decision support systems. The presented method is experimentally verified for 7 different currency pairs.

The objectives of the research presented in the paper include investigation and comparison of different market indicators and construction of a rule-based trading system generating signals on the basis of the most effective market indicators. A special interest is focused on a new module in the system for automatic selection of such indicators using information from the market. Assuming that the efficiency is measured as the ratio of profitable orders to an overall number of orders taken on the basis of the generated signals, we take into account that the number of signals generated for different indicators may differ significantly. Our research hypothesis is as follows: there does not exist single market indicator (or set of indicators) which constantly dominate another indicator in the sense of the efficiency, and that the efficiency of the selected indicator is closely related with the time window used in the analysis. We assume, that for the long enough time window the similar efficiency of all the considered indicators may be observed and that the differences may relate rather to the short time windows. To confirm the above hypothesis an experimental rule-based trading system was developed and numerical experiments have been made using real data from the Forex market for six currency pairs, ten technical indicators and different time windows in the period 2015 – 2018. The obtained numerical results are presented and discussed in this paper.

This article is organized as follows: in Sect. 2 we introduce some related works. Section 3 describes a general modular construction of the trading information system. It includes the proposed in the paper new module for selection of the most effective indicators. Section 4 presents short information about the market indicators and rules used in the system. In Sect. 5 we describe the procedure for selecting the market indicators deriving the most efficient signals. The results of experiments conducted on the real-world data along with a short discussion are presented in Sect. 6.

2   Related Works

The bibliography on trading systems is very rich. Different methods and approaches are applied. Only selected papers are referred below. One of the first approaches including the expert knowledge in trading systems was proposed in [6], where the concept of fuzzy logic representation along with the Mamdani's form of rules was described. The methods of rough sets theory were used in [7]. Nowadays the main area of interest is focused on general concepts related directly to the hybrid systems (including the neural networks) and systems in which new rules are discovered. There are some papers dealing with neural networks like [8]. One of the most promising approaches involves the concept of the Psi Sigma Neural Network (PSI) and the Gene Expression algorithm (GEP) described in [10]. A very interesting connection between technical analysis indicators and the fundamental indicators like interest rates and gross domestic products can be found in [9].

The second group of articles is related to the systems based on the new trading rules formulated with the application of the genetic programming [11] or grammar evolution [12]. This approach is often based on approximate mechanisms like metaheuristics.

A text mining expert system on the Forex market can be found in [13]. A concept of ensemble methods, when multiple approaches combining a different set of rules gave promising results is presented in [14]. However, there are no articles directly related to the problem of selection of efficient indicators solved by a module included in the trading system.

3   General Schema of the Trading System

A general modular construction of the system is presented in Fig. 1. The "Opening the position" phase includes all actions required to generate the BUY signal and the BUY order for a given currency pair. In this paper, we propose to extend the typical trading system by the module presented by the dotted rectangle in the middle of the first phase. We assume, that a large spectrum of different market indicators could be defined by the decision maker at the system initialization. A further selection of the indicators used to generate signals is derived within the additional module of the first phase – called the "selection". Selection of the efficient indicators is the core idea of this article. It includes verification of efficiency of the indicators on the basis of the historical data and selection of the best ones. The proposed approach includes preliminary investigation of all considered indicators, checking their effectiveness. Only selected indicators having the highest effectiveness are used in the system. Finally, only a limited set of indicators along with the respective indicator rules is used to generate the signals and finish the first phase.

The second is the "managing the position" phase. It involves methods to properly manage the position in such way, that eventually losses are minimized, and profit is maximized. There are many approaches that include opening additional positions which allow increasing the potential profit. The last phase is "the closing position", in which on the basis of some predefined conditions the order (or orders) is closed and the account balance is updated.

In this article, we are particularly concerned with the proposed additional module selecting indicators in the "opening the position" phase. The module investigates the preliminary effectiveness of all considered indicators and selects indicators with the highest final score. The selected indicators will be used in the system to generate trading signals.

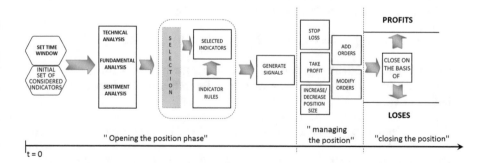

Fig. 1. General schema of the trading system

4   Overview of Rules in the Proposed System

There are numerous indicators which can be used to generate BUY or SELL signals. In this research, we investigated 10 different technical analysis indicators which can be included in the trading system. For every indicator, we derive details about the rule implemented in the trading system. We focused only on the BUY signals as examples. The detail information about some of the formal definitions of the indicators and the rules for the BUY and SELL signals can be found in [16]. For all considered indicators we calculated their values on the basis of the $n = 14$ last readings. The rules for the BUY signals for the considered indicators are presented below:

- CCI (Commodity Channel Index) – the oscillator used to estimate potential overbought and oversold levels possibly the most effective during the side trend. Described by the following binary activation function:

$$cond_{CCI_{Buy}} = true \text{ if } (CCI_n(r-1) < -100) \wedge (CCI_n(r) > -100), \quad (1)$$

where $CCI_n(r)$ is the value of the CCI indicator in the present reading r, while $r - 1$ means the previous reading;
- RSI (Relative Strength Index) – oscillator with assumptions similar to the CCI indicator with the following activation function:

$$cond_{RSI_{Buy}} = true \text{ if } (RSI_n(r-1) < 30) \wedge (RSI_r(r) > 30), \quad (2)$$

where $RSI_n(r)$ is the value of the RSI indicator in the present reading r;
- OsMA (Oscillator – Moving Average) – an indicator used to estimate the difference between MACD indicator and moving average calculated on the basis of the MACD indicator:

$$cond_{OsMA_{Buy}} = true \text{ if } (OsMA_n(r-1) < 0) \wedge (OsMA_n(r) > 0), \quad (3)$$

where $OsMA_n(r)$ is the value of the OsMA indicator in the reading r;
- Bulls power – simple indicator used to measure the strength of the rising trend with binary activation rule described as follows:

$$cond_{bulls_{Buy}} = true \text{ if } (bulls_n(r-1) < 0) \wedge (bulls_n(r) > 0), \quad (4)$$

where $bulls_n(r)$ is the value of the bulls indicator in the reading r;
- Bears power – indicator similar to the Bulls indicator used to measure the falling trend:

$$cond_{bears_{Buy}} = true \text{ if } (bears_n(r-1) < 0) \wedge (bears_n(r) > 0), \quad (5)$$

where $bears_n(r)$ is the value of the bears indicator in the reading t;
- Parabollic SAR – indicator used to estimate the direction of the actual trend:

$$cond_{SAR_{Buy}} = true \text{ if } (SAR_n(r-1) < Close(r-1)) \wedge (SAR_n(r) > Close(r)), \quad (6)$$

where $SAR_n(r)$ is the value of the SAR indicator in the reading r and Close(r) is the instrument price at the end of the reading r;

– Moving Average – the classical indicator used for the price noise reduction:

$$cond_{MA_{Buy}} = true \text{ if } (MA_n(r-1) < Close(r-1)) \wedge (MA_n(r) > Close(r)), \tag{7}$$

where $MA_n(r)$ is the value of the moving average in the reading r;
– DeMarker – indicator used to estimate potential overbought and oversold levels:

$$cond_{DMBuy} = true \text{ if } (DM_n(r-1) < 0.3) \wedge (DM_n(r) > 0.3), \tag{8}$$

where $DM_n(r)$ is the value of the DeMarker indicator in the reading r;
– Force Index – combines price and volume to assess the force behind price movements and spot eventual trend changes:

$$cond_{FIBuy} = true \text{ if } (FI_n(r-1) < 0) \wedge (FI_n(r) > 0), \tag{9}$$

where $FI_n(r)$ is the value of the force index in the reading r;
– Williams – momentum indicator measuring the overbought and oversold levels:

$$cond_{WilliamsBuy} = true \text{ if } (Williams_n(r-1) < -80) \wedge (Williams_n(r) < -80), \tag{10}$$

where $Williams_n(r)$ is the value of the Williams indicator in the reading r. One should note, that in the case of the last indicator we assume, that the BUY signal is generated when the value of the indicator is below the -80 during two successive readings.

5   Market Indicators Selection

There is a relatively large set of indicators which can be used to generate BUY or SELL signals. At a given time window some of them generate the signal but other ones don't. The question arrives which of the indicators generates the most beneficial signals/orders. In this paper selection of the most effective indicators is performed on the basis of their past behavior. We assume, that the historical performance of each considered indicator can be used as a suitable evaluator for further results achieved by this indicator for the given currency pair.

Let's assume, that for a given time window T, set $U_{i,curr}$ includes all transactions opened on the basis of rules from the indicator i in time window T. The number of profitable transactions for the indicator i and the currency pair $curr$ can be measured as:

$$|U_{i,curr}| = \{u_i \in U_{i,curr} : order(u_i, \delta) > \epsilon\}, \tag{11}$$

where $order(u_i, \delta)$ is the result of the transaction u_i after time window δ, while ϵ is a threshold used to exclude situations, for which the price of the instrument was in a very narrow price channel smaller than ϵ. This situation basically corresponds to the case when no significant price movement was observed. In the proposed approach only indicators with the highest efficiency are included in the trading system. Such indicators have to be selected from the overall number of the indicators considered initially. The $|U_{i,curr}|$ is calculated for all $i = 1, 2, ..., n$ and represents the number of successful transactions to overall number of transactions. Finally, the efficiency of each indicator is the ratio of the number of profitable transactions, to overall number of transactions:

$$eff(i) = \frac{|U_{i,curr}|}{U_{i,curr}}. \tag{12}$$

We assume, that the set of market indicators selected on the basis of the $eff()$ is profitable in the next time window t. The dependency between T, t and δ is presented in Fig. 2.

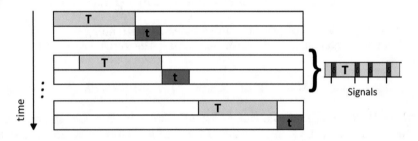

Fig. 2. The windows T, t, δ moving in the algorithm actions

The following pseudocode of the procedure is proposed:

Algorithm 1. Derivation of the most efficient indicators

---
begin
1   Initialize the set of indicators $U_{i,curr}$ for all currency pairs
2   Set the length of the time period T, t, δ and parameter m
3   **for** *each currency pair curr* **do**
4   Calculate the $U_{i,curr}$ for all indicators i
5   Calculate the efficiency $|U_{i,curr}|$ for all indicators i
6   Sort indicators descending, according to their efficiency
7   Select top m indicators and include them in "Opening the transaction" module

---

The initial set of indicators defined in line 1 could include as many indicators as possible. All calculations are derived on the basis of time period T and the results included in the list of the best indicators are valid through the time window t, while every transaction profitability is checked after the time window δ. The procedure is repeated after time window t. The core of the procedure can be seen in lines 4–7, where for every currency pair the following steps are performed: calculation of all transactions made in time window T for all indicators according to the crisp trading rules defined in Sect. 4; calculation of the efficiency for all indicators; ranking and selection of the best indicator (or top m indicators having the best efficiency). Only one, or m selected indicators will be further used in the trading system for generation of the trading signals.

6 Numerical Experiments

We performed experiments on 6 different currency pairs: AUDUSD, EURJPY, EURUSD, GBPUSD, USDCHF and USDJPY for 10 technical indicators. For every indicator we used the classical crisp rules described in details in Sect. 4. The results are presented in Tables 1 and 2. The following notations for indicators are used in the tables: CCI – Commodity Channel Index; RSI – Relative Strength Index; OsMA – Oscillator of Moving Average; Bears – Bears Indicator; Bulls – Bulls Indicator; SAR – Parabolic SAR Indicator; MA – Moving Average; DM – DeMarker Indicator; FI – Force Index; % W – Williams Indicator. The analyzed time window covered IX 2015 to IV 2018 with 4000 readings divided into four separate time windows T. For every time window the effectiveness of all indicators was investigated.

In Table 1 one can see the ratio of the successfully predicted transactions for all considered indicators. The number of generated signals for some indicators (like RSI or DM) is very low, especially, when compared with the number of signals generated for the Williams indicator. In general, it can be observed, that the considered indicators achieved effectiveness greater than 50%. The differences between a number of successful and unsuccessful orders are very small.

To investigate these observations in details, we calculated percentage effectiveness for all analyzed indicators for every currency pair separately. The results can be observed in Table 2. One of the most important observations is that the effectiveness for all analyzed indicators is not constant, and fluctuations easily exceed 15%. Since in the most cases effectiveness does not exceed threshold between 40% – 60%, such differences can be crucial for overall trading system

Table 1. The ratio of the profitable orders to non-profitable orders

	AUDUSD	EURJPY	EURUSD	GBPUSD	USDCAD	USDCHF	USDJPY
CCI	127/97	125/99	116/108	120/104	98/126	111/113	115/109
RSI	38/34	38/34	37/35	39/33	36/36	36/36	36/36
OsMA	94/90	96/88	101/83	92/92	102/82	84/100	88/96
Bears	135/141	147/129	134/142	136/130	155/121	129/147	131/145
Bulls	148/132	146/134	148/132	145/135	135/145	134/146	134/146
SAR	102/102	105/99	104/100	110/94	109/95	93/111	93/111
MA	126/118	124/120	131/113	132/112	129/115	110/134	111/133
DM	76/48	62/62	65/59	61/63	56/68	63/61	67/57
FI	123/121	132/112	129/115	131/113	131/113	117/127	119/125
%W	366/306	349/323	307/365	326/346	315/357	313/359	329/343

Table 2. The percentage of successfully predicted transactions for all currency pairs; I – covers period from IX 2015 – V 2016; II period – V 2016 – I 2017; III period – I 2017 – VIII 2017; IV period – VIII 2017 – IV 2018

AUDUSD	CCI	RSI	OsMA	Bears	Bulls	SAR	MA	DM	FI	%W
AUDUSD I	68%	44%	54%	57%	59%	47%	59%	61%	52%	52%
AUDUSD II	50%	44%	50%	49%	49%	51%	49%	48%	51%	53%
AUDUSD III	55%	61%	48%	45%	51%	53%	44%	74%	49%	60%
AUDUSD IV	54%	61%	52%	45%	53%	49%	54%	61%	49%	54%
EURJPY I	63%	50%	61%	52%	54%	49%	54%	45%	52%	48%
EURJPY II	64%	56%	50%	61%	56%	53%	56%	52%	66%	50%
EURJPY III	52%	50%	59%	58%	51%	49%	51%	45%	56%	56%
EURJPY IV	45%	56%	39%	42%	47%	55%	43%	58%	43%	54%
EURUSD I	46%	56%	48%	43%	44%	59%	41%	42%	46%	35%
EURUSD II	43%	50%	48%	46%	57%	49%	54%	61%	54%	54%
EURUSD III	55%	33%	65%	54%	51%	41%	59%	55%	46%	45%
EURUSD IV	63%	67%	59%	51%	59%	55%	61%	52%	66%	50%
GBPUSD I	59%	56%	54%	54%	56%	55%	48%	48%	56%	43%
GBPUSD II	41%	50%	37%	42%	39%	49%	44%	45%	39%	42%
GBPUSD III	63%	50%	48%	51%	57%	51%	61%	45%	61%	55%
GBPUSD IV	52%	61%	61%	51%	56%	61%	64%	58%	59%	54%
USDCAD I	34%	56%	63%	55%	44%	51%	61%	58%	51%	46%
USDCAD II	55%	61%	50%	52%	51%	55%	48%	58%	52%	57%
USDCAD III	38%	44%	48%	55%	44%	49%	49%	32%	48%	33%
USDCAD IV	48%	39%	61%	62%	53%	59%	54%	32%	64%	51%
USDCHF I	39%	28%	39%	45%	36%	45%	39%	52%	38%	33%
USDCHF II	54%	50%	46%	52%	44%	49%	51%	52%	59%	50%
USDCHF III	54%	61%	50%	39%	57%	49%	39%	58%	39%	54%
USDCHF IV	52%	61%	48%	51%	54%	39%	51%	42%	56%	50%
USDJPY I	43%	33%	50%	42%	41%	29%	34%	52%	39%	43%
USDJPY II	52%	44%	43%	48%	54%	53%	43%	61%	49%	52%
USDJPY III	57%	44%	52%	52%	50%	53%	54%	61%	56%	48%
USDJPY IV	54%	78%	46%	48%	46%	47%	51%	42%	51%	52%

effectiveness. At the same time, none of the indicators achieved more than 80% and less than 20% of successful transactions.

We also investigated, which indicator was mostly selected as the best – these results presented as the histogram can be seen in Fig. 3. It can be seen, that two indicators: RSI and DM achieved a visible advantage over remaining indicators. The Bulls indicator was not selected even once, and there is a group of indicators, which achieved the best results only for a single analyzed time period.

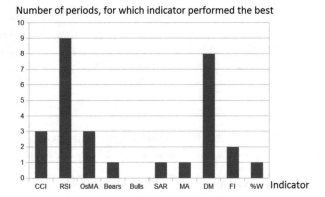

Fig. 3. The most efficient indicators

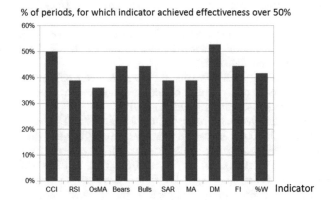

Fig. 4. The overall number of time periods (each period including approx. 1000 readings) for which given indicator had the highest effectiveness

To estimate the overall efficiency of all indicators, we calculated the number of time periods, for which indicators achieved at least 50% efficiency. The results can be seen in Fig. 4.

7   Conclusions and Future Works

In this article a novel approach is presented, in which a trading system including the module selecting market indicators on the basis of their past effectiveness is proposed. The efficiency of frequently used indicators is verified in the experimental studies. The studies show, that in general, all the market indicators have some predictive capabilities, which are slightly above the "luck guess" level. However, when analyzing small trading periods, there are some indicators which are far more effective than others. What more, the set of the most effective indicators changes over time. It confirms that the proposed approach is promising.

Such effective indicators should be sequentially selected and recommended for further analysis in the trading systems - manual and automatic. In further works the proposed module will be added to the trading systems constructed with the use of the fuzzy multicriteria approach presented in [15].

References

1. Ramasamy, R., Mohd, H.M.H.: Chaotic behavior of financial time series an empirical assessment. Int. J. Bus. Soc. Sci. **2**(3), 77–83 (2011). https://doi.org/10.30845/ijbss

2. Nuti, G., Mighaemi, M., Treleaven, P., Yingsaeree, C.: Algorithmic trading. Computer **44**(11), 61–69 (2011)

3. Mabu, S., Hirasawa, K., Obayashi, M., Kuremoto, T.: Enhanced decision making mechanism of rule-based genetic network programming for creating stock trading signals. Expert Syst. Appl. **40**(16), 6311–6320 (2013). https://doi.org/10.1016/j.eswa.2013.05.037

4. Chatrath, A., Miao, H., Ramchander, S., Villupuram, S.: Currency jumps, cojumps and the role of macro news. J. Int. Money Financ. **40**, 42–62 (2014). https://doi.org/10.1016/j.jimonfin.2013.08.018

5. Baccianella, A., Esuli, S., Sebastiani, F.: SentiWordNet 3.0: an enhanced lexical resource for sentiment analysis and opinion mining. In: Proceedings of European Language Resources Association (ELRA) (2010)

6. Dourra, H., Siy, P.: Investment using technical analysis and fuzzy logic. Fuzzy Sets Syst. **127**, 221–240 (2002). https://doi.org/10.1016/S0165-0114(01)00169-5

7. Wang, Y.F.: Mining stock price using fuzzy rough set system. Expert Syst. Appl. **24**(1), 13–23 (2003). https://doi.org/10.1016/S0957-4174(02)00079-9

8. Thawornwong, S., Enke, D., Dagli, C.: Neural networks as a decision maker for stock trading: a technical analysis approach. Int. J. Smart Eng. Syst. Des. **5**(4), 313–325 (2003). https://doi.org/10.1080/10255810390245627

9. Eng, M.H., Li, Y., Wang, Q.G., Lee, T.H.: Forecast forex with ANN using fundamental data. Int. Conf. Inf. Manag. **1**, 279–282 (2008)

10. Sermpinis, G., Laws, J., Karathanasopoulos, A., Dunis, C.L.: Forecasting and trading the EUR/USD exchange rate with gene expression and psi sigma neural networks. Expert Syst. Appl. **39**, 8865–8877 (2012). https://doi.org/10.1016/j.eswa.2012.02.022

11. Lee, C.S., Loh, K.Y.: GP-based optimisation of technical trading indicators and profitability in FX market. Neural Inf. Proc. **3**, 1159–1163 (2002)

12. Brabazon, A., O'Neill, M.: Evolving technical trading rules for spot foreign-exchange markets using grammatical evolution. Comput. Manag. Sci. **1**(3), 311–327 (2004)

13. Nassirtoussi, A.K., Aghabozorgi, S., Wah, T.Y., Check, D., Ngo, L.: Text mining of news-headlines for FOREX market prediction: a multi-layer dimension reduction algorithm with semantics and sentiment. Expert Syst. Appl. **42**(1), 306–324 (2015). https://doi.org/10.1016/j.eswa.2014.08.004

14. Mabu, S., Obayashi, M., Kuremoto, T.: Ensemble learning of rule-based evolutionary algorithm using multi-layer perceptron for supporting decisions in stock trading problems. Appl. Soft Comput. **36**, 357–367 (2015). https://doi.org/10.1016/j.asoc.2015.07.020

15. Juszczuk, P., Kruś, L.: Supporting multicriteria fuzzy decisions on the forex market. Multi. Criteria Decis. Mak. **12**, 60–74 (2017)
16. Kirkpatrick II, Ch.D, Dahlquist, J.R.: Technical Analysis. Complete Resource for Financial Market Technicians. FT Press, New Jersey (2010)

Impact of the Size of Equity
on Corporate Liquidity

Sebastian Klaudiusz Tomczak[(✉)]

Faculty of Computer Science and Management,
Wroclaw University of Science and Technology,
Ul. Wybrzeże Wyspiańskiego 27, 50-370 Wrocław, Poland
sebastian.tomczak@pwr.edu.pl

Abstract. In the market economy correct determination of the financial structure is immensely important. The right equity form can contribute to its competitive edge, whereas the improper one may lead up to many problems, among other things, the loss of corporate ability to pay off its financial obligations.

This paper covered the analysis of fifteen thousand manufacturing companies, among them the ones, marked by high quantities of equity in the total assets, have been selected along with intervals of values. Next, the analysis of financial liquidity has been carried out within the group in question – the most typical indicators of liquidity have been computed. Namely, current ratio, quick ratio, cash ratio, size of working capital I and II ratio.

The comparative analysis of the indicator of the size of equity with the indicators of liquidity makes it possible to answer the question: Are the companies having a high share of equity in the total assets marked by overliquidity? ANOVA test has been applied to analyse the influence of the size of equity on the corporate liquidity. Based on the results of the test, there is a significant impact on the value of liquidity ratios of manufacturing companies. The companies having a high share of equity in the total assets were marked by overliquidity. In addition, the standard values for liquidity ratios were calculated.

Keywords: Equity capital · Liquidity ratios · Manufacturing sector
ANOVA

1 Introduction

The proper balance of equity and debt in a company is important especially for managers. The proper balance allows managers to regulate the cost of capital and financial risk. In turn, this enable to achieve a long-term financial objective, which is to maximize the company's value through a higher return on capital than the cost of its acquisition and maintenance.

The loss of corporate capability to pay its debts is one of the most frequent given reasons for corporate financial condition which may lead to corporate bankruptcy [3]. It should be added that apart from the loss of liquidity [15] there are in literature of the subject many other reasons also related or not to erroneous management of corporate financial resources, for instance improper corporate financial structure [2] or misguided

© Springer Nature Switzerland AG 2019
Z. Wilimowska et al. (Eds.): ISAT 2018, AISC 854, pp. 134–144, 2019.
https://doi.org/10.1007/978-3-319-99993-7_13

corporate strategy [13]. Therefore incorrect management of liquidity as well as financial structure contributes towards its annihilation [10]. In the light of this fact, determination of the proper financial structure as well as reasonable management of liquidity seem to be the necessary condition for corporate correct workings. It worth adding that the size of equity influences on the value of profitability ratios of manufacturing companies but the impact was statistical irrelevant [14].

In this paper analysis of the influence of increase in equity in the total assets on liquidity of production companies was made. It will allow to give an answer to question: are the companies with a huge share of equity in the total assets characterized by overliquidity?

2 Research Methodology

The focus in this paper is on manufacturing sector. Research time-scale is set to 16 years (2000–2015). The choice of this period is determined by data availability. In the examined interval of time over 15 thousand firms from the sector in question were flirted drawn from the EMIS (86 thousand financial reports were analysed). The gathered data enabled computation of the following financial indicators for the itemized companies: equity ratio (ER); indicator of the size of working capital (SWC1), (SWC2) current ratio (CR); quick ratio (QR), cash ratio (CaR).

Then selection of companies was carried out to choose these with a high share of equity in its total assets. The proportion was regarded to be high if it reaches fifty one percent. Companies with quantities below this threshold have been excluded from the sample.

The next step defines intervals (classes) for the equity ratio. The following were chosen: 0,51–0,60; 0,61–0,70; 0,71–0,80; 0,81–0,90; 0,91–1,00. Depending on the value of equity, the remaining firms were assigned to their corresponding class. For instance a particular company with the value of its indicator of equity at the level of sixty five percent is going to be assigned to the interval 0,61–0,70. Also other computed indicators of this company will be allotted to the interval representing the size of equity. For example the company is described by the averaged values of indexes in the examined period: ER = 65%; SWC1 = 0,32; SWC2 = 22,24; CR = 2,03; QR = 1,61; CaR = 0,16. The values of these indexes will point to the class 0,61–0,7 the quality of equity. Each firm will be assigned to its corresponding interval of the size of equity. Thanks to this one can investigate how the increase in equity will influence the levels of liquidity indexes.

According to the size of the equity ratio different companies have been assigned to different classes. There are 1131 firms in the first class 0,51–0,60. The second 0,61–0,70 contains 1976 and in the third 0,7–0,8 there are 1847 companies. The penultimate 0,81–0,9 and the last comprise 1452 and 598 companies accordingly.

The aim of setting intervals of the equity ratio is to check its influences on corporate liquidity. Making use of these intervals is to answer the questions: does corporate liquidity augment with the increase in equity in the total assets? Are the firms with a high percentage of equity in the total assets (nearing one) characterized by overliquidity.

This paper is written with the following hypothesis in mind:

H10: A high share of equity in the total assets has statistically significant impact on the liquidity of manufacturing companies.

H1A: A high share of equity in the total assets has no statistically significant impact on the liquidity of manufacturing companies.

In order to analyse the impact of the size of equity on the level of liquidity ANOVA test has applied (the hypothesis will be tested against the significance of 0.05). This test is a frequently used method for analysing dissimilarities between the averages of two or more intervals (in this particular case differences between five defined intervals of values of the ratio). This test one can estimate whether the difference in means of analysed intervals is statically significant. The ANOVA test was applied by the use of the Statistica suite.

3 Results

The reference indicator for liquidity ratios to be analysed with is the size of equity ratio. This index states to what degree company assets is covered by equity capital. Equity capital is needed for setting up a business. The values of this ratio circulate from zero to one, but also can be less than zero. If the value of this indicator are negative the company is in danger of bankruptcy. There is, however, some regularity, the higher value of this ratio the better financial standing of a company is. Thanks to a high value of index of equity capital in financing the assets companies feel more independent of lenders of external sources of financing [1].

In order to verify importance of impact of the size of equity capital on liquidity the following intervals of this ratio has been defined: 0,51–0,60; 0,61–0,70; 0,71–0,80; 0,81–0,90; 0,91–1,00. In the research sample there have been companies with the share of at least 51% equity in the total assets such businesses have been chosen. Each firm is assigned to its corresponding interval of the size of equity. The analysed liquidity ratios are going to be compared with equity. The results are going to be presented on Figs. 1, 2, 3, 4, and 5.

The first examined indicator of liquidity associated with the index of the size of equity is the index of the size of working capital I (Fig. 1). One assumes that high level of this index means huge involvement of working capital in the total assets, and also high corporate financial liquidity. This indicator returns values oscillating around 0,5 [12]. Nevertheless, it should be noted that the size of this index is also influenced by the involved assets and its degree of usage. The value of the indicator will be relatively high in a case of low assets or a significant degree of its usage. In this case it would not mean a favorable situation of this particular entity [9]. There are diverse strategies to manage that capital, however, not every yields the desired effect [18].

Having analysed the first index of liquidity, one can conclude that the value of the indicator of the-size-of-working-capital for the firms with a high share of equity in the total assets is greater than the median of the values of the indicator for the whole research sample which resides in the interval 0,13–0,21. The highest values of this index are generated by these companies whose shares of equity in the total assets are very near one.

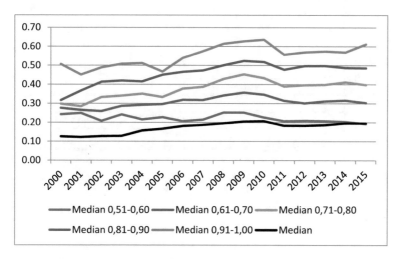

Fig. 1. Yearly values of the indicator of the size of working capital I within the division into classes.

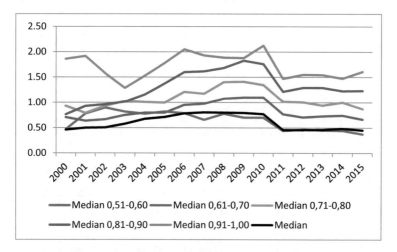

Fig. 2. Yearly values of the indicator of the size of working capital II within the division into classes.

The second examined liquidity indicator with the connection to the size of equity is the size-of-working-capital II (Fig. 2). It is similar to the already mentioned indicator of the size of working capital I. There is only one difference in the formula. Here the value of working capital is to be divided by fixed assets.

By analysing the second indicator, it can be concluded that the median of the indicator of the size of working capital II is four times higher than the level of working capital I and it takes values within the interval 0,45–0,80 in the period in question. It should be mentioned that the value of the median for the whole sample coincides with

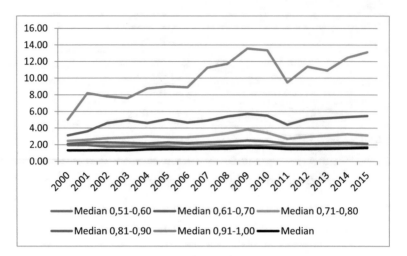

Fig. 3. Yearly values of the current ratio with the division into classes.

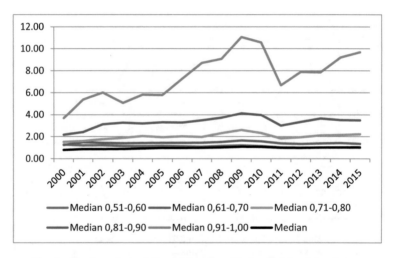

Fig. 4. Yearly values of the quick ratio with the division into classes.

the one for companies with equity in the assets assigned to the interval 0,51–0,60. In addition, one notices lower differences in the values of indicator between the intervals of the share of equity in the total assets, which may suggest smaller relevance of its influence on working capital.

The third examined index with reference to the indicator of the size of equity is the current ratio (Fig. 3). According to literature values of this indicator should oscillate between 1,2 and 2,0 [4]. Unfortunately, adopted norms of the index seem to be little precise, since individual authors present varied intervals of the optimal values of this index [5, 8]. Therefore the value of this index should be compared with the general one representing the sector in which the firm operates [6].

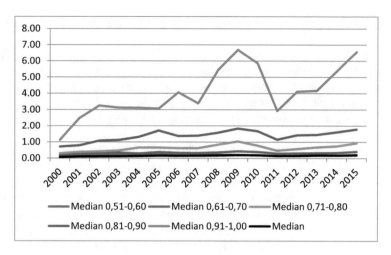

Fig. 5. Yearly values of the cash ratio with the division into classes.

From the drawing extra posed research results it follows that also in this case the median of the values of the indicator for the whole sample coincides with the one for the companies possessing equity in the total assets in the interval 0,51–0,60. In the examined period it oscillates between 1,32–1,63. However, the values of the current ratio for firms with equity ratio nearing one are even eight times higher than the value of the median for the whole sample, which denotes corporate overliquidity.

The next indicator being investigated with relation to the index of the size of equity is the quick ratio (Fig. 4). One takes it that the indicator at the level of 1,0 or so is correct and sufficient for maintenance of financial liquidity [11]. Whereas the index lower than 1,0 denotes huge difficulties in dealing with current liabilities. However, this regularity does not hold in case there are relevant dissimilarities between the period of drainage of short-term receivables and the period of payment of the short-term liabilities. The level of this index is strongly correlated with the difference of the cycles of receivable and liabilities [17].

The research results presented in Fig. 4 indicate that the median of values of the quick ratio for the whole sample in the period in question oscillates between 0,79 and 1,08. In addition, also here the median of the values of the indicators for the whole sample coincides with the one for the firms possessing equity in the total assets in the interval 0,51–0,60 and it differs significantly from the values of the index for the companies preferring to make use of equity such firms are characterized by overliquidity.

The last index that is analysed in connection with the size of equity is the cash ratio (Fig. 5). This indicator shows capability of an organization to perform payments in the right-away mode. It indicates which parts of the short-term financial obligations will immediately be finalized assuming that the possessed cash is the source of their cover. One takes is that the value of this index should be in the interval 0,1–0,2 [7]. The higher index the greater capability to finalize short-term obligations and vice versa. It worth emphasizing that values of this indicator lower than 0,06 indicate a lack of liquidity [16].

Carrying out analysis of the results presented in Fig. 5, it can be observed that the median of the cash ratio in the period in question resides in the interval 0,06–0,20. Moreover, in this case the values of the median of this index show similar dependencies to the median for the whole sample of companies with the highest share of equity, and of the firms with the share of equity equaling to at least the value of the defined threshold.

The provided research results prove that values of the median of analysed indicators for the firms with equity share nearing one in the total assets are considerably higher than the median of values for the whole research sample that equals to 15 thousand firms. It can thus be concluded that the higher equity in the total assets the greater liquidity. As for companies employing only equity there is substantial overliquidity.

On the basis of the carried out research one cannot determine any level of relevance of the individual classes of the median. Thus the level of relevance of the intervals of the median of intervals is going to be verified by means of the ANOVA test for independent samples (see Table 1). Before the ANOVA test was applied, the values of the ratios were cleared from outliers in order to obtain a normal distribution.

By analysing the following table, one can conclude that initially these differences between intervals were irrelevant. However, from 2007 through 2015 there dissimilates were already statistically relevant with just one exception as far as the cash ratio is concerned, which stood out from the others.

The presented research results show that the higher share of equity in the assets the grater corporate liquidity will be. Moreover, one should affirm that companies making use of only equity are marked by huge overliquidity.

Table 1. Results of the ANOVA test for the investigated intervals of values of median for the itemized financial indicators (2000–2015).

Year	Class/Ratio	SWC1	SWC2	CR	QR	CaR
2000	1	5	5	4, 5	4, 5	3, 4, 5
	2	No	5	4, 5	4, 5	3, 4
	3	No	No	5	No	1, 2, 4
	4	No	No	1, 2	1, 2	1, 2, 3
	5	1	1, 2	1, 2, 3	1, 2	1
2001	1	4, 5	No	3, 4, 5	4, 5	3, 4
	2	4, 5	No	4, 5	4	3, 4
	3	5	No	1, 5	No	1, 2
	4	1, 2	No	1, 2	1, 2	1, 2
	5	1, 2, 3	No	1, 2, 3	1	No
2002	1	4, 5	No	3, 4, 5	3, 4, 5	3, 4
	2	4, 5	3, 4	3, 4, 5	4, 5	3, 4
	3	5	2	Yes	1, 4, 5	1, 2
	4	1, 2	2	Yes	1, 2, 4	1, 2
	5	1, 2, 3	No	Yes	1, 2, 3	No

(*continued*)

Table 1. (*continued*)

Year	Class/Ratio	SWC1	SWC2	CR	QR	CaR
2003	1	3, 4, 5	No	3, 4, 5	3, 4, 5	No
	2	4, 5	No	3, 4, 5	3, 4, 5	No
	3	1, 4	No	Yes	Yes	No
	4	1, 2, 3	No	1, 2, 3	1, 2, 3	No
	5	1, 2	No	1, 2, 3	1, 2, 3	No
2004	1	3, 4, 5	No	3, 4, 5	3, 4, 5	No
	2	4, 5	No	3, 4, 5	3, 4, 5	No
	3	1, 4	No	Yes	1, 2, 4	No
	4	1, 2, 3	No	Yes	1, 2, 3	No
	5	1, 2	No	Yes	1, 2	No
2005	1	3, 4, 5	No	3, 4, 5	3, 4, 5	No
	2	4, 5	No	3, 4, 5	3, 4, 5	No
	3	1, 4	No	1, 2, 4	1, 2, 5	No
	4	1, 2, 3	No	1, 2, 3	1, 2	No
	5	1, 2	No	1, 2	1, 2, 3	No
2006	1	Yes	3, 4, 5	Yes	Yes	No
	2	Yes	4	Yes	Yes	No
	3	Yes	1, 4	1, 2, 3	1, 2, 4	No
	4	1, 2, 3	1, 2, 3	1, 2, 3	1, 2	No
	5	1, 2, 3	1	1, 2	1, 2	No
2007	1	Yes	Yes	Yes	Yes	No
	2	Yes	1, 4, 5	Yes	Yes	No
	3	Yes	1, 4, 5	Yes	Yes	No
	4	Yes	1, 2, 3	1, 2, 3	1, 2, 3	No
	5	Yes	1, 2, 3	1, 2, 3	1, 2, 3	No
2008	1	Yes	Yes	Yes	Yes	No
	2	Yes	1, 4, 5	Yes	Yes	No
	3	Yes	1, 4	Yes	Yes	No
	4	Yes	1, 2, 3	1, 2, 3	Yes	No
	5	Yes	1, 2	1, 2, 3	Yes	No
2009	1	Yes	Yes	Yes	Yes	No
	2	Yes	Yes	Yes	Yes	No
	3	Yes	1, 2, 4	Yes	Yes	No
	4	Yes	1, 2, 3	1, 2, 3	1, 2, 3	No
	5	Yes	1, 2	1, 2, 3	1, 2, 3	No
2010	1	Yes	Yes	Yes	Yes	No
	2	Yes	1, 4, 5	Yes	Yes	No
	3	Yes	1	Yes	Yes	No
	4	Yes	1, 2	1, 2, 3	Yes	No
	5	Yes	1, 2	1, 2, 3	Yes	No

(*continued*)

Table 1. (*continued*)

Year	Class/Ratio	SWC1	SWC2	CR	QR	CaR
2011	1	Yes	Yes	Yes	Yes	No
	2	Yes	Yes	Yes	Yes	No
	3	Yes	1, 2, 4	Yes	Yes	No
	4	Yes	1, 23	Yes	Yes	No
	5	Yes	1, 2	Yes	Yes	No
2012	1	Yes	Yes	Yes	Yes	No
	2	Yes	Yes	Yes	Yes	No
	3	Yes	1, 2, 4	Yes	Yes	No
	4	Yes	1, 23	Yes	Yes	No
	5	Yes	1, 2	Yes	Yes	No
2013	1	Yes	Yes	Yes	Yes	No
	2	Yes	Yes	Yes	Yes	No
	3	Yes	Yes	Yes	Yes	No
	4	Yes	1, 2, 3	1, 2, 3	Yes	No
	5	Yes	1, 2, 3	1, 2, 3	Yes	No
2014	1	Yes	Yes	Yes	Yes	No
	2	Yes	Yes	Yes	Yes	No
	3	Yes	1, 2, 4	Yes	Yes	No
	4	Yes	1, 2, 3	1, 2, 3	Yes	No
	5	Yes	1, 2	1, 2, 3	Yes	No
2015	1	Yes	4, 5	2, 3, 4	1, 2, 3	No
	2	1, 4, 5	4	1, 3, 4	1, 3, 4	5
	3	1	No	1, 2, 4	1, 2, 4	No
	4	1, 2	1, 2	1, 2, 3	1, 2, 3	5
	5	1, 2	1	No	Yes	2, 4

Legend: *SWC1* - the size of working capital ratio I, *SWC2* - the size of working capital ratio II, *CR* - current ratio, *QR* - quick ratio, *CaR* - cash ratio;
1 = 0,51–0,60; 2 = 0,61–0,70; 3 = 0,71–0,80; 4 = 0,81–0,90; 5 = 0,91–1,00;
No = ddifferences between the compartments are statistically insignificant;
Yes = ddifferences between the compartments are statistically significant.

4   Conclusions

This paper examines over 15 thousand manufacturing companies. Time-span incorporated sixteen years (2000–2015). The firms that maintained a high percentage of equity in the total assets were chosen among all firms being examined for further consideration. The threshold value was set to 51%. Then the following intervals (classes) of equity ratio were determined 0,51–0,60; 0,61–0,70; 0,71–0,80; 0,81–0,90; 0,91–1,00.

In order to compare the relevance of the influence of equity ratio on liquidity several financial indicators were computed for the itemized companies: indicator of the size of working capital I, II; current ratio; quick ratio, cash ratio. The statistical relevance of the influence of the size of equity on the corporate liquidity was checked by using ANOVA test.

Presented results of this research back up the relevance of equity in the total assets for liquidity of companies. The higher percentage of equity in the total corporate assets, the higher its liquidity, and marked overliquidity comes into being in these companies that make use of only equity. Therefore, the null hypothesis should be confirmed.

It should be pointed out that the median values of ratios for the whole research sample can be treated as the reference value for the selected indexes of the manufacturing sector. The reference values of ratios for the selected sector oscillate in the range:

- 0,12–0,21 for the size of working capital I,
- 0,45–0,80 for the size of working capital II,
- 1,32–1,63 for current ratio,
- 0,79–1,03 for quick ratio,
- 0,06–0,21 for cash ratio.

References

1. Walczak, M. (ed.): Analiza finansowa w zarządzaniu współczesnym przedsiębiorstwem. Difin, Warszawa (2007)
2. Bruno, A.V., Leidecker, J.K.: Causes of new venture failure, 1960s vs. 1980s. Bus. Horiz. **31** (6), 51–56 (1988)
3. Coface report: Upadłości firm w Polsce w 2015 r. http://www.coface.pl/Aktualnosci-i-Media/Biuro-prasowe/. Accessed 1 Feb 2015
4. Czekaj, J., Dresler, Z.: Zarządzanie finansami przedsiębiorstw. Podstawy teorii, PWN, Warszawa (2005)
5. Dudycz, T.: Siła diagnostyczna wskaźników płynności. In: Kuciński, K., Mączyńska, E. (eds.) Zagrożenie upadłością, Materiały i prace IFGN, SGH, Warszawa, pp. 63–76 (2005)
6. Dudycz, T.: Uwarunkowania interpretacyjne wskaźników płynności. In: Sierpinska, M. (ed.) Zarządzanie finansami we współczesnych przedsiębiorstwach, T. 1., pp. 11–26. VIZJA PRESS&IT, Warszawa (2009)
7. Dudycz, T., Hamrol, M., Skoczylas, W., Niemiec, A.: Finansowe wskaźniki sektorowe - pomoc przy analizie finansowej i ocenie zdolności przedsiębiorstwa do kontynuacji działalności. Rachunkowość **3**, 1–6 (2005)
8. Hodun, M., Żurakowska-Sawa, J.: Płynność finansowa przedsiębiorstw przemysłowych w zależności od strategii gospodarowania kapitałem obrotowym. Zeszyty Naukowe SGGW - Ekonomika i Organizacja Gospodarki Żywnościowej **97**, 67–80 (2012)
9. Kusak, A.: Płynność finansowa Analiza i sterowanie. WWZ, Warszawa (2006)
10. Mączyńska, E: Bankructwa przedsiębiorstw. Wymiar teoretyczny, statystyczny i rzeczywisty, "Biuletyn PTE" **1**, 7–35 (2013)
11. Michalski, G.: Strategiczne zarządzanie płynnością finansową w przedsiębiorstwie. CeDeWu, Warszawa (2010)
12. Skowrońska-Mielczarek, A., Leszczyński, Z.: Analiza działalności i rozwoju przedsiębiorstwa. PWE, Warszawa (2008)

13. Szczerbak, M.: Przyczyny upadłości przedsiębiorstw w świetle opinii syndyków i nadzorców sądowych. In: Kuciński, K., Mączyńska, E. (eds.) Zagrożenie upadłością, Materiały i prace IFGN, SGH, Warszawa, pp. 36–45 (2005)

14. Tomczak, S.K.: Influence of the size of equity on corporate efficiency. Oeconomia Copernicana 8(2), 239–254 (2017)

15. Tomczak, S.: Comparative analysis of the bankrupt companies of the sector of animal slaughtering and processing, equilibrium. Q. J. Econ. Econ. Policy 9(3), 59–86 (2014)

16. Tomczak, S.: Comparative analysis of liquidity ratios of bankrupt manufacturing companies. Bus. Econ. Horiz. 10(3), 151–164 (2014)

17. Wasilewska, E.: Model szybkiego testu płynności finansowej. Zeszyty Naukowe SGGW - Ekonomika i Organizacja Gospodarki Żywnościowej 89, 109–122 (2011)

18. Wasilewski, M., Zabolotnyy, S.: Sytuacja finansowa przedsiębiorstw o odmiennych strategiach zarządzania kapitałem obrotowym. Zeszyty Naukowe SGGW - Ekonomika i Organizacja Gospodarki Żywnościowej 78, 5–20 (2009)

Estimation of the Probability of Inversion in the Test of Variable Dependencies

Mariusz Czekała[1(✉)], Agnieszka Bukietyńska[1],
and Agnieszka Matylda Schlichtinger[2]

[1] Wroclaw School of Banking, Wrocław, Poland
{mariusz.czekala,
agnieszka.bukietynska}@wsb.wroclaw.pl
[2] Department of Physics, University of Wrocław, Wrocław, Poland
mathilde.schlichtinger@gmail.com

Abstract. In the present work, the distribution of the number of inversions has been considered. This constitutes a generalization of the distribution used in the Kendall test. Explicit formulas have been given for the calculation of the first two moments of the distribution of the number of inversions. In addition, proposal for the estimation of the probability of inversion has been provided. The proposed estimator, based on the monotonicity of the expected value, is an unbiased estimator.

Keywords: Inversion · Kendall test · Expected value · Variance
Unbiased estimator

1 Introduction

The destination of the paper is to consider the distribution of inversions. The mentioned distribution was earlier presented in the paper [1]. The specified case of this distribution is identical with the distribution which is used in the classical Kendall's test. However, the classical Kendall's test is used to test a hypothesis about independence versus dependence. The power of this test is unknown, because of lack of commonly accepted distribution, when the alternative hypothesis holds.

The mentioned inversion distribution will be analyzed in detail in the paper. Moreover, the alternative method of computing of inversion number will be presented. In the paper the arithmetical formulas, based on inversions number for the samples that were smaller, than the currently analyzed ones, were applied. In the present paper the sequence A008302 [10] will be generated thanks to the use of the algebra of matrices. There turns out that both ways are equivalent. The sequence A008302 describes the number of permutations with specified number of inversions. This corresponds with situation, when probability of the inversion is equal to 0.5.

The probability 0.5 may also occur in one of the formulations of hypothesis about independence that constitutes a tested hypothesis in Kendall's test. Nevertheless, there is a difficulty in interpretation the case of rejecting null hypothesis. In this case, however, did not exist a proposition of alternative distribution. Evoking the proposition from the paper [1] the analysis of two first moments of inversions distribution will be

© Springer Nature Switzerland AG 2019
Z. Wilimowska et al. (Eds.): ISAT 2018, AISC 854, pp. 145–156, 2019.
https://doi.org/10.1007/978-3-319-99993-7_14

added here. In the mentioned work expected values and variances for chosen values of p and n were computed. Here, the explicit formulas of expected values and variances for any n and p will be presented.

Subsequently, problem of estimation of the probability of the inversion will be considered. In this order, the formula for the expected value of inversions number will be applied. Estimation will be based on the method of maximal likelihood. The paper will be finished with two empirical examples. The first of them (for n = 4) will constitute the illustration of the estimation method, while the second one is intend to present the applying for analysis of expert's notes. This will be a new example of applying. Till now, the presented method was applied to the analysis of the rankings of investments funds and to the analysis of surveys associated with innovations in enterprises.

2 Number of Inversions – Algebra of Matrices

Before starting the consideration, the definition of the inversion will be mentioned.

Definition 1.
There is introduced the sequence of observations $(X_1, \ldots, X_n)$. The inversion for i and j occurs, if and only if $i > j$, but $X_i < X_j$.

There is assumed that the considered distribution is continuous. Therefore, there are no ties. In Table 1. the number of inversions for chosen values of n is presented. Continuous values may be computed thanks to the methods of matrices algebra.

Table 1. Sequences A008302 for chosen n and a number of inversions.

n/Inversion	0	1	2	3	4	5	6	7	8	9
1	1									
2	1	1								
3	1	2	2	1						
4	1	3	5	6	5	3	1			
5	1	4	9	15	20	22	20	15	9	4
6	1	5	14	29	49	71	90	101	101	90
7	1	6	20	49	98	169	259	359	455	531
8	1	7	27	76	174	343	602	961	1415	1940
9	1	8	35	111	285	628	1230	2191	3606	5545
10	1	9	44	155	440	1068	2298	4489	8095	13640
11	1	10	54	209	649	1717	4015	8504	16599	30239

Source: own work

For $n = 1$ there exists only one permutation without inversion. In this case, number of inversion can be described as $C_1 = (1)$, where C_1 is the number of permutations without inversion.

For $n = 2$ there are two permutations.

This is easy to note that $C_2 = \begin{pmatrix} 1 \\ 1 \end{pmatrix} = \begin{pmatrix} 1 \\ 1 \end{pmatrix} \cdot C_1$.

Let $M_{k,k-1}$ be a band matrix with bandwidth k. Particularly:

$$M_{2,1} = \begin{pmatrix} 1 \\ 1 \end{pmatrix},$$

$$M_{3,2} = \begin{pmatrix} 1 & 0 \\ 1 & 1 \\ 1 & 1 \\ 0 & 1 \end{pmatrix},$$

$$M_{4,3} = \begin{pmatrix} 1 & 0 & 0 & 0 \\ 1 & 1 & 0 & 0 \\ 1 & 1 & 1 & 0 \\ 1 & 1 & 1 & 1 \\ 0 & 1 & 1 & 1 \\ 0 & 0 & 1 & 1 \\ 0 & 0 & 1 & 1 \end{pmatrix}.$$

Let $N_n = \frac{n(n-1)}{2}$ be the maximal number of inversions for n observations. With respect to this, $M_{n,n-1}$ is characterized by $N_n + 1$ rows and $N_{n-1} + 1$ columns. Miscellaneous rows of the sequence A008302 may be computed according the formulas:

$$C_3 = M_{3,2} \cdot C_2$$
$$C_4 = M_{4,3} \cdot C_3$$
$$\vdots$$
$$C_n = M_{n,n-1} \cdot C_{n-1}$$

Especially:

$$C_3 = \begin{pmatrix} 1 & 0 \\ 1 & 1 \\ 1 & 1 \\ 0 & 1 \end{pmatrix} \cdot \begin{pmatrix} 1 \\ 1 \end{pmatrix} = \begin{pmatrix} 1 \\ 2 \\ 2 \\ 1 \end{pmatrix},$$

and

$$C_4 = \begin{pmatrix} 1 & 0 & 0 & 0 \\ 1 & 1 & 0 & 0 \\ 1 & 1 & 1 & 0 \\ 1 & 1 & 1 & 1 \\ 0 & 1 & 1 & 1 \\ 0 & 0 & 1 & 1 \\ 0 & 0 & 1 & 1 \end{pmatrix} \cdot C_3 = \begin{pmatrix} 1 \\ 3 \\ 5 \\ 6 \\ 5 \\ 3 \\ 1 \end{pmatrix}.$$

One may notice that the vectors C_3 and C_4 are consistent with the subsequent rows of the terms of the sequence A008302 for $n = 3$ and $n = 4$.

The above considerations show that the matrices algebra may constitute a convenient tool for computing the miscellaneous terms of the considered sequence. The terms of this sequence are used for computing the probabilities in the inversion test with an assumption that null hypothesis is valid. To obtain he mentioned probabilities the terms in n-th row ought to be divided by $n!$.

3   Inversions Distribution Under Dependence

Below there is presented an inversions distribution for inversions probabilities that are different than 0.5.

For particular values of p and q, this is quite easy to compute expected value and variance. However, in the estimation process, explicit formula of the at least estimation value is required. It is implied by the assumption which is presented below that observations results are consistent with expected results.

Additional number of inversions is presented in Table 2.

Table 2. Distribution of Y_k (the case of associations).

Y_k	0	1	2	3	....	$k-1$
Probability	q^{k-1}/w_k	$q^{k-2}p/w_k$	$q^{k-3}p^2/w_k$	$q^{k-4}p^3/w_k$	.....	p^{k-1}/w_k

Source: own work
where $w_k = (q^k - p^k)/(q - p)$ for $q \neq p$
In case of $p = 0.5$ distribution of Y_k has the form

Table 3. Distribution [5] of Y_k for $p = 0.5$

Y_k	0	1	2	3	...	...	$k-1$
Probability	1/k	1/k	1/k	1/k	...	...	1/k

Source: own work

Due to the fact that the number of inversions is a sum of an additional values of inversion for any n, the following formula holds:

$$I_n = \sum_{i=1}^{n} Y_i.$$

In general case, the distribution of the number of inversions I_n has the following form (Table 4):

Distribution of the inversions number is a consequence of the assumption about the distribution of the additional inversions number. The correctness of this fact was proved in the paper [4].

Table 4 Distribution of I_n (the case of associations).

I_n	0	1	...	k	...	N_n
Probability	$\left\{\begin{matrix} N_n \\ 0 \end{matrix}\right\} \frac{q^{N_n}}{W_n}$	$\left\{\begin{matrix} N_n \\ 1 \end{matrix}\right\} \frac{q^{N_n-1}p}{W_n}$	$\cdots$	$\left\{\begin{matrix} N_n \\ k \end{matrix}\right\} \frac{q^{N_n-k}p^k}{W_n}$	$\cdots$	$\left\{\begin{matrix} N_n \\ N_n \end{matrix}\right\} \frac{p^{N_n}}{W_n}$

Source: own work

In Table 3. $\left\{\begin{matrix} N_n \\ k \end{matrix}\right\}$ described the number of permutations of n-elements having exactly k inversions. At the same time:

$$W_n = \sum_{k=0}^{N_n}\left\{\begin{matrix} N_n \\ k \end{matrix}\right\}q^{N_n-k}p^k.$$

Basing on the distribution of Y_n there exists a possibility to compute an expected value of Y_n :

$$EY_n = \frac{1}{W_n}\left[1\cdot q^{n-2}p + 2q^{n-3}p^2 + \cdots + (n-1)p^{n-1}\right]$$

$$= \frac{p}{W_n}\left[q^{n-2} + 2q^{n-3}p + 3q^{n-4}p^2 + \cdots + (n-1)p^{n-2}\right]$$

$$= \frac{p}{W_n}\frac{\partial}{\partial x}\left[\left(q^{n-2}x + q^{n-3}x^2 + q^{n-4}x^3 + \cdots + x^{n-1}\right)\right]_{|x=p}$$

Using a differential calculus one may apply the formula of the sum of geometrical sequence

$$EY_n = \frac{p}{W_n}\frac{\partial}{\partial x}\left[\left(q^{n-2}x\cdot\frac{1-\left(\frac{x}{q}\right)^{n-1}}{1-\frac{x}{q}}\right)\right]_{|x=p} = \frac{p}{W_n}\frac{\partial}{\partial x}\left[\left(\frac{q^{n-1}x - x^n}{q-x}\right)\right]_{|x=p}$$

$$= \frac{p}{W_n}\left[\frac{(q^{n-1}-nx^{n-1})(q-x)-(-1)(q^{n-1}x-x^n)}{(q-x)^2}\right]_{|x=p}$$

$$= \frac{p}{W_n}\left[\frac{q^n - nqx^{n-1} + nx^n - x^n}{(q-x)^2}\right]_{|x=p} = \frac{p}{W_n}\left[\frac{q^n - nqx^{n-1} + (n-1)x^n}{(q-x)^2}\right]_{|x=p}$$

Subsequently, after substitution of p in the place of x the value of expected value will be obtained. This is a function of variables p, n (because $q = 1 - p$).

$$EY_n = \frac{p}{W_n}\left[\frac{q^n - nqp^{n-1} + (n-1)p^n}{(q-p)^2}\right] = \frac{p(q-p)}{q^n - p^n}\cdot\left[\frac{q^n - nqp^{n-1} + (n-1)p^n}{(q-p)^2}\right]$$

$$= \frac{p}{q-p} \cdot \frac{q^n - nqp^{n-1} + (n-1)p^n}{q^n - p^n} = f_n(p)$$

The mentioned function is signed with a symbol $f_n(p)$ (Fig. 1).

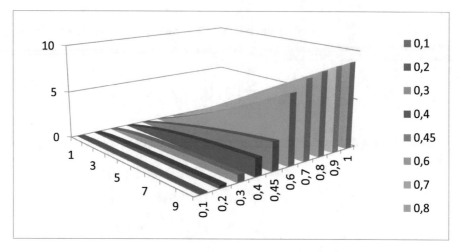

Fig. 1. Expected values of EY_n for chosen values of n and p. Source: own work

Since

$$I_n = \sum_{i=1}^{n} Y_i,$$

therefore,

$$EI_n = \sum_{k=1}^{n} f_k(p).$$

Analogically, the variance of the random variables I_n and Y_n can be obtained. The value of the variance completes the characteristics of the distribution of inversions. In this paper, the idea of computing variance will be presented, despite the fact, that it is not necessary in the estimation process. The identity $EY_n^2 = EY_n(Y_n - 1) + EY_n$ will be applied during computing the second moment of these random variables (Table 5).

The method of computing the variance is similar, but it requires finding the second derivative of the proper function.

Let $Z = [(Y_n)(Y_n - 1)]$

Table 5. Expected values of Y_n and I_n, and example values of p (0.6 and 0.52)

n	$EY_n, p = 0.6$	$EI_n, p = 0.6$	$EY_n, p = 0.52$	$EI_n, p = 0.52$
1	0.00	0.00	0.00	0.00
2	0.60	0.60	0.52	0.52
3	1.26	1.86	1.05	1.57
4	1.98	3.85	1.60	3.17
5	2.76	6.61	2.16	5.33
6	3.58	10.18	2.73	8.07
7	4.44	14.62	3.32	11.38
8	5.32	19.94	3.92	15.30
9	6.24	26.18	4.53	19.83
10	7.18	33.36	5.15	24.98
11	8.13	41.49	5.79	30.77

Source: own work

$$E(Z) = \frac{1}{w_n} \left[1 \cdot 0 \cdot q^{n-2}p + 2 \cdot 1 \cdot q^{n-3}p^2 + 3 \cdot 2 \cdot q^{n-4}p^3 + \cdots + (n-1)(n-2)q^0 p^{n-1} \right]$$

$$= \frac{p^2}{w_n} \left[2 \cdot 1 \cdot q^{n-3} + 3 \cdot 2q^{n-4}p + 4 \cdot 3q^{n-5}p^2 + \cdots + (n-1)(n-2)p^{n-3} \right]$$

$$= \frac{p^2}{w_n} \frac{\partial^2}{\partial x^2} \left[(q^{n-3}x^2 + q^{n-4}x^3 + q^{n-5}x^4 + \cdots + q^{n-n}x^{n-1}) \right]_{|x=p}$$

Using differential calculus twice, one may apply the formula of the sum of geometrical sequence to get the second derivative of the above function.

$$E(Z) = \frac{p^2}{w_n} \left(\frac{2q^n - n(n-1)q^2 x^{n-2} + 2n(n-2)qx^{n-1} - (n-1)(n-2)x^n}{(q-x)^3} \right)_{|x=p}$$

Subsequently, after substitution of p in the place of x the value of expected value will be obtained. This is a function of variables p, n (because $q = 1 - p$).

$$E(Z) = \frac{p^2}{w_n} \left(\frac{2q^n - n(n-1)q^2 p^{n-2} + 2n(n-2)qp^{n-1} - (n-1)(n-2)p^n}{(q-p)^3} \right)$$

The variance of $V(Y_n) = EY_n^2 = EY_n(Y_n - 1) + EY_n$ and the variance of the number of inversions $V(I_n) = \sum_{k=1}^{n} V(Y_k)$.

4   Estimation of Probability of Inversion

It is easy to note that for $p = 0.5$, the expected value EI_n is equal to $\left(\frac{n(n-1)}{4}\right)$, what means $EI_n = N_n p$. Therefore, it seems that the natural estimator has the following form:

$$\hat{p} = \frac{I_n}{N_n}.$$

Unfortunately, the estimator $\hat{p}$ is unbiased, if and only if $p = 0.5$. Thus, the attempt of codification the observed value of inversions number with expected value is natural.
Since

$$EY_n = f_n(p),$$

thus

$$EI_n = \sum_{k=1}^{n} f_k(p) = F_n(p)$$

Estimator of the form $\hat{p} = F_n^{-1}(I_n)$ may be proposed. This is significant to stress that the functions $f_n(p)$ and $F_n(p)$ (as the sum of increasing functions) with constant n are increasing function of p. That is why, the inverse function always exists.

To obtain the estimator value one ought to solve the equation

$$F_n(p) = \hat{I}_n,$$

where $\hat{I}_n$ is the inversions number that was obtained basing on the sample.

For single ordering there exists only finite number of the inversions number that is possible to observe.

Remark 1. The case of $n = 3$

There will be considered a specified case, when $n = 3$.

In this case, the number of inversions that is possible to observe is equal to 0 or 1 or 2 or 3. Solving four equations $F_n(p) = \hat{I}_n$ for $\hat{I}_n \in \{0, 1, 2, 3\}$ allows to obtain four values of estimators $\hat{p} = 0$ or $\hat{p} = 0.3611$ or $\hat{p} = 0.6389$ or $\hat{p} = 1$. The considered estimator is unbiased. This is easy to show that in any of these cases. It can be shown:

$$\frac{0 \cdot \hat{q}^3 + 1 \cdot 2\hat{q}^2\hat{p} + 2 \cdot 2\hat{q}\hat{p}^2 + 3\hat{p}^3}{W_3} = \hat{I}_3$$

Where $\hat{q} = 1 - \hat{p}$

The estimator that was presented above is characterized of the property of unbiasedness. Proposition 1 says about it.

Proposition 1.

Estimator $\hat{p} = F_n^{-1}(I_n)$ is unbiased.

Proof.

$$E\hat{p} = EF_n^{-1}(I_n) = E\left(F_n^{-1}(F_n(p))\right) = p$$

This implies that after the substitution $\hat{p}$ in the place of p in the distribution of inversions number, the expected value of inversions number will be equal to the observed number of inversions.

Remark 2. Purchase decisions

In Table 6. there are presented amounts of orders as a whole and an order of an expert, who has been chosen for the analysis.

Table 6. Rank for selected product groups

Product group	Total from sale/quantity	Purchase of the expert	Total from sale/quantity – rank	Purchase of the expert – rank
MP	92 452	26 430	1	2
MX	76 717	30 690	2	1
LP	67 159	20 160	3	3
LH	26 914	19 380	4	4
TB	24 133	2 115	5	5
MN	15 933	1 320	6	6

Source: own work based on data [9] from MG4 Company Ltd

Specified orders have been assigned with ranks. In the two-rows description, the permutation $\begin{pmatrix} 1 & 2 & 3 & 4 & 5 & 6 \\ 2 & 1 & 3 & 4 & 5 & 6 \end{pmatrix}$ was obtained. Therefore, in result there exists one inversion (alteration of order for numbers 1 and 2). In the case of the lack of inversion there would be a perfect consistency between expert's forecasting and the real sales.

Because, one inversion was observed, for obtaining the value of estimator of p parameter, one ought to solve the equation

$$\hat{p} = F_n^{-1}(1).$$

This was necessary to use the numerical analysis. The value of estimator is presented in the Table 7. in bold. This value is equal to 0.1517. As it was noted above, there is a value of an unbiased estimator. Not only this value may be found in the table, there is also values of estimator for other numbers of inversions. One can observe that $\hat{p}(I_n) + \hat{p}(N_n - I_n) = 1$.

Pragmatically, this is sufficient to compute about the half of the searched values.

In the analysis of sales of the company MG4 there exists a possibility of multiple checking of any expert. In the mentioned company, data are related to miscellaneous assortments. In each of these cases there is a possibility to create the function of

Table 7. Number of inversion, number of permutations and probability

I_n	P_n	$\hat{p}$
0	1	0
1	5	0.1517
2	14	0.2457
3	29	0.3123
4	49	0.3641
5	71	0.4075
6	90	0.4462
7	101	0.4823
8	101	0.5177
9	90	0.5538
10	71	0.5925
11	49	0.6359
12	29	0.6877
13	14	0.7543
14	5	0.8483
15	1	1

Source: own work

distribution. Subsequently, for any expert, mixture of distributions may be made. This is a method to obtaining the distribution of inversions (interpreted as mistakes of experts) for any of analyzed experts.

One of such distributions with estimated p parameter on the level 0.1517 has been presented in the Fig. 2.

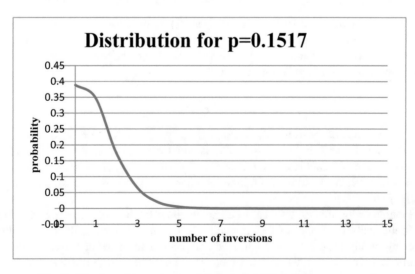

Fig. 2. Number of inversions and their probabilities. Source: own work

5 Summary

The proposed method was applied in the financial issues. The quality of investing in Open Pension Funds (OFE) and investments companies [3]. The results were presented in the works [1]. There was also given a speech about applications of the above method to analysis of surveys related to innovation [2]. Now, the application work that is connected with the note of experts is prepared.

What is peculiarly interesting, the described method may also have many wider applications. Some of them concern genetics, in particular, the issue of nucleotide order [8], which is associated with genetic mutations, and in the future they may be used with theoretical genetics. The method may also be useful in the physical sciences, in particular in matters related to quantum mechanics or elementary particle physics, because these theories are based on probability. However, it is also possible to find application in statistical physics, especially when describing individual states of the system, where inversions and permutations can be crucial.

The concept of using the method described in this work in the theory of music and composition may also be interesting. The method could be a way to study various sound sequences or be used in the composition as an alternative way of obtaining certain characteristic sound sequences.

However, this is not necessarily a question of the universality of the discussed method. It is rather a consequence of the structure of the Universe, which is reflected in mathematical formulas. They are present in strict sciences as well as humanities and arts, even if this is not fully disclosed to observers.

References

1. Bukietyńska, A.: The test of inversion in the analysis of investment funds. Cent. East. Eur. J. Manag. Econ. 5(3), 277–289 (2017)
2. Bukietyńska, A., Czekała, M., Kłosowski, J.: Analiza ankiet o innowacyjności z cechami na skali porządkowej – test istotności. In: Presented speech, III Konferencja Naukowa "W kierunku zarządzania procesowego organizacją – trendy i wyzwania", Uniwersytet Szczeciński (2018)
3. Bukietyńska, A., Czekała, M., Wilimowska, Z., Wilimowski, M.: The inversion test of the investment funds efficiency measures. In: Information Systems Architecture and Technology: Proceedings of 38th International Conference on Information Systems Architecture and Technology – ISAT 2017 – Part III, Springer (2017)
4. Czekała, M., Bukietyńska, A.: Distribution of inversions and the power of the Kendall's test. In: Information Systems Architecture and Technology: Proceedings of 37th International Conference on Information Systems Architecture and Technology – ISAT 2016 –Part III, Springer (2017)
5. Feller, W.: An Introduction to Probability Theory and Its Application. Wiley, New York (1961)
6. Kendall, M.G., Buckland, W.R.: A Dictionary of Statistical Terms. Oliver and Boyd, Edinburgh (1960)
7. Netto, E.: Lehrbuch der Combinatorik, 2nd edn, p. 96. Teubner, Leipzig (1927)

8. Schlichtinger, A.M.: On the Epigenetic Factors of Ageing of Organisms; Noumenology of Biological Changes; Theoretical Model of Ageing. Doctoral dissertation. State of New Jersey (2017)
9. Data related to sales obtained from MG4 Company
10. The On-Line Encyclopedia of Integer Sequences, sequence A008302

Models of Organization Management

The Role of Business Intelligence Tools in Harvesting Collective Intelligence

Khaled Saleh Al Omoush[1]([⊠]), Raed M. Alqirem[1],
and Sabri R. Alzboon[2]

[1] Al Zaytoonah University of Jordan, Amman, Jordan
k.Alomoush@zuj.edu.jo
[2] Argosy University, Los Angeles, USA

Abstract. The present study aims to develop and empirically validate a framework for exploring the role of business intelligent tools in shaping the dimensions of collective intelligence. A questionnaire survey was developed to collect data from 9 firms across all industries with a sample of 89 respondents. Structural Equation Modeling, using smart PLS was conducted to analyze the data. The results indicated that business intelligent tools play a significant role in harvesting the dimensions of collective intelligence, including collective cognition, shared memory, knowledge sharing, and collective learning.

Keywords: Business intelligent · Collective intelligence · Collective cognition
Shared memory · Collective learning

1 Introduction

In today's dynamic and changing environment, business organizations need to innovate, make complex decisions, develop creative solutions and behave as human beings to serve their survival and superiority. The more organizations are able to be intelligent the more they will be able to ensure their survival and success in a highly competitive environment. This perspective establishes a new approach of re-examining organizations as intelligent entities that are evolving in the same manner as biological entities that compete for survival and growth in an ecological system using their intelligence [1]. In such a paradigm, the contemporary organizations are visualized as intelligent entities possessing an orgmind with high Collective Intelligence (CI) and not merely as economic production machines [2].

The basic notion of CI stands on two complementary pillars, which together constitute the concept of CI bridged by the collaboration among group of individuals [3–5]). The first pillar is related to the limited capabilities of individuals, including individual's bounded rationality, cognitive limits, cognitive bias, and nobody knows everything. The second pillar represents the power of collective collaboration through the ability of a group to find better solutions to the same problems, solve more problems than its individual members, evolve toward higher order complexity thought, and engage in intellectual collaboration in order to create, innovate, and invent.

CI intends and deserves to become a full discipline with its formal framework, tools, measuring instruments, and practical applications by virtue of the advances in

Z. Wilimowska et al. (Eds.): ISAT 2018, AISC 854, pp. 159–172, 2019.
https://doi.org/10.1007/978-3-319-99993-7_15

collaborative technological applications [5, 6]. It is an emerging field that seeks to merge human and technological intelligences with the aim of achieving results that are unattainable by either of these entities alone [5]. The new Information and Communication Technologies (ICTs) has effectively changed the way intelligence is collectively developed [7]. The advances in collaborative Internet applications have given impetus to the emergence, dissemination, and application of CI.

In terms of value creation, business intelligence (BI) appears to be among the most promising technologies in recent years [8]. According to McHenry [9], a true BI system should provide collaboration tools that allow for the capture and reuse of knowledge that is created in the process of collaboration, with links to any and all supporting materials that were used in making the decision. Imhoff and White [10] investigated how BI results in the discovery, access, integration, and management of information that are published and analyzed, and used by workgroups in a collaborative process. The integration of collaborative systems and BI enhance the decision-making ability of companies by leveraging the ability to manage data from the collaborative system and the analytical capabilities of BI systems [11]. Imhoff and White [10] demonstrate that organizations using collaborative systems integrated with BI tools achieve higher levels of decision-making performance compared to organizations that only use collaborative systems. Previous studies revealed that using BI tools is an important aspect of organizational intelligence. According to Imhoff and White [10], BI techniques have mainly been developed to detect hidden structures of the considered problems by CI and to leverage the advantages of qualitative and quantitative intelligence.

Recently, literature has been focusing on business value of BI tools. Although many efforts have been made to capture how BI generates business value, it is safe to conclude that there is much to learn about the value creation induced by this dominant IT domain in the context of harvesting CI. Despite the IT revolution and the continuous growth in its role in harvesting CI, little attention has been paid to the ways in which BI tools can contribute to harvesting CI. Therefore, the present study aims to develop and empirically validate a framework for exploring the role of BI tools in shaping the dimensions of CI.

The rest of this paper is organized as follows. Section 2 reviews related literature. Then we propose the research model and hypotheses in Sect. 3. Section 4 reports instrument development and data collection. Section 5 presents the results followed by a discussion of these results in Sect. 6. Finally, we conclude and highlight the limitations and future work of research in Sect. 7.

2   Literature Review

Intelligence can be defined as a very general mental capability that, among other things, involves the ability to reason, use memory, solve problems, think abstractly, comprehend complex ideas, and learn from experience [12]. Another useful definition of intelligence was provided by the American Psychological Association as the ability to use memory, knowledge, experience, understanding, reasoning in order to solve problems and adapt to new situations. According to Yick [13], intelligence is spread over a spectrum, with proto-intelligence at one end and CI at the other. However, Many definitions of CI have been built around the idea of intelligence. For example, Pór [14] defined it as the capacity of a human community to evolve toward higher order complexity thought, problem-solving, and integration through collaboration and innovation. According to Gan and Zhu [16], CI is the ability of a group, a team, an organization, a community, and the whole society to learn, solve problems, and plan the future in order to understand and adapt to the to environmental conditions based on the convergence of individual or distributed intelligence and the integration of the whole strength and unity. Schut [17] also defined CI as a shared intelligence that emerges from the collaboration of individuals.

Recently, much work has been done to develop different frameworks for describing different issues of CI. This interest can be seen across a wide range of research in many fields, including the computer science and artificial intelligence, psychology, sociology, collective problem solving, knowledge management, learning, democracy, collaboration, social media and others. The partnership between IT and CI is an emerging field of study and is still in its early stages. A large amount of research (e.g., [6, 14]) has developed theoretical and conceptual frameworks for understanding the phenomenon and foundation of IT-enabled CI. A review of the literature reveals that much effort (e.g., [18, 19]) has been devoted to examining the technical aspects and development of CI systems. Particular attention has been paid to developing algorithms for optimizing the capabilities and quality of CI systems. With the growing needs of collective problem-solving and group decision-making, CI has attracted increasing attention. Given that, many studies (e.g., [20, 21]) were conducted on different aspects of the group decision support systems through harvesting CI. Recently, a considerable amount of research (e.g., [18, 19]) was conducted on integrating efforts from the computer science and artificial intelligence communities into the study, design and development of computational approaches that are inspired by the CI of biological populations. The review of the literature indicates that most previous research has sought to model or suggest different applications of collaborative web-based CI as solutions to specific problems in specific areas outside the business context (e.g., [7, 16, 21]).

In today's challenging and competitive environment, BI systems have become key strategic tools, which directly impact on understand business situation and the success of any project implementation [22]. They are considered in today's business environment as a promising source of IT business value [23]. Some studies have emphasized the organizational impact of BI, suggesting that the introduction of BI systems implies not only technological enhancement, but also a revolution in the way that business activities and decision-[22] having a comprehensive knowledge of all factors that affect a business, such as customers, competitors, business partners, economic environment, and internal operations, therefore enabling optimal decisions to be made at all levels of organization. A typical BI infrastructure consists of data storage, processing, and delivery [8]. A typical BI system includes components such as a data warehouse, automated Extract-Transform-Load (ETL) utilities for transferring and transforming data within the system, and software platforms for developing end-user tools such as reports, On-Line Analytical Processing (OLAP) utilities for on-line investigation of data, digital dashboards, data mining tools, and possibly others. The combination of these infrastructural technologies and tools creates a technological environment that enables organizations to acquire better BI capabilities, which can lead to better decision making and to improved organizational performance [8, 23].

The literature (e.g., [8, 11]) provides a great deal of evidence supporting the idea that the rise of collaborative systems has accompanied with the advances of intelligence mechanisms, techniques, and tools. According to Nofal and Yusof [22], BI systems Facilitate collaboration between departments and business stakeholders. Imhoff and White [10] demonstrate that organizations using collaborative systems integrated with BI tools achieve higher levels of decision-making performance compared to organizations that only use collaborative systems. In addition, previous studies (e.g., [8, 15, 22]) revealed that using BI tools is an important aspect of organizational intelligence. According to Imhoff and White [10], BI techniques have mainly been developed to detect hidden structures of the considered problems by CI and to leverage the advantages of qualitative and quantitative intelligence. BI solutions enable an intelligent usage and analysis of company's business data [11]. They provide a large functionality with a vast collection of applications and tools that enables delivery of consistent and high quality information through gathering, providing access to, and analyzing data.

The review of literature shows that many efforts (e.g., [8, 11, 22, 24]) have been made to capture how BI generates business value. These studies frequently distinguish between the operational and strategic impacts of BI resources, where the former impacts included efficiency improvement, process optimization, and cost reduction, whereas the latter impacts included improvements in effectiveness, profitability, market share, and customer satisfaction. The common denominator between previous studies is that little attention has been given for the integration of BI and CI. More specifically, the aforementioned literature review indicates a lack of empirical research on the role of BI tools in harvesting the dimensions of CI has not been previously conducted. Therefore, the present research aims to develop and validate a framework to study the contribution of BI tools in establishing CI dimensions.

3   Research Model and Hypotheses

The research model (Fig. 1) proposes that BI tools contribute significantly to harvesting CI by supporting the achievement of its dimensions. The dimensions of CI was extracted based on a review of the literature on the concept of intelligent, CI, and intelligent organizations. These dimensions are collective cognition, shared memory, collective problem-solving, knowledge sharing, and collective learning.

Below, each construct of the research model is discussed, followed by the related hypotheses.

3.1   Collective Cognition

Most of the previous research (e.g., [2, 5, 13, 16, 25]) has paid considerable attention to the collective cognition as one of the very important features of CI. Atlee [25] indicated that the concept of CI needs to co-exist with the concept of cognition. According to Fadul et al. [26], the field of CI should be seen as a human enterprise in which mindsets, willingness to share, and openness to the value of distributed intelligence for the

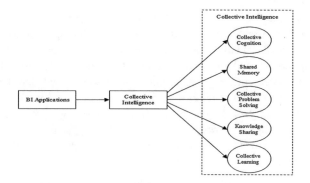

Fig. 1. Research model and hypotheses

common good are paramount. In this case, people make shared meaning of their diverse perspectives and experiences by surfacing, testing, and improving the collective thinking [27]. According to Atlee [25], most of CI aspects can be readily explained in terms of cognitive synergies among the participants.

Previous research (e.g., [8, 9, 22, 29]) confirmed the role of BI tools in providing cognitive appropriation mechanisms to facilitate the collective cognition features that enable decision makers to unify all streams of knowledge, share meaning, adjust their understanding, perform collective awareness, reveal concrete expressions of shared metacognition, and other cognitive tasks. BI is as an overarching term for decision support systems that are based on the intelligent exploration, integration, aggregation and analysis of data from various resources toward understanding business situation and improving decision making [22]. BI tools offer a comprehensive view of the entire organization, permit the analysis of business activities from multiple perspectives [8]. Davenport [24] highlighted the transition toward a shared culture of fact-based decision making that is associated with the use of business analytics and the adoption of BI systems. Ramamurthy et al. [29] also confirmed the role of BI tools in establishing a data-sharing culture. Therefore, this study proposes the following hypothesis:

H1. BI tools play a significant role in achieving collective cognition.

3.2 Shared Memory

Shared memory is central to the idea of CI (e.g., [13, 30, 31]). It can be described as a type of collective memory and thought of as a repository for historical information, an organization's accumulated knowledge and generative intelligence that results from collaboration among organizational members [30]. The information and knowledge memory stores previous cases that have been solved either as a success or a failure [31]. Mačiulienė [15] demonstrate that the intelligent use of stored information and knowledge ensures adequate decision support and leads to effective problem-solving practices as well as the development of new ideas, prototypes, competencies, and activities.

The previous research (e.g. [8, 11, 22, 23]) confirmed the critical role of BI tools in aggregating, sorting, organizing, refining, storing, integrating, and accessing organizational memory at all organizational levels. BI tools provide the memory functionality through data warehouse and data marts that are able to ensure a uniform and consistent storage and depth accessing of all data, information, and knowledge that are relevant for the organization and the environment it is acting in [22, 32]. Data storage in BI systems includes a large-scale repository of integrated organizational data and the hardware for managing and storing it. Such a repository typically includes a centralized data warehouse that covers a broad range of business activities, and a collection of departmental data marts [8]. Therefore, this study proposes the following hypothesis:

H2. BI tools play a significant role in achieving shared memory.

3.3 Collective Problem Solving

A considerable stream of research (e.g., [5, 7, 16, 33]) claims that better understanding of CI can be achieved through examining it as a collective problem-solving capability. Staškevičiūtė et al. [15] suggest that CI is the motivation behind all forms of group problem solving since the birth of collaboration. Dumas [27] asserted that the CI of any group increases the capacity for effective action in pursuit of common aims and finding emergent and sustainable solutions to the complex problems and challenges faced by organizations and communities. Suárez et al. [7] stated that CI often displays remarkable aggregate abilities to solve different classes of complex problems, more quickly and efficiently than experts or dedicated organizations.

BI solutions provide a valuable collection of applications and tools that support the collaborative problem-solving. BI tools help users to identify and solve their problems and discover business risks and opportunities, providing historical reports, data analyses and alerts that signalize problems and possible threats where managers can react faster to the problems that appear [11]. Fink et al. [8] described data mining as a complex problem-solving system. According to Tang [33], data mining techniques mainly developed to detect hidden structures of the concerned problems by CI and leverage advantages of qualitative and quantitative intelligence. Fink et al. [8] also demonstrated the role of organizational data-warehouse in utilizing data and models to solve semi-structured and unstructured organizational problems. Therefore, this study proposes the following:

H3. BI tools play a significant role in achieving collective problem solving.

3.4 Knowledge Sharing

The accelerated transformation of the knowledge revolution is propelling the human world rapidly into the intelligence era [2]. Previous studies (e.g. [2, 13, 15]) revealed that knowledge sharing is a highly significant function of intelligent organizations. Previous studies (e.g. [26, 34, 35]) also indicated that CI is fostered through knowledge sharing. Schut [17] argue that CI is the knowledge shared by people. Mosia and Ngulube [34] demonstrated that the management of CI in organizations has to facilitate knowledge transfer and sharing. Liu et al. [34] explained that, by sharing knowledge, each agent coevolves with other agents to adapt to the rapidly changing environment, resulting in augmented knowledge workers and improved organizations' CI. Fisch et al. [35] explained how the exchange of knowledge or meta-knowledge can certainly be seen as a higher-level form of CI.

Previous studies (e.g., [9]; Bach et al., 2016) confirmed that BI has contributed to the generation and sharing of a massive amount of knowledge content. Nofal and Yusof [22] mentioned that BI is a term introduced as a tool that represents a set of concepts and methods originated to enhance decision-making in business through knowledge sharing and systems usages. Moffett et al. [32] concluded that Web-based collaborative intelligent tools, such as data warehousing, data-mining, Knowledge directories, Knowledge-based systems, and intelligent support systems play a crucial role in utilization of knowledge sharing. Yogev et al. [23] indicated that operational BI capabilities include the amount of cooperation and knowledge sharing. Therefore, the study hypothesizes the following:

H4. BI tools play a significant role in achieving knowledge sharing.

3.5   Collective Learning

An extensive body of literature (e.g. [2, 6, 16, 26]) confirmed the relationship between CI and learning. According Gan and Zhu [16], CI is a sustainable human ability for individual and collective learning in which emergent patterns of meaning, coordination flows, insights, and inspiration interact and grow on each other to deepen the under-standing of knowledge and to progressively integrate the individual intelligences. Staškevičiūtė et al. [15] identified learning as intelligent ability to develop new knowledge and skills, and to learn from experience. Likewise, Chujfi and Meinel [1] described intelligence as a mental ability and the power of learning, understanding, and knowing. The collective learning represents an important aspect of organizational intelligence [2]. Yick [13] demonstrated that, in the intelligent organization, learning represents the biggest pool of intense intelligence sources. Gruber [6] suggested that true CI can emerge if the data collected from all those people is aggregated and recombined to create new ways of learning that individual humans cannot do by themselves.

Organizational learning is an important theoretical lens for understanding how BI creates business value [8]. BI systems also offer facilities of discovering trends and patterns that can be used for enhancing the collective learning toward unknown problem solving [11]. Yogev et al. [23] suggest that operational BI capabilities include the ability to acquire new knowledge. Fisch et al. [35] concluded that BI systems enable the collective learning by dissemination of the new knowledge and learn from experiences of other to act efficiently in a very dynamic environment. Given that, this study proposes the following:

H5. BI tools play a significant role in achieving collective learning.

4   Research Methodology

4.1   The Measurement and Instrument Development

The measurement instruments were derived based on an extensive review of the related literature. The scale items of constructs' variables were adapted to the context of the study from the literature in the areas of intelligence, CI, intelligent organizations, decision support systems, and BI tools.

Empirical data for this study were gathered via a self-administered questionnaire. As shown in Table 1, the questionnaire includes a total of 24 questions that represent the research model constructs. Respondents were asked to rate a set of statements on a 5-point Likert-scale, ranging from "not at all" (represented as one) to "to a great extent" (represented as five).

Table 1. Constructs and measurements of the research model

Constructs	Code	Measurement items
BI tools		*Indicate the extent to which the following BI tools are used in your organization*
	BI1	Data warehouses, including data marts;
	BI2	Data integration;
	BI3	Real-time reporting;
	BI4	Interactive reports (ad hoc);
	BI5	Intelligent agent;
	BI6	Online analytical processing (OLAP);
	BI7	Analytical applications, such as trend analysis and "what if " scenarios;
	BI8	Data mining techniques;
	BI9	Dishoarding, including dishoarding of metrics, key performance indicators and alerts
The dimensions of CI		*To what extent do you think that business intelligent applications contribute to the following*
Collective cognition	CC1	Facilitating a deep collective understanding and shared meaning of information and concepts;
	CC2	Enabling people to share their diverse perspectives and insights;
	CC3	Facilitating the aggregation of individual's perceptions into a collective perception;
Shared memory	SM1	Enabling the gathering, organization, storage, and access of all information assets;
	SM2	Working as a repository for information and knowledge for future use;
	SM3	Involving capabilities of recalling relevant information from long-term memory;

(continued)

Table 1. (*continued*)

Constructs	Code	Measurement items
Collective problem solving	CPS1	Assisting in identifying and formulating problems;
	CPS2	Providing mechanisms to aggregate participants' judgments about problems and solutions;
	CPS3	Providing a rich set of techniques to compare alternative solutions to problem;
Knowledge sharing	KS1	Facilitating transfer and exchange of best practices;
	KS2	Enhancing the ability to share and integrate the diversity of knowledge;
	KS3	Enabling collaborative groups to share their work reports and official documents;
Collective learning	CL1	Supporting the continuous learning ability that enables the collaborative groups to consume new information and evolve with time;
	CL2	Enabling the individuals to collectively learn, renew capabilities, and put new learning to use;
	CL3	Facilitating the development of new skills and learning from others' experience;

4.2 Sampling and Questionnaire Distribution

The study was conducted on a sample drawn from Jordanian firms that are currently adopting BI tools for at least one year. The author selected 6 companies that provide IT solutions and obtained a list of their client companies. The author also benefited from lists of satisfied customers that are provided by IT solution vendors on their websites and brochures. The obtained list constituted an initial sample of 17 companies across all industries. Exploratory telephone interviews were carried out to provide the final list. Finally, 9 firms that agreed to participate in the study were included in the final sample. After 40 days of preparation, 148 questionnaires were distributed, and a total of 96 questionnaires were collected. By examining the cases, 7 responses were found to be incomplete and were thus eliminated from the data analysis, resulting in 89 valid responses, which represents a response rate of 60.1%.

Based on the assumption that participants must be well informed or actively engaged in BI adoption and use in their firm, The targeted respondents were top-level executives, including individuals in such roles as Chief Executive Officer, Chief Financial Officer, Chief Information Officer, Chief Operating Officer, and Vice President. Furthermore, the sample included IT directors, sales and marketing directors, manufacturing directors, finance/accounting directors, human resource directors, procurement/purchasing directors, logistics/supply chain directors, customer service directors, and research and development directors.

5 Data Analysis and Results

The Partial Least Squares (PLS) method was employed to analyze a complete survey dataset. PLS is very appropriate when theoretical information is low, as in the area of the role of BI tools in harvesting CI. It is also robust in that it does not require a large sample or normally distributed multivariate data. Smart PLS 2.0 software was used for the data analysis in this study.

5.1 Reliability and Validity of Measures

Convergent validity was assessed by comparing the confirmatory factor loadings of each item with its intended construct. To purify scales with the goal of improving their measurement, Confirmatory Factor Analysis (CFA) was applied. The results of CFA indicated that some items were not satisfied with factor loadings of less than 0.50 on their own constructs and had to be removed from the scale. Specifically, one item (BI4) was removed from BI tools and one more (CL3) was removed from the collective learning scale. Regarding the model's measurement reliability, Table 2 lists the loading average, composite reliability (CR), and Average Variance Extracted (AVE) for each construct. All item loadings are greater than 0.6, and all values of AVE are larger than 0.5. All constructs exhibited acceptably high scores of composite reliability, exceeding the 0.70. In addition, Cronbach's alpha was employed to evaluate the reliability of the model constructs and their internal consistency. As presented in Table 2, all constructs exhibited acceptably high reliabilities, with values of Cronbach's coefficient alpha exceeding 0.70.

Table 2. Validity and reliability estimates of the constructs

Constructs	AVE	Loading average	CR	Cronbach's Alpha
BI tools	0.535	0.834	0.873	0.825
Collective cognition	0.727	0.870	0.889	0.816
Shared memory	0.715	0.743	0.838	0.724
Collective problem solving	0.787	0.723	0.879	0.861
Knowledge sharing	0.641	0.830	0.877	0.814
Collective learning	0.565	0.751	0.819	0.714

To examine the discriminant validity, the study has compared the square root of AVE for each construct and factor correlation coefficients. As shown in Table 3, the square root of each construct's AVE value is larger than its correlations with other constructs, which suggests good discriminant validity.

Table 3. The square root of AVE and factor correlation coefficient

No.	Constructs	1	2	3	4	5	6
1	BI tools	**0.731**					
2	Collective cognition	0.380	**0.853**				
3	Shared memory	0.652	0.376	**0.846**			
4	Collective problem solving	0.196	0.401	0.122	**0.887**		
5	Knowledge sharing	0.372	0.704	0.305	0.419	**0.800**	
6	Collective learning	0.453	0.519	0.332	0.367	0.515	**0.752**

5.2 Assessing the Structural Model and Testing Research Hypotheses

The results of the structural modeling analysis are illustrated in Fig. 2.

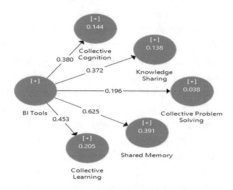

Fig. 2. Tested research model

Table 4 presents the results of the hypothesis tests. It is recommended that the t-value be greater than 2.0. Furthermore, the strengths of the hypothesized paths and whether each path is significant are evaluated by the standardized path coefficient.

Table 4. Significant paths and hypothesis results

H#	Standardized coefficient	T statistics	P value	The result
H1	0.380	4.067	0.000	Supported
H2	0.372	8.111	0.000	Supported
H3	0.196	1.342	0.180	Not supported
H4	0.625	3.357	0.001	Supported
H5	0.453	4.948	0.000	Supported

As shown in Table 4, four paths were found to be significant; accordingly, their four associated hypotheses were supported (H2, H4, H5, H6). In contrast, H3 was not supported because the path of the variable was not found to be significant.

6 Discussion

In terms of value creation, BI appears to be among the most promising technologies in recent years. Although many efforts have been made to capture how BI generates business value, there is much to learn about the value creation induced by this dominant IT domain in the context of harvesting CI. Therefore, the present study aimed to develop a framework for exploring the role of BI tools in shaping the dimensions of CI. Drawing on literature review, five dimensions of CI were proposed in the research model: collective cognition, shared memory, collective problem-solving, knowledge sharing, and collective learning.

The empirical results indicated that BI tools contribute to the harvesting of collective cognition. This finding is consistent with the previous studies (e.g., [8, 9, 22, 28]) that confirmed the role of BI tools in providing cognitive appropriation mechanisms to facilitate the collective cognition features that enable decision makers to unify all streams of knowledge, share meaning, adjust their understanding, reveal concrete expressions of shared metacognition, and other cognitive tasks. The results also revealed that BI tools play a significant role in achieving shared memory. Previous research (e.g., [8, 11, 23]) revealed the critical role of BI in aggregating, sorting, organizing, storing and accessing organizational memory at all organizational levels.

The results of the present study indicated a significant role of BI tools in achieving knowledge sharing. This finding is in agreement with previous research. For example, Nofal and Yusof [22] mentioned that BI is a term introduced as a tool that represents a set of concepts and methods originated to enhance decision-making in business through knowledge sharing and systems usages. Moffett et al. [32] concluded that Web-based collaborative intelligent tools, such as data warehousing, data-mining, Knowledge-based systems, and intelligent support systems play a crucial role in utilization of knowledge sharing. Finally, The results revealed that BI tools play a significant role in achieving collective learning. According to Fink et al. [8], organizational learning is an important theoretical lens for understanding how BI creates business value, especially given that BI systems are deployed to facilitate decision support, environmental adaptation, and organizational innovation. This finding is also consistent with the previous studies (e.g., [23, 35]) that BI systems offer facilities of discovering trends and patterns that can be used for enhancing collective learning.

7 Conclusions and Implication

The present study has sought to develop a framework under which to study the contribution of BI tools in establishing the dimensions of CI. The findings can be useful for research and in practice, as the need to re-examine organizations as evolving intelligent entities is becoming increasingly prevalent and the transition towards CI forms of synergies in an amalgamation of humans and technology is advancing. Investigating the role of BI tools in in harvesting CI provides a deeper understanding of the business value of these applications and enriches the scientific literature in the area of BI adoption by specifying new directions of theoretical and empirical research. In addition, research findings imply that CI initiatives must consider investment not only in primary collaborative technology but also in complementary BI applications. The results can be used to help BI systems and algorithm developers gain a better understanding of the requirements of CI and build specialized collaborative BI applications to achieve its dimensions.

Despite its contributions, as with any research study, there are some limitations, which can serve as directions for future research. Although the design of the research model builds on the theoretical insights resulting from the literature analysis, the study did not reflect the theoretical factors that can impact the role of BI tools in achieving the dimensions of CI, such as the organization's characteristics, human capital, and top-management support.

References

1. Chujfi, S., Meinel, C.: Patterns to explore cognitive preferences and potential collective intelligence empathy for processing knowledge in virtual settings. J. Interact. Sci. 3(5), 1–16 (2015)
2. Ng, P., Liang, T.: Educational institution reform: insights from the complexity–intelligence strategy. Hum. Syst. Manag. 29(1), 1–9 (2010)
3. Heylighen, F.: Collective intelligence and its implementation on the web. Comput. Math. Organ. Theory 5(3), 253–280 (1999)
4. Lévy, P.: From social computing to reflexive collective intelligence. Inf. Sci. 180(1), 71–94 (2010)
5. Lykourentzou, I., Vergados, D.J., Kapetanios, E., Loumos, V.: Collective intelligence systems: classification and modeling. J. Emerg. Technol. Web Intell. 3(3), 217–226 (2011)
6. Gruber, T.: Ontology of folksonomy: a mash–up of apples. Int. J. Semant. Web Inf. Syst. 3 (1), 1–11 (2008)
7. Suárez, V.E., Bucheli, V., Garcia, A.: Collective intelligence: analysis and modelling. Kybernetes 44(6/7), 1122–1133 (2015)
8. Fink, L., Yogev, N., Even, A.: Business intelligence and organizational learning: an empirical investigation of value creation processes. Inf. Manag. 54(1), 38–56 (2017)
9. McHenry, W.: Linking decision artifacts: a means for integrating business intelligence and knowledge management. Electr. J. Knowl. Manag. 14(2), 91–102 (2016)
10. Imhoff, C., White, C.: Collaborative BI: theory becomes reality. Bus. Intell. J. 18(2), 40–46 (2013)
11. Matel, G.: A collaborative approach of business intelligence systems. J. Appl. Collab. Syst. 2 (2), 91–101 (2010)
12. Barlow, J., Dennis, A.: Not as smart as we think: a study of collective intelligence in virtual groups. J. Manag. Inf. Syst. 33(3), 684–712 (2016)
13. Yick, T.: Intelligence Strategy: the evolution and co–evolution dynamics of intelligent human organizations and their interacting agents. Hum. Syst. Manag. 23(2), 137–149 (2004)
14. Pór, G.: Augmenting the collective intelligence of the ecosystem of systems. Syst. Res. Behav. Sci. 31(5), 595–605 (2014)
15. Staškevičiūtė, I., Neverauskas, B., Čiutienė, R.: Applying the principles of organisational intelligence in university strategies. Eng. Econ. 3(48), 63–72 (2006)
16. Gan, Y., Zhu, Z.: A learning framework for knowledge building and collective wisdom advancement in virtual learning communities. Educ. Technol. Soc. 10(1), 206–226 (2007)

17. Schut, M.: On model design for simulation of collective intelligence. Inf. Sci. **180**(1), 132–155 (2010)
18. Gonzalez-Pardo, A., Palero, F., Camacho, D.: An empirical study on collective intelligence algorithms for video games problem-solving. Comput. Inform. **34**(1), 233–253 (2015)
19. Asgari, A., Lee, W.: Simulating collective intelligence of bio-inspired competing agents. Expert Syst. Appl. **56**, 56–67 (2016)
20. Bundzel, M., Lacko, J., Zolotová, I., Kasanický, T., Zelenka, J.: Artificial intelligence aggregating opinions of a group of people. Comput. Inform. **35**(6), 1491–1514 (2016)
21. Mačiulienė, M., Skaržauskienė, A.: Emergence of collective intelligence. J. Bus. Res. **69**(5), 1718–1724 (2016)
22. Nofal, M., Yusof, Z.: Integration of business intelligence and ERP. Proc. Technol. **11**, 658–665 (2013)
23. Yogev, N., Fink, L., Even, A.: How business intelligence creates value. In: ECIS, p. 84, June 2012
24. Mannino, M., Hong, S.N., Choi, I.J.: Efficiency evaluation of data warehouse operations. Decis. Support Syst. **44**(4), 883–898 (2008)
25. Atlee, T.: Co–intelligence, collective intelligence, and conscious evolution. In: Tovey, M. (ed.) Collective Intelligence: Creating a Prosperous World at Peace, Virginia, pp. 5–14 (2008)
26. Fadul, J.: Collective learning: applying distributed cognition for collective intelligence. Int. J. Learn. **16**(4), 11–20 (2009)
27. Dumas, C.: Hosting conversations for effective action. J. Knowl. Global. **3**(1), 99–116 (2010)
28. March, S., Hevner, A.: Integrated decision support systems: a data warehousing perspective. Decis. Support Syst. **43**(3), 31–43 (2007)
29. Ramamurthy, K., Sinha, A.: An empirical investigation of the key determinants of data warehouse. Decis. Support Syst. **44**(4), 817–841 (2008)
30. Jacko, J.A., Salvendy, G., Sainfort, F.: Intranets and organizational learning: a research and development agenda. Int. J. Hum. Comput. Interact. **14**(1), 93–130 (2002)
31. Lancieri, L.: Relation between the complexity of individuals' expression and groups dynamic in online discussion forums. Open Cybern. Syst. J. **2**(1), 68–82 (2008)
32. Moffett, S., Parkinson, S.: Technological utilization for knowledge management. Knowl. Process Manag. **11**(3), 175–184 (2004)
33. Tang, X.: Towards meta-synthetic support to unstructured problem solving. J. Inf. Technol. Decis. Mak. **6**(3), 91–108 (2007)
34. Liu, P., Raahemi, B., Benyoucef, M.: Knowledge sharing in dynamic virtual enterprises. Knowl.-Based Syst. **24**(3), 427–443 (2011)
35. Fisch, D., Jänicke, M., Kalkowski, E., Sick, B.: Learning from Others: exchange of classification rules in intelligent distributed systems. Artif. Intell. **187–188**, 90–114 (2012)

The Concept of Conjoined Management

Marian Hopej[1] and Marcin Kandora[2]([⊠]) [ID]

[1] Wrocław University of Science and Technology, Wrocław, Poland
[2] Głębinów 43, 48-300 Nysa, Poland
mj.kandora@gmail.com

Abstract. In contemporary businesses substantial efforts are made to promote workforce diversity. Some leading organizations publically announce the introduction of I&D (inclusion and diversity) strategies and train their management staff in diversity management. Those initiatives most often build implicitly on the assumption that workforce diversity is positively correlated with organizational performance which, as research results show, is somewhat debatable. Consequently, the gap between the key assumptions of diversity management and business practice calls for further research. This article is an attempt to address it by presenting the concept of conjoined management drawing on observations of management routines in the best European football clubs, i.e. highly dynamic organizations which deal quite well with diverse workforce. In essence, conjoined management consists of five dimensions: talent focused recruitment, sense of unity for a common goal, collective group identity, transparency of operations and inside-out leadership. To demonstrate application of the concept in practice a case study has been presented on an application integration department within a multinational FMCG company.

Keywords: Conjoined management · Diversity · Integration

1 Introduction

The members of every organization always differ from one another. They have different age, marital status, level of education, gender, hierarchical position, values and beliefs. This raises the question: how to manage diverse staff?

The answer is sought within diversity management, which presupposes the need to create awareness and acceptance of differences between organizational participants and to initiate actions allowing to benefit from their (differences) presence [11]. However, as the results of several scholarly inquiries show, the effects of such management approach are not indisputably positive. In general, the point is that:

- Although substantial efforts are being made to promote diversity and equal treatment of people, the outcomes are significantly different from expectations. For example, the effectiveness of training programs aiming to promote diversity is not satisfying; it doesn't seem to alter neither the behavior nor the beliefs of organizational participants [5].
- The research results on the benefits that the organization derives from the diversity of its staff are unclear. Some scholars, including Adler and Gundersen [1], confirm

© Springer Nature Switzerland AG 2019
Z. Wilimowska et al. (Eds.): ISAT 2018, AISC 854, pp. 173–187, 2019.
https://doi.org/10.1007/978-3-319-99993-7_16

the positive relationship between the staff diversity (in terms of their gender) and the (financial) effectiveness of the organization, others do not [3]. Some of them point to the relationship between diversity and various mediating variables (e.g. innovation or strategy), which in turn are to affect effectiveness, though the interrelations between mediating variables and organizational efficiency are not verified in advance but a priori assumed[1].

So there is a clear gap between the key assumptions of diversity management and business practice. This article is an attempt to address this discrepancy by presenting the concept of conjoined management drawing on observations of management activities in the best European football clubs, largely because of their far-reaching agility. Additionally:

- In a football club, similar management challenges are basically to be addressed as in contemporary dynamic business organizations [4].
- These clubs are largely free of prejudices associated with employee diversity.

In other words, the aim of the article is to present the concept of conjoined management, i.e. focused on equal treatment of people in organizations and elimination of various types of prejudices. It draws on the experience of highly dynamic organizations which deal quite well with diverse workforce.

To demonstrate its application in practice a case study has been presented on an integration competence center within a multinational FMCG (FTSE100) company.

2   Workforce Diversity and Organizational Performance – Research Results

The research covered 166 clubs from 10 European leagues, both the top and the mediocre ones (selected based on the UEFA ranking). The relation between two independent variables, players' nationality and age (indicators of the so-called primary identity) and the dependent variable, points scored by these clubs in the season 2016/2017 and place in the final table of competitions, was analyzed. The results of the calculations performed (with the Kolmogorov-Smirnov test) show that the hypothesis about the normal distribution cannot be rejected both for diversity of nationality and age[2].

Two consecutive tables show correlations between differences in nationality (Table 1) and age (Table 2) and club performance.

The results generally show that there is no significant correlation between the diversity of players measured in terms of nationality and club results. One should also accept the lack of correlation between the diversity of players' age and the club performance even though it was important for two of the ten leagues, but contrary - once positive, other time negative.

---

[1] It concerns among others the research of Stahl et al. [13] and Joshi and Roh [10].

[2] The measurement of age diversity was conducted with the kurtosis statistic which is a measure of the concentration of results around the average. The nationality diversity was measured through dividing the number of countries of players' origin through the total number of players employed in a club.

Table 1. Correlation between national diversity and club performance and club market value (Spearman correlation test) (The interpretation of the r-Spearman factor: 0–0,3 means that the correlation is weak or non-existent, 0,3–0,5 moderate, 0,5–0,7 strong, 0,7–1,0 very strong.)

National diversity		Club points	Club position
LaLiga (Spain)	Correlation coefficient	0,163	− 0,163
	Significance	0,492	0,494
	N	20	20
Premier League (England)	Correlation coefficient	0,292	− 0,266
	Significance	0,211	0,257
	N	20	20
Bundesliga (Germany)	Correlation coefficient	− 0,182	0,165
	Significance	0,471	0,512
	N	18	18
Serie A (Italy)	Correlation coefficient	0,368	− 0,368
	Significance	0,110	0,110
	N	20	20
Ligue 1 (France)	Correlation coefficient	0,138	− 0,144
	Significance	0,562	0,546
	N	20	20
First Division (Cyprus)	Correlation coefficient	0,279	− 0,279
	Significance	0,334	0,334
	N	14	14
Ekstraklasa (Poland)	Correlation coefficient	0,481	− 0,462
	Significance	0,059	0,071
	N	16	16
Liga I Bergenbier (Romania)	Correlation coefficient	0,383	− 0,380
	Significance	0,177	0,180
	N	14	14
Premyer Liquasi (Azerbaijan)	Correlation coefficient	0,143	− 0,143
	Significance	0,736	0,736
	N	8	8
Allsvenskan (Sweden)	Correlation coefficient	0,300	− 0,312
	Significance	0,259	0,239
	N	16	16

**Significant correlation on the level of 0.01 (two-tailed); *Significant correlation on the level of 0.05 (two-tailed).
Source: [14]

Although the study involved only two dimensions of diversity, its results are consistent with the findings of earlier mentioned research, fortifying the skepticism about the positive impact of workforce diversity on organizational performance. Obviously, this calls for particularly cautious introduction of diversity-enhancing initiatives that are supposed to positively influence organizational performance. On the

Table 2. Correlation between age diversity and club performance (Spearman correlation test)

		Club points	Club position
LaLiga (Spain)	Correlation coefficient	0,190	− 0,205
	Significance	0,421	0,387
	N	20	20
Premier League (England)	Correlation coefficient	0,028	− 0,033
	Significance	0,907	0,890
	N	20	20
Bundesliga (Germany)	Correlation coefficient	0,264	− 0,286
	Significance	0,290	0,250
	N	18	18
Serie A (Italy)	Correlation coefficient	0,035	− 0,035
	Significance	0,885	0,885
	N	20	20
Ligue 1 (France)	Correlation coefficient	0,452*	− 0,462*
	Significance	0,045	0,040
	N	20	20
First division (Cyprus)	Correlation coefficient	− 0,574*	0,574*
	Significance	0,032	0,032
	N	14	14
Ekstraklasa (Poland)	Correlation coefficient	0,071	− 0,176
	Significance	0,794	0,513
	N	16	16
Liga I Bergenbier (Romania)	Correlation coefficient	0,262	− 0,253
	Significance	0,366	0,383
	N	14	14
Premyer Liquasi (Azerbaijan)	Correlation coefficient	− 0,524	0,524
	Significance	0,183	0,183
	N	8	8
Allsvenskan (Sweden)	Correlation coefficient	− 0,296	0,303
	Significance	0,266	0,254
	N	16	16

**Significant correlation on the level of 0.01 (two-tailed); *Significant correlation on the level of 0.05 (two-tailed).
Source: [14]

other hand, it would be incorrect not to realize that, in contemporary organizations, perhaps especially in football clubs, the diversity of players is an indisputable fact, therefore it is necessary to limit prejudiced behaviors that negatively impact organizational performance. Consequently, the assumption on which diversity management is based needs to be adjusted. It is not so much about increasing diversity and profiting from it, but more about building an integrated and prejudice-free organization. Management based on such an assumption can be labelled as a "conjoined", as it emphasizes equal treatment of all employees.

The dimensions of conjoined management are elaborated on in the next part of this paper. Although there is no simple analogy between the best European football clubs and other types of organizations, especially business organizations, it seems possible to identify its most important aspects based on the experiences of these clubs, which are largely free from prejudice, which is confirmed regularly in UEFA reports.

3   The Dimensions of Conjoined Management

1. In football, the success depends primarily on whether the club managed to acquire talented players. Why is FC Paris-Saint-Germain paying over 200M EUR for Neymar and 180M EUR for K. Mbappe? Why does FC Barcelona buy P. Cutinho for over 150M EUR, and why does J. Mourinho hire R. Lukaku in Manchester United paying 75M GBP to Everton? The answer is simple: such players make the difference between winning and losing. It is therefore not surprising that the best clubs compete fiercely for talent, which sometimes resembles a war, and this game is planned in all aspects. As can be seen in Fig. 1, each phase of this process overlaps with the next one, meaning that, for example, the sale involves some activities related to the search for successors, while in the hold phase one has to take care of the sale. Although many European clubs are doing well in some phases, only the best master the whole cycle.

Knowing this, it can be assumed that conjoined management:

- concerns mainly people.
- involves recruitment of employees not conducted under the banner of diversity. The most important criterion for assessing candidates for work are their talents, including problem solving skills, initiative and independence, interpersonal skills and preferences for teamwork.
- relies only to a small extent on the classic principle of personnel stability. More precisely, its use requires far-reaching prudence.
- requires the ability to see potential in possible candidates, and after recruitment, their continuous and dynamic development.

2. The sense of unity for a common goal integrates the players of the best European clubs on the pitch and outside it. G. Graham, a Scottish football player and country representative, as well as a coach, writes about it in the following way: "Only a person practicing a team sport who participated in some important matches can understand this sense of unity that warms up the atmosphere in the locker room before the first whistle. This feeling cannot be described, but you can almost touch it. It is a combination of harmony and desire to present yourself not only individually but with regard to the entire team" [4, pp. 223–224]. This point of view is also supported by A. Fergusson: "As a football manager, I have never been interested in sending a jumble of players out to the pitch. There is no substitute for talent, but on the pitch talent without a sense of unity of purpose is a grossly overvalued advantage" [4, p. 225].

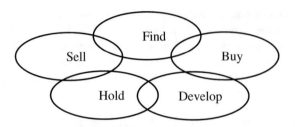

Fig. 1. The talent cycle Source: [4, p. 244]

The sense of unity for a common goal is another dimension of conjoined management. It involves primarily:

- Definition of achievable, free from contradictions, quantitative strategic and operational objectives. It should also be ensured that they are, if possible, jointly determined rather than arbitrarily set.
- Stimulation and preservation of a strong need for achievements, so that the desire to achieve a goal becomes one of the main driving forces for a team.
- Following the classic principle of subordinating the personal interest to the overarching purpose. As it is known, it says that in the organization the interest of its member, or of a group of people, should not prevail over the interest of the whole organization, while it is always preferable to seek the possibility of reconciling both, being sometimes different but deserving respect [7].
- Supporting the formation of effective self-organization, which is the embodiment of the idea that the system should be organized in such a way that it organizes and regulates itself [12]. A prerequisite for self-organization is the joint pursuit of organizational participants to achieve something that goes beyond individual goals.

3. Many managers of leading football clubs are interested in the personal development and non-football life of players. For example, former Leeds United manager and England's national team coach D. Revie kept close relationships with his players and their families, as did B. Paisley, former manager of Liverpool FC. The human side of the relationship with the players was also not neglected by the well-known Manchester United's manager M. Briesby: "I led the team as I believed the players wanted it, above all, I wanted to create more human relations than those I knew from my professional career. Sometimes the boys were left alone with their problems, they were not given enough attention. From the beginning I tried to make every employee feel like an important element in the club" [4, p. 295].

The managers of the best clubs also focus a lot on what connects the players. An expression of this approach is the belief of A. Ferguson that Manchester United's players had to be mentally strong. He wanted them to never to give up. "I kept saying: if you let go for the first time, you let go the second time as well. Additionally, there was as well the work ethics and energy that seemed to emanate from me to the whole team. (…) I kept telling the players that talent is nothing but hard work throughout the whole life. I demanded even more from the stars. I was telling them: you have to show that you are the best players. And they showed. That's why they are stars - because they can work harder" [6, p. 91].

Another kind of similarity connects the players of P. Guardiola. His football philosophy, the essence of which is the high mobility of players, many short, low-risk passes, creating many goal situations, requires players who keep well on the ball, i.e. with a low center of gravity. Therefore, it is not surprising, that the average height of FC Barcelona players he trained a few years ago was the lowest in Europe (1.77 m), and the one of his current players (Manchester City) is the lowest in the Premier League.

Based on the above remarks, it can be assumed that the next dimension of conjoined management is a collective group identity, which is related to the classic principle of human treatment of employees. It also indicates the need for human and equal treatment of all organizational members, but the main emphasis is on making efforts to make them feel like in a family. It involves:

- The focus of managers on what is common for, and not differentiating, the organizational participants. In other words, it is not necessary to increase diversity but rather foster similarities.
- The need to manifest common values and team identity. Research carried out by social psychologists reveals that people are more cordial and trusting to those who resemble their appearance or dress similarly [8]. This implies that existing differences between people (e.g. racial) may be, to some extent erased, sunk in the sea of similarities.

4. Football is extremely transparent. One former CEO of Manchester United once said that there are about 40 general shareholder meetings in each football season, involving 40,000 people each. In fact, everything can be seen easily, and what is hidden is being revealed by inquisitive media [4]. This also applies, perhaps even above all, to prejudiced behaviors, which are almost immediately disclosed and stigmatized, also by UEFA. For example, recently one of the activists of the English club West Ham United scandalously commented on footballers from Africa, stating that he no longer wants to hire them in the club because of the problems they cause, mainly off the pitch. He was immediately suspended by the club's authorities with the following justification: "West Ham have terminated the contract of director of player recruitment, (…), with immediate effect following his unacceptable comments that were widely reported in the press" (sport.tvp.pl/35844917/nie-chcemy-pilkarzy-z-afryki-rasistowski-skandal-w-anglii, 25/02/2018). The proceedings have also been initiated by the English football federation. It is impossible to disagree that the lack of transparency and openness is major impediment in business, blocking the potential of talented people, their cooperation, rapid decision making, and thus achievement of competitive advantage [15]. Moreover, low transparency was identified in the study of Ashkenas, Siegal and Spiegel [2], as one of the key factors contributing to the organizational complexity.

Increased transparency and openness, facilitates sound relationships between organizational participants, and is therefore necessary. It requires in particular:

- A simplification of organizational structure, implying a reduction of organizational rules, which is synonymous with far-reaching organizational flexibility. However, its outcome should not be the creation of opaque, multi-line structural solutions, or the frequent application of various forms of teamwork to the DNA of all applicable rules [9].

- Shaping the culture of openness and honesty. In general, it is important for organizational participants to be convinced that speaking openly and admitting the truth is their duty and is always safe. Although such a culture may inevitable for success, its inculcation in people is extremely difficult and time-consuming, but as Welch and Welch [15] emphasize, "for a company to be open, you have to reward those who live it, propagate it, talk about it loud. Most importantly, you should demonstrate openness enthusiastically, even exaggerate, even if you are not the boss".
- Employees who trust in openness and transparency, which is probably the key mechanism to prevent the accumulation of complicated interpersonal relations. In order to build such trust, senior management must convince organizational participants, also through own example, that it has real value and is not just a lip service.

4   Conjoined Management in a Diverse Integration Competence Center – A Case Study

The following case study[3] refers to an application integration department of a leading FMCG company. The company operates globally and is present in 160 markets. Its origins date back from eighteenth century but in its current form it was shaped in the 90s when it became listed on the London Stock Exchange as a FTSE 100 company. Over the past 20 years the company has developed very rapidly, largely through mergers and acquisitions of other companies from the same industry.

ESB facilitated integration plays a vital role in the analyzed company. It ensures harmonization and monitoring of enterprise data exchange inside a diverse application architecture where particular functional domains are supported by dedicated stand-alone systems. Without integration, redundant data processing followed by its subsequent and effortful alignment would seriously impact organizational performance and impede the company's responsiveness to environmental changes (mainly related to frequent introduction of new regulatory requirements). The advanced and mature integration platform allows for quick response to new requests for integration (independently from the technology to be supported) through relatively fast delivery of new interfaces ensuring correct validation, transformation, enrichment and routing of messages. In the analyzed company, integration is predominantly asynchronous.

The company used originally many different, but rather less advanced, application integration patterns. The main characteristics of the evolved (and not deliberately designed) integration architecture were: the usage of asynchronous integration patterns, such as flat and XML file exchange by means of FTP scripts and based on batch processing, usage of ETL tooling and shared databases, low degree of data standardization and interface reuse, tight coupling of systems through point-to-point interfaces.

The situation was aggravated around 2010–2012, when the European Commission introduced legal obligations (within the industry the company is operating) to track and trace the company's finished goods throughout the supply chain, from their

---

[3] Parts of this case were also included in M. Kandora, 2017, In search for effective methods of routine formation, Management. The Journal of the University of Zielona Góra, 21(1), pp. 20–39.

manufacturing at the plant until the sale to the first wholesaler. The consequence of these new regulations were: an enormous increase in data exchange (both in terms of volume and frequency) between the different information systems and the need for extensive adaptation work in the field of application integration. To address the challenge a decision was taken to implement an Enterprise Service Bus (ESB), a highly performant message oriented middleware, enabling a more loosely-coupled information system landscape through the use of publish-subscribe integration patterns and canonical data models, establishing a single point of control for interfaces, ensuring their better transparency and facilitating their reuse. The ESB selection and deployment decision was followed by a half year pilot project, which was successfully delivered in late 2013.

The introduction of advanced middleware software, together with an increasing demand for integration, strongly influenced the decision to set-up a dedicated application integration department. Along with the establishment of the department, external consultants were also hired to address the growing resource requirements. In addition, to ensure knowledge retention, recruitment of integration engineers and administrators (for internal positions) started. The positions were to be located both in the Netherlands and in Germany.

The selection of suitable candidates quickly became a serious inhibitor of the recruitment process. The Dutch and German labor markets turned out to be quite limited and one could notice that true "war for talents" was taking place there. Therefore, the search for candidates promptly expanded to whole Europe. Occasionally even candidates from outside Europe were contacted and interviewed.

Eventually, after two years of recruitment, about 15–20 people (20 people during the largest project load) were hired, forming a rather diverse team (Table 3).

Table 3. Dimensions of application integration department's diversity

Diversity criteria	Degree	Details
Nationality	High	Germany, Netherlands, Spain, Poland, Portugal, India, Slovakia, Turkey
Gender	Low	During the greatest highest delivery load in the 20-people department only 2 women worked
Age	Medium	5 people below 30; 8 people between 30 and 40; 7 people above 40
Localization	High	Netherlands, Germany, Spain, Slovakia, Poland
Application integration experience	Medium	3 people < 5 years of experience 10 people between 5 and 10 years of experience 7 people > 10 years of experience
Personality profiles (DISC method)	High	2 people with a profile close to D 3 people with a profile close to I 5 people with a profile close to S 5 people with a profile close to C
Organizational awareness	Medium	2 people > 10 years in the company 5 people between 2 and 4 years in the company 13 people < 1 year in the company

Source: own study

It consisted of an integration architect, integration consultants and administrators of the integration platform. It included both internal employees and external consultants, representatives of various nationalities, located in different locations (cooperating with each other remotely), with different levels of integration and industry knowledge, various levels of analytical, communication and technical skills.

As one can guess, the high diversity of the team naturally created the necessity of its far-reaching integration.

In order to create a high-performant and cohesive team, every employee was firstly made familiar the main objective of the department which was "the delivery of integration solutions which will enable [the company] to perform its operations fast, secure and well-aligned". To this end, a series of workshops was held in which all employees were discussing, how can this general objective be achieved best. During these discussions, a kind of vision emerged that assumed that in four years the application integration department should transform from an enabling function into a value generating unit and become a trusted business partner which proactively initiates improvements in business processes through innovative integration solutions. The team agreed that to achieve this, in each of the following 4 years different detailed objectives should be reached. In the first year the goal was to set the standards for requirements analysis, programming (i.e. development guidelines), platform administration, code delivery, cost estimations. The ambition was also to define a scalable integration architecture, develop the basis of work organization and focus on delivering current projects within time and budget to build credibility. For this purpose, a 1 year project (called "foundation build") was launched. The objective for the second year was to achieve high predictability and scalability in providing integration services and become a trusted internal supplier. In the third year, the goal was to transform the department into an integration competence center, widely recognized in the organization. In the fourth year the department was expected to be mature enough to proactively trigger integration initiatives which would help the value-generating business functions to increase added value.

For each stage, specific deliverables and KPIs were defined in terms of operating costs, maturity and reliability of integration solutions, process standardization and the availability of specific artifacts associated with them.

Such vision, detailed in a road map, developed in a team and achievable, has become the main bond of team members' activities. It defined commonly accepted framework for actions and, laid down in the jargon of workshop participants, created the foundations for effective internal communication. The vision focused the attention of team members on what is truly important and what is secondary, simplifying work organization. Because it was developed by the team (the head of the department was mainly facilitating the discussions during the workshops) it did not have to be intensively promoted and was easily accepted.

During the workshops there were naturally opposite points of view, resulting from prior experiences, participant's personalities, varying degrees of organizational awareness and beliefs on what directions of action are preferred and which counterproductive. Nevertheless, a strong desire to achieve the set goals and manager's moderation allowed eventually to work out commonly accepted solutions. Although the team members worked day-to-day in 3 different locations, all of them were in one place during the workshop, which helped to build the foundations of personal relationships. The joint development of action plans for the next years ensured initial integration of the team. A well-known saying, that in order to unite people one should "constrain them to join in building a tower" (and oppositely if they are made to "hate each other" food should be thrown amongst them - A. St. Exupery, Fortress), turned out to be true again in this case. Differences between participants moved to the background. The construction of the proverbial tower could have been started because there was a basic premise - the majority of team members shared a strong need for achievement.

The acceptance of set objectives and their joint implementation was not enough to effectively provide integration services and continue the journey through subsequent stages of the roadmap. It was important to ensure that, the large variety of ideas on how to provide services, does not induce unnecessary complexity and chaos.

Over time, it has been noticed that the available standards do not harmonize suf-ficiently team member's activities. Frequent conflicts and the intensive managerial involvement in solving them inhibited team development. In response, a set of values which should direct the actions of each individual was jointly defined. These values were meant to ensure that team members always act in accordance with the vision and contribute to its implementation. These values were:

- **simplicity** – always choose what is simpler, composed of a smaller number of elements and interactions between them; for example, when designing an interface, choose solutions that are easier and cheaper to maintain afterwards, rather than the most technologically advanced ones;
- **internal customer focus** – we are here to solve business problems; a guideline was adopted to deliver for each business request for integration a cost estimation within two weeks;
- **effectiveness** and **independence** – recognition that achieving the goal is the most important in every situation; not stopping in the event of adversity, but continually searching for alternative ways to achieve the goal, asking others, looking for new solutions, if known approaches do not seem to lead to the expected result; high independence - in the absence superiors or advices from others, decide on your own;
- **impossible doesn't exist** – the adoption of a principle that it is forbidden to use the word "impossible", instead it should be demonstrated that every problem can be solved in many ways, which differ however in complexity, effort and risk;
- **mutual respect** – recognition that each member of the team has talents and drawbacks but deserves respect, the lack of which will not be accepted;
- **continuous development** – ambitious goals (in the perspective of four years) require continuous development of each team member, otherwise they become unattainable;
- **keep high standards** – manifesting, among others, in: not being satisfied with incomplete solutions or partial explanations of issues that do not clearly indicate their root-cause, delivery of fully documented integrations and thorough knowledge transfers to the support department, emphasis on precise communication and keeping promises (e.g. due dates),
- **honesty and openness** – a principle has been adopted that team members should always present their reasons directly, without taking care about being "diplomatic"

or violating an existing and comfortable group consensus; openly criticize if a given solution conflicts with the established rules of action; undermine the standards themselves if it is seen that they no longer correspond to reality. However, it has been agreed that each criticism should be supplemented with well-justified (cost, risk, scope change and plan for its implementation, dependencies) proposals for solutions and expressed with respect for the authors of the criticized solutions. People are expected to speak clearly, precisely and simply and with respect for others. A critique which is only an expression of temporary dissatisfaction was not accepted.

The degree to which a given team member was demonstrating these values was made, together with the level of achieving individual goals, an assessment criterion in the employee's annual appraisal, and became linked to the bonus system.

Not all values were easy to accept and implement. Openly telling the truth turned out to be uncomfortable for team members originating from cultures with a large power distance. The pressure on effectiveness and independence did not fully suit employees from cultures in which group decision making predominated. In such cases, a lot of coaching was required to make people understand why the value driven behaviors are necessary, what problems are they solving and how to best deal with situations requiring action in accordance with a given value.

The strong need for achievement and the presence of mutually accepted values facilitated the emergence of collective group identity. It is worth emphasizing that it was more an evolutionary process than a designed one. It was accelerated by collective project involvement, collective problem solving and collectively achieving success. The final success was the eventual proof that the adopted solutions, established work organization and collectively shared values worked in practice, fortifying mutual trust and confidence in having chosen the right values.

It should also be emphasized that the team members also maintained relationships outside of work. They met each other and with their families. Together, they practiced sport and participated in various events. They helped each other privately. Even if one of the employees left the team, contact was kept with him and occasionally was invited to some of the team's events. Relationships established in this way proved to be more durable than only labor relations. This often resulted in new ideas of solutions that the former employees observed with the new employer and passed on to the current team members during private meetings.

The preservation of collective group identity was, paradoxically, also facilitated by the great cultural diversity of the team. The main point was here that there was not one dominant national culture in the team, whose representatives could already at the start bias (even unconsciously) the perception of organizational life.

The awareness of team members that all of them, without exceptions, adhere to the same rules (or at least sincerely try to do so) was fundamental to keep a high degree of cohesion. It was not easy in remote work settings (employees were spread over three main and many other project locations). Direct, face-to-face interactions were often impossible. The remote work settings allowed to "hide" many aspects of activities, and the lack of face-to-face communication caused that frequently during discussions (conducted by means of ICT tooling) significant nuances were overlooked. To maintain a high degree of transparency in such conditions, it was decided:

- to conduct daily half-hour stand-ups during which each participant was explaining what is he currently working on, what work he completed the day before and what problems is he facing;
- to hold weekly one-and-a-half hour meetings on which the project portfolio was discussed, employees were assigned to new projects, technical details of planned solutions were elaborated on, cost estimation were shared and aspects of continuous improvement addressed. An overview of the project portfolio was maintained, which showed who is working with whom on which project and who its client was;
- to make the MS Outlook calendars fully visible to each other which gave everyone insight on the accessibility of all team members and visibility about their occupancy;
- to adopt a tool in which everyone had an insight into the current location of another team member at a given day, existing travel arrangements, as well as temporary unavailability (due to e.g. vacation or illness);

The leader and the way how he managed the team influenced its (team's) cohesion. This was not only about the type of opaque leadership style (democratic or instructive) or specific leadership skills, but about the degree of intensity with which he participated in the everyday life of the team. The higher it was, the more nuances the manager was able to (1) identify and correct, (2) detect, mediate and resolve conflicts, (3) recognize talents and areas for improvement of individuals and, consequently assign them to projects which suit them best and/or gave them the opportunity to develop even more. He was able to tie up a diverse group on a current basis, steer its transformation into a team and then develop it in the desired direction.

A high degree of participation in the team's everyday life allowed to build good relationships with employees, which were more than relations of superiority and subordination. The manager became a friend of his direct reports. However, he had to constantly emphasize what is important for the team, what is he aiming at and watching over the observance of rules and values. He had to ensure that comprehension and friendliness does not become more important than striving for the goal and that the high involvement in everyday life of the team does not turn into micromanagement. A manager strongly involved in the everyday life of the team must finally have the ability to synthesize the details observed at the micro level into more general rules and patterns at the macro level and to assess whether they support or hinder their overall goals. In other words, such a manager must be able to constantly carry out analysis and synthesis, move (mentally) between the macro and micro levels, engage in details and regularly step out of the operational matters. It seems that such a way of managing the team can be labeled as "inside-out leadership".

In football the presence of inside-out leadership seems to be more likely than in business[4] (primarily due to its nature, i.e. daily training, locker-room conversations,

---

[4] Similar examples can be found in military history where some of the great commanders (see e.g. Alexander the Great) were leading their armies this way.

joint trips, camps, tournaments) where many managers still tend to preserve a high power distance and maintain a high degree of impersonality in managing their direct reports. On the other hand, a coach or football manager in the best clubs is supported by specialized staff (physiotherapists, data analysts, specialists in tactics, fitness trainers, psychologists) who assist him in leading the team, including the aforementioned detail analysis and its synthesis into patterns, which makes it easier for him to manage his subordinates effectively. In business, a team leader usually does not have such facilities and can often count only on the support of his line manager, assuming that he is vitally interested in its development and the results of his department. A business leader must therefore work on his self-awareness and develop the ability to lead inside-out individually.

It seems that a high involvement in team's everyday life, enriched with the ability to synthesize details into more general patterns, can be seen as another element facilitating the integration of diverse groups into efficiently functioning teams. It requires more effort and talent than managing less diverse staff, but it seems to be a promising proposition for managers who want to benefit from diversity through applying conjoined management or who are faced with diverse staff and want to form a cohesive team.

5 Summary

The presented concept of conjoined management has five dimensions:

- Recruitment focused on talent acquisition, not carried out under the banner of diversity,
- A sense of unity for a common goal,
- Collective group identity,
- Transparency of operations,
- Inside-out leadership.

It seems that it will be the more effective, i.e. more free from any kinds of prejudices (always hindering organizational performance):

- the more effective the recruitment of talented staff,
- the more far-reaching sense of unity for a common goal,
- the stronger the collective group identity,
- the greater the transparency of operations,
- the more frequently inside-out leadership is being enacted.

Conjoined management it is not oriented on increasing staff diversity, because, as research shows (first part of the article), the relationship between the workforce diversity and company performance appears to be highly debatable.

The presented concept seems to be applicable, as it turns out, not only football clubs. This is for at least two reasons:

- contemporary organizations, as already mentioned, face principally the same problems that solve football clubs,

- the dimensions of conjoined management strongly refer to the classic principles of management, applicable primarily, as it is known, to business organizations.

It should be noted, however, that in the case of organizations larger than football clubs, or the organizational unit considered in the article, conjoined management will be more complex and more challenging.

References

1. Adler, N.J., Gundersen, A.: International Dimensions of Organizational Behavior. Cengage Learning, Boston (2007)
2. Ashkenas, R., Siegal, W., Spiegel, M.: Mastering organizational complexity: a core competence for 21st century leaders. In: Woodman, R., Pasmore, W., Shani, A.B. (eds.) Research in Organizational Change and Development. Emerald Group Publishing Limited, Bingley (2014)
3. Bell, S.T., Villado, A.J., Lukasik, M.A., Belau, L., Briggs, A.L.: Getting specific about demographic diversity variable and team performance relationships: a meta-analysis. J. Manag. **37**(3), 709–743 (2011)
4. Bolchover, D., Brady, C.: 90-minutowy menedżer. Lekcje z pierwszej linii zarządzania. Wydawnictwo Zysk S-ka, Poznań (2007)
5. Burrel, L.: Nadal ulegamy stereotypom. HBR Polska 11 (2016)
6. Elborse, A., Fergusson, A.: Poradnik Fergussona. HBR Polska 06 (2016)
7. Fayol, H.: Ogólne zasady administracji. In: Kurnal, J. (ed.) Twórcy naukowych podstaw organizacji. PWE, Warszawa (1972)
8. Haidt, J.: Prawy umysł. Dlaczego dobrych ludzi dzieli religia i polityka (2014)
9. Hopej-Kamińska, M., Zgrzywa-Ziemak, A., Hopej, M., Kamiński, R., Martan, J.: Simplicity as a feature of an organizational structure. Argum. Oecon. **34**(1), 259–276 (2015)
10. Joshi, A., Roh, H.: The role of context in work team diversity research: a meta-analytic review. Acad. Manag. J. **52**(3), 599–627 (2009)
11. Koźmiński, A.K., Jemielniak, D., Latusek-Jurczak, D.: Zasady zarządzania. Oficyna a Wolters Kluwer business (2014)
12. Malik, F.: Kieruj, działaj, żyj. Zawód menedżer, MT Biznes Sp. z o.o., Warszawa (2015)
13. Stahl, G.K., Maznevski, M.L., Voigt, A., Jonsen, K.: Unraveling the effects of cultural diversity in teams: a meta-analysis of research on multicultural work groups. J. Int. Bus. Stud. **41**(4), 690–709 (2010)
14. Tworek, K., Zgrzywa-Ziemak, A., Hopej, M., Kamiński, R.: Workforce diversity and organizational performance. Submitted to the International Journal of Human Resource Management
15. Welch, J., Welch, S.: Winning znaczy zwyciężać. Wydawnictwo Studio Emka (2015)

On-line sources

1. sport.tvp.pl/35844917/nie-chcemy-pilkarzy-z-afryki-rasistowski-skandal-w-anglii. 25 Feb 2018

The Simplification of Organizational Structure: Lessons from Product Design

Marian Hopej[1] and Marcin Kandora[2](✉) (iD)

[1] Wrocław University of Science and Technology, Wrocław, Poland
[2] Głębinów 43, 48-300 Nysa, Poland
mj.kandora@gmail.com

Abstract. The organizational structure of an enterprise is continuously exposed to increasing complexity induced by growing dynamics and complexity of its environment. It is therefore not surprising that the simplification of structural solutions is becoming increasingly important and implies challenging Ashby's law. In the literature on the structuring of organizations, the principle of simplicity appears to be ambiguous and not well articulated. Moreover, the following research problem seems to be left unaddressed: can the simplification of an organizational structure be solely narrowed down to its complexity reduction or should it be complemented by other measures, such as an increase in employee participation (in the simplification process), to be effective? The concept proposed in this paper represents an attempt to unravel it. It is based on J. Maeda's ten laws of simplicity of product design which, as it turns out, have significant transfer potential into the field of organizing. It stresses that an effective simplification of an organizational structure cannot be based solely on complexity reduction but requires as well an active development of the so-called bottom-up organizational culture and continuous improvement of the self-management system.

Keywords: Organizational structure · Simplification · Complexity
Simplicity principle

1 Introduction

It is assumed that an organizational structure consists of a set of rules governing the behavior of organizational members [22]. In general, by limiting discretion it makes their behavior more predictable. It should, on the one hand, shape organizational order (by structuring the elements and activities of an organization) and, on the other, foster organizational agility in its relations with the environment. It is thus one of the key management tools used for complexity reduction.

The structuring of organizations is one of the most fundamental themes in management science, so it is no surprise that many scholars, as well as managers, formulate different principles for it. One of them is the **principle of simplicity** which includes as well the notion of aesthetics. That is mainly because what is appreciated is simultaneously in accord with the aspect of organizational harmony [20]. Although striving for simplicity of organizational structures is not new, scholars stress it more strongly

© Springer Nature Switzerland AG 2019
Z. Wilimowska et al. (Eds.): ISAT 2018, AISC 854, pp. 188–200, 2019.
https://doi.org/10.1007/978-3-319-99993-7_17

recently, particularly in the context of globalization and digitalization of economy. Consequently, Ashby's law [1, p. 207] stating that "only variety can destroy variety", is being challenged. Moreover, practitioners like J. Welch, a long-time CEO of GE, articulate that "for a large organization to be effective, it must be simple" [24, p. 114]. Adapting to environmental complexity through complication of organizational rules ends up with bureaucracy taking over an organization which will, sooner or later, impede its adaptation to changing environmental conditions [5, 12]. Additionally, as Ghoshal and Nohria [7, p. 24] state, "complexity is costly and difficult to manage, and simplicity, wherever possible, is a virtue".

In the rich body of literature on the structuring of organizations, the principle of simplicity appears to be ambiguous and not well articulated. Consequently, the following research problem arises: can the simplification of organizational structure be solely brought down to its complexity reduction or does it need to be complemented by other measures associated with more employee participation in this regard? The approach to structural simplification, presented in this paper, attempts to fill this research gap. It builds on approaches to simplification applied in engineering and technical sciences, where simplicity is primarily associated with product design. More specifically, it is based on the guidelines for simplifying product design formulated by J. Maeda in "The Laws of Simplicity" [16]. They stem from a careful observation of the work of experienced designers, focused on customer experience, and form a conceptual framework for the analysis and interpretation of simplicity (Table 1). As pointed out by Griesbach, Kaudela-Baum and Wyss [10], these laws can be considered as good problem-solving guidelines in managing of organizations. They can be seen as a promising vantage point when analyzing and assessing approaches to the simplification[1] of management systems and organizational structures (always entangled in a specific context) in particular.

2   The Simplification of Organizational Structure

How could the attempts to simplify an organizational structure, following the first law, i.e. **reduction**, look like? As already mentioned, an approach consistent with Ashby's law would result in increasing bureaucratization of a structural solution and would lead to a far-reaching limitation of organizational participants' discretion. J. Welch was well aware of this and urged GE employees to reject and fight bureaucracy. He argued, "you will not release the potential of people if you think for them (…), leave them free and do not sit on their neck, remove the bureaucrats and the barriers of hierarchy" [14, p. 13].

One cannot disagree with this view. The best way to reduce complexity is to provide people with more freedom. As Crozier [5, p. 47] states, "structures and procedures are not a response to complexity. They lead to stiffness and complication. Only people can absorb complexity because only they can find solutions by thinking forward, redefining problems, investing in knowledge, formulating policies".

---

[1] In this paper "simplification" is understood in accordance with the interpretation of Gregory and Rawling [9] as the removal of sources of complexity and waste in the organization.

Table 1. The ten laws of simplicity by J. Maeda

Law	Content
Reduce	The simplest way to achieve simplicity is through thoughtful reduction
Organize	Organization makes a system of many appear fewer
Time	Savings in time feel like simplicity
Learn	Knowledge makes everything simpler
Differences	Simplicity and complexity need each other
Context	What lies in the periphery of simplicity is definitely not peripheral
Emotions	More emotions are better than less
Trust	In simplicity we trust
Failure	Some things can never be made simple
The One	Simplicity is about subtracting the obvious, and adding the meaningful

Source: [16, p. ix]

This involves reducing the number of organizational rules, which is inextricably related to increasing decision-making freedom. Consequently, the notion that "organizing is the art of doing extraordinary things with the help of ordinary people" seems to be losing significance and the slogan "professionalization of people instead of complicating structures and procedures" is gaining it [5].

What does it mean to **organize**? In general, the essence of organizing is, on the one hand, the division (differentiation) of work and on the other, its integration. From organizational structure's simplicity perspective, two aspects of organizational differentiation are most important, i.e., the way work is grouped on the second (from the top) hierarchical level, which determines the overall nature of the structural solution, and the size of the core organizational units.

As it is known, during organizing the functional criteria might be followed, resulting in a functional structure, or the object one (market, product, customer, etc.), which characterizes a divisional structure. The two forms can co-exist in parallel, which somehow seems to be common as structures of contemporary organizations, especially of the relatively bigger ones, are most often hybrid, i.e. two grouping criteria are used at the second (from the "top") hierarchical level. However, it is important to remember that diversity induces complexity, and thus such solutions cannot be positively evaluated for simplicity. Conversely, it can't be excluded that for some organizations a hybrid form is inevitable. Such structure should however not be the first but last choice [17].

The principle of simplicity needs to be considered too when determining the number and size of the core organizational units (created on the second, from the "top", hierarchical level). The dilemma to be addressed here is if there should be less larger ones or relatively more but rather smaller ones? As demonstrated by Tworek, Hopej and Martan [25] using fractal calculus, the larger the number of core organizational units, the greater the complexity of the structure. Thus, simplicity of organizational structure is negatively correlated to the number of core organizational units [25].

It's widely known that each division of work increases complexity and that the creation of specialized roles and organizational units generates demand for integration which can be accomplished through hierarchy (inseparable from direct supervision), planning and programming, and mutual adjustment [22]. The last one is predominantly used in matrix structures where the principle of unity of command isn't followed. There might be circumstances in which a matrix solution is inevitable, because a better alternative cannot be temporarily found. It hinders, however, focusing on business priorities, and thus, threatens efficient management [17]. Good results can be achieved when focusing on the most important matters which is particularly difficult in matrix structures. They might be perceived modern but they do not represent true progression [17, p. 12].

Another law (**time**) requires commenting on the model of a simple organizational structure, proposed by Mintzberg [18] and commonly referred to in literature. The main part of it is the so called strategic apex, while the primary coordination instrument is direct supervision.

It is characterized by a two-level hierarchy, a low degree of specialization, formalization, and a high degree of centralization, as one person usually sets the rules which everyone else has to follow. Consequently, there is little room for maneuver which could be greater, if the activities wouldn't be coordinated by means of direct supervision but through mutual adjustment (between employees). Hence, the question arises: is such a structure, which provides people with substantial decision-making freedom, a simpler one? An analysis performed using fractal calculations indicates that structural complexity increases with increasing intensity of direct supervision [12]. Thus, a simple structure should not be a centralized, but rather a decentralized one, undoubtedly accelerating decision-making processes.

A simple structure fits small organizations. In others, paraphrasing Einstein's words ("everything should be done as simple as possible but not simpler"), the structure should be as simple as possible, but not simpler. Hence, as decentralized as possible. However, decentralization must be complemented by the centralization of information flows. "For this purpose, coordination nodes should be created in the same way as those used in aviation. They must be organized so that at every point one can tell whether something has gone out of control, and if so, to intervene immediately" [17, p. 371].

The structure with the least possible degree of specialization, centralization, formalization, and standardization, as well as the least developed hierarchy, is a flexible solution that agrees with the next law (**learning**). It should facilitate uninterrupted exchange of information inside an organization and the communication with its environment aiming to reduce the natural boundaries in place. In other words, it is necessary to increase the permeability of both vertical and horizontal organizational boundaries.

From an organizational learning perspective vertical and horizontal boundaries negatively impact the acquisition of knowledge associated with the work accomplished in other organizational areas. They hinder the collection of feedback about the outcomes of activities conducted in other organizational units. Tight boundaries decrease transparency and reduce the intensity of cooperation which, in the study of Ashkenas, Siegal and Spiegel [3, p. 45], have been identified as key drivers of complexity.

Vertical boundaries originate from hierarchical layers. The more permeable they are, the more significant capabilities become (as opposed to formal positions and roles). This facilitates decision-making, getting closer to emerging issues, and allows for better use of other people's ideas. In turn, increasing permeability of horizontal boundaries, between the core organizational units, can limit conflicts and consequently enable employees to concentrate on achieving organization's goals [2].

According to the next law (**differences**), simplicity and complexity need each other. The organizational rules prevalent in a complex structure can be seen as useful reference when designing a simpler structure (as their contrary). Thus, because of the interdependence of simplicity and complexity (the greater the simplicity, the smaller the complexity and vice versa, i.e., the greater the complexity, the smaller the simplicity), a potential structural simplification may build to some extent on the identification of the symptoms of complexity growth (Table 2). Moreover, it has to be complemented by well-prepared and thorough change management. That is mainly because the implementation of changes aiming to simplify an organizational structure is usually more difficult than of those increasing structural complexity (e.g. it is easier to create a new position than to eliminate one).

Although people can adapt to changes meant to simplify organizational structures and are doing it better, it cannot be forgotten that such adjustments can be compared to surgical interventions on a living organism without anesthesia. A good surgeon knows that if a surgery is not necessary, it should not be performed. Effective managers are doing the same. They avoid continuous restructuring and reorganizing and agree to alter the structure only when it is really necessary, ensuring well-prepared change management and shielding [17].

The law "**context**" complements the law "**differences**". It provides a guideline on how to shift between complexity and simplicity. The simplicity of an organizational structure can be theoretically shaped independently (to some degree) from the context, i.e. through referring to the structural complexity symptoms (presented in Table 2) and forming the to-be structure as their opposite. However, the simultaneous preservation of structural simplicity and a good degree of fit between an organizational structure and its context requires empirically grounded knowledge about relations between the degree of simplicity of structural solutions and its contingencies.

A study of 100 companies operating in Poland revealed that none out of 11 structure-forming factors[2] was fully predictive (a stepwise regression analysis has been applied) for all structural characteristics, although for each of them statistically significant models were proposed (Table 3).

Although the identified dependencies are neither unavoidable nor certain, they lend credence to the influence of a few factors on organizational structure's simplicity, including: the size of the organization (the smaller the organization, the simpler the structure), the degree of diversification (the smaller the degree of diversification, the simpler the structure), manufacturing technology (the more non-routine the manufacturing technology, the simpler the structure), organizational culture (the more open the culture, the simpler the structure), the dependence on the environment (the lower the

---

[2] They were: environment uncertainty, organizational dependence on the environment, degree of diversification, organizational culture, professionalism of the employees, leadership, management aspiration to simplify the organization, manufacturing technology, use of IT, organization history and its size (number of employees, revenues).

Table 2. Some symptoms of organizational structure complexity growth

Structure complexity growth	Symptoms
Increase of the degree of centralization, specialization, formalization, standardization, and the expansion of the organizational hierarchy	Growth of inertia, friction, and sluggishness in the organization
	"Noise" in the information channels, distorted information
	Focus on irrelevant matters or wrong priorities
	Declining rationality of decisions
	Lower motivation to work
	Monotony, physical and mental fatigue
	Reduction of staff's involvement in pursuing common goals
	Smaller ability to react to new problems
	Skilled incompetence
	Vicious circle of bureaucracy
	Many meetings with many people
	Necessity to hire specialists and coordinators

Source: own work

Table 3. Structure-forming factors and structural characteristics

Structural characteristic	Corrected R^2	Variables explaining structural characteristics
Hierarchy	0.488	Number of employees ($\beta = 0.485$; $p < 0.001$), revenue ($\beta = 0.179$, $p < 0.05$), degree of diversification ($\beta = 0.166$, $p < 0.05$), management aspiration to simplify the organization ($\beta = -0.28$, $p < 0.01$)
Centralization	0.191	Number of employees ($\beta = 0.300$, $p < 0.01$), manufacturing technology ($\beta = -0.270$, $p < 0.05$), organizational culture ($\beta = 0.263$, $p < 0.05$), organization history ($\beta = -0.763$, $p < 0.001$)
Specialization	0.649	Manufacturing technology ($\beta = -0.763$, $p < 0.001$)
Formalization	0.391	Number of employees ($\beta = 0.479$, $p < 0.001$), organizational dependence on the environment ($\beta = 0.196$, $p < 0.01$), management aspiration to simplify the organization ($\beta = -0.185$, $p < 0.01$)
Standardization	0.203	Manufacturing technology ($\beta = -0.408$, $p < 0.001$), number of employees ($\beta = 0.198$, $p < 0.05$)

Source: own work based on [12]

dependence, the simpler the structure) and the management's aspiration to simplify the organization (the greater the aspiration to simplify the organization, the simpler the structure).

The recalled guidelines for the simplification of structural solutions, related to the "differences" and "context" laws, together with the presented dependencies should be included in the **mechanism of structural ballast reduction**. It is grounded on:

- An **emotional** (another Maeda's law) attitude of all organization members towards redundant and complex organizational rules. It is key to treat them as a personal enemy and fight them (as did J. Welch in GE; see [14]).
- A simple idea to ask the following questions: What shouldn't I do? What can be simplified without lowering the quality of work? Those questions should be asked about the structure and its contingencies, e.g. the used forms, the created reports or the production program in place. Asking them should not only serve the organization as a whole but also support each employee [17]. Moreover, it should be complemented by a regular review of historically grown routines and procedures to determine their appropriateness [3, pp. 43–44].
- The simplification of the structural solution by the employees themselves, in the scope of their responsibilities. This should be preceded by liaising with the relevant stakeholders, including the line manager, and other subject matter experts. The more important the change, the greater the number of mutual alignments to be made.

Trust, precisely trust in simplicity and mutual trust, is the essence of the eighth Maeda's law. It is crucial for the functioning of the structural ballast reduction mechanism and key to prevent structural complexity accumulation.

Trust in simplicity is essential for the process of persisting emotions associated with either complexity or simplicity, i.e. the sense of discomfort when encountering complexity and the relief, sense of comfort and enthusiasm experienced when facing simplicity. To develop it, senior management has to convince the employees (through leading by example, keeping promises and demonstrating that simplicity leads to good results) that simplicity has a real value in the organization and is truly used as one of the key decision criteria and not just a lip service. Decisive and consistent leadership, emphasizing the value of simplicity, is the key to build such trust.

For the structural ballast reduction mechanism to function well, the trust in simplicity needs to be complemented by mutual trust. If the management succeeds to develop and maintain it, it will positively impact the quality of workplace life and cooperation making structural simplification more effective. When people trust each other, they risk more and are more innovative. They require less control and less detailed procedures which can usually be replaced by general guidelines. As emphasized by Sztompka [23], mutual trust is a fundamental element of good living in all dimensions. When there is no foundation of trust, any restructuring is harder to complete. Moreover, the degree of employees' involvement in simplification initiatives will further decline.

Mutual trust is difficult to sustain. It is a fragile resource that is part of the so-called bottom-up organizational culture which can also be defined as the culture of bottom-up collaboration. It is based on the assumption that people have a natural predisposition to cooperate and refers to the latest experiments and discussions on the unselfish gene [4] or to cooperation itself, seen as the third pillar of evolution, apart from mutation and natural selection [19]. According to Benkler [4], the adoption of such an assumption should lead to the rejection of bureaucratic rules of supervision and control in the interaction with subordinates. They should be rather replaced with engagement and the sense of common purpose. In contrast to the so-called top-down culture, a bottom-up culture is conducive to the motivation to experiment and to the formation of the belief that a change aiming to simplify an organizational structure isn't something worse than maintaining the status quo [11].

The interpretation of the ninth law (**failure**), adopted in this article, is based on the assumption that an organizational structure free of problems and frictions doesn't exist. All structural solutions are imperfect. They are prone to conflicts and tensions due to the continuous pressure on organizational differentiation and organizational integration. They exhibit antagonisms between people, create ambiguities and other complications. There is also no ideal simplification of the structure, as it usually requires to compromise with the challenges associated with the ever-increasing environmental uncertainty. Therefore, the simplification of a structural solution can't simply be perfect. It can be only made as imperfect as possible, which is not a negative thing, unless for the purists of management [17].

Several scholars support the above findings. The manifestation of modern management is often associated with the introduction of different forms of teamwork, like project teams, into traditional structural solutions (e.g. functional) which constitutes the DNA of the organizational rules in place [15, 21]. However, such structures are relatively complex, because of their hybrid nature, and often contradict the principle of unity of command. Their complexity is even greater when the teamwork is poorly organized, and unfortunately this is often the case. Many teams are in fact groups that are created thoughtlessly. They are too big, their objectives are ambiguous and their members' roles and responsibilities are imprecisely formulated [17]. To be able to achieve any results, ambiguity and chaos are eliminated by introducing supplementary, mostly more detailed, rules.

Teamwork requires good cooperation. If this is the case, the introduction of different types of teamwork into traditional, hierarchical structures can be seen as management's consent for employees' entrepreneurial initiatives. However, the adaptation to environmental complexity through self-organization and mutual adjustment requires not only exceptional skills but as well, perhaps even above all, ownership and accountability, not only for own results but also for those of others. Also, it requires behaviors that do not go beyond the norms of decency and propriety. This seems to imply the need for extreme caution about the modern, but relatively complex structural solutions.

Finally, the last law (**the one – framework law**) which can be seen as the general guideline for the simplification of organizational structures which are exposed to the continuous regeneration of complexity. This is mainly due to:

- The ever-increasing complexity of the environment which the structure should reduce through a system of rules specifying expected behaviors [22].

- Automatic renewal of structural complexity. The point is that organizational differentiation and integration cause tensions which are easiest eliminated through the introduction of new rules that further detail expected behaviors. Hence, if simplification measures are not taken, the complexity of the structural solution continues to grow which is in line with the second principle of thermodynamics - as time goes by, all things strive for less simplicity and greater disorder [13].

Therefore, an organizational structure should be regularly evaluated by the management on the basis of two criteria:

- Rationality, i.e., the goodness of contextual fit. A total fit is not possible in practice since contextual elements often create contradictory requirements for the formation of structural dimensions. It is therefore necessary to use simplified models of both sets of characteristics when aligning an organizational structure with its context. In consequence, a particular context can be addressed well not only by one structural solution, but by a set of acceptable solutions limited by their goodness of contextual fit [12].
- The simplicity of the structural solution acceptable in a given context (the simplest solution should be chosen from the set of acceptable solutions).

The results of such analysis should support senior management in the simplification of organizational structure, including the correction of mistakes made during the functioning of the structural ballast reduction mechanism, not free, after all, of both subjective and objective limits of the decisions' rationality.

A well-functioning structural ballast reduction mechanism seems to be a possible way to institutionalize the process of "subtracting the obvious and adding the meaningful" at the tactical and operational level of the organization. Together with regular reviews of company's structure, aiming to evaluate its rationality and simplicity, it constitutes a relatively consistent and holistic implementation of the tenth Maeda's law concerning organizational structures. It shall be stressed though, that it is primarily aiming to reduce excess but already existing structural ballast which Giroud and Karim [8, p. 130] compare to "organizational cholesterol". The functioning of such a structural ballast reduction mechanism can be compared to the treatment of symptoms of diseases whereas it is well known (in medicine) that it is cheaper and more effective to implement preventive measures. The actions which can be taken in order to prevent the structural ballast to accumulate absorb less management energy and are easier to institutionalize than the *ex-post* corrections as they result in less resistance to change. However, they require significantly more imagination and organizational design capabilities, as they must be planned *ex-ante*, in the initial phase of structural design.

Seen through the lens of organizing, this mechanism represents an element of a self-management system which needs to be developed and continuously improved.

The concept of a self-management molecule, as shown in Fig. 1, illustrates its essence: the ability of each member of the organization to influence its management, including the structure as a basis for management actions.

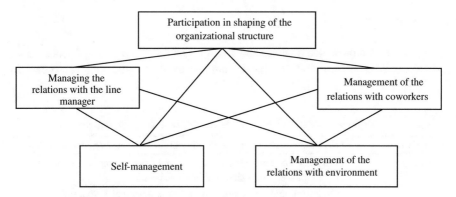

Fig. 1. Self-management molecule Source: own work based on [17]

This molecule can be replicated because everyone has their superiors, department co-workers, cooperates with the outside world, and manages her/himself [17]. By spreading over the whole organization it is used in all organizational units, also in the traditional, hierarchical structures.

Such a self-management molecule can function particularly well in relatively independent organizational units and on the lowest hierarchical layers. It is supposed that at this level the effect of increased empowerment on employees' motivation, their innovativeness and their line managers' span of control will be the strongest, enabling the managers to focus more on staff development and coaching than on their super-vision[3]. Moreover, the more independent from each other the organizational units are, the lower will be the impact of decisions taken in one of them on the others, and thus, the lower the number and frequency of necessary alignments between particular molecules which will ease their replication. For more interdependent units the neces-sary integration has to be ensured. This can be most easily accomplished through the before-mentioned mutual adjustment process or regular collective problem solving sessions (in case of more strategic decisions), resembling to some extent the Work-Out meetings introduced by J. Welch in GE [24].

The replication of the self-management molecule should, however, not transform into a spontaneous, uncontrolled process. Senior management should promote this concept among the employees, as the leading management approach, reducing potential resistance.

---

[3] This is not a stand-alone view, as in some management concepts like lean management particular attention is given to this level of hierarchy (see key lean management guidelines as "go and see", "respect the front line employee" or "teamwork").

3 Summary

The results of the presented analysis show that Maeda's laws have significant transfer potential and, as it turns out, all can be applied as well in the field of organizing (Table 4).

Table 4. Transfer of Maeda's laws into the field of organizing

Maeda's law	Recommendation for the simplification of organizational structure
Reduce	Limiting the number of organizational rules, which means increasing the employees' decision-making freedom
Organize	Organizational differentiation and organizational integration
Time	Acceleration of decision-making by increasing decentralization
Learn	Increasing permeability of vertical and horizontal organizational boundaries
Differences	When simplifying an organizational structure, a complex structure can be used for reference
Context	The simplicity of the organizational structure cannot be considered separately from the characteristics of other elements of the organization and its environment
Emotions	Mobilizing employees' energy in simplifying everything that is possible in the organization
Trust	The simplification of organizational structure builds on the trust in simplicity and the bottom-up culture a key element of which is mutual trust
Failure	There is no ideal simplification of organizational structure because it usually requires to compromise, among others, with the requirements of rapidly growing environmental complexity
The One (framework law)	Due to the continuous regeneration structural complexity, it is crucial to institutionalize a mechanism for prevention and elimination of structural ballast, which is part of the self-management system

Source: own work

From the analysis of the presented approach to the simplification of organizational structure several conclusions can be drawn:

- **The simplification of organizational structure is not purely about complexity reduction.** It also requires the introduction of complex measures for building and continuously improving the self-management system, which is a mechanism intended to prevent the creation and to remove accumulated structural ballast. It is an attempt to embody the idea that one needs to trust the innovative potential of organization's members in simplifying structural solutions. It is characterized by a high degree of empowerment and employee participation who act in a relatively informal system. Under such conditions, the execution of activities is governed more by organizational routines than by formal organizational rules.

- **The effective functioning of the structural ballast prevention and reduction mechanism requires a special ability - proficiency in identifying increases of structural complexity.** The employees should therefore develop a good sense for identifying initiatives leading to potential complexity increase (to be able to prevent it) and symptoms of its excessive accumulation (to be able to reduce it). To do so they should practice the simplification of structural solutions and be actively coached by the management.
- The incubation of organizational routines that co-create the mechanism of prevention and reduction of structural ballast requires **management's determination in fighting complexity.** Management should also **encourage others to challenge existing rules** if there are good reasons for doing so.
- **The application of the principle of simplicity to organizational structures means that they should be as simple as possible,** which is tantamount to their greatest possible flexibility. It should not result in the overlap of different forms of teamwork with the essentials of organizational rules in place, e.g. in the form of classic, hierarchical structures.
- **The simplification of organizational structure should build on a bottom-up organizational culture** that preserves a kind of civic community in the organization [6]. In general, it refers to the extent to which the employees help each other, suppress egoism and observe the norms of decency and propriety.
- The conducted analysis indicates that **the simplification process itself is more vital to the simplification of organizational structures than a detailed description of their simplicity**. This is because employees' innovativeness, inseparable from the simplicity principle, may result in various structural solutions characterized by low degree of specialization, centralization, formalization, standardization and low hierarchy.

Some of the issues discussed in this paper require further refinement to increase the application potential of the presented concept. In particular, it would be useful to analyze and *describe the approach to the implementation and replication of the elements of the self-management system*, including the mechanism for prevention and reduction of structural ballast. Furthermore, from the management's perspective, it would be highly beneficial to develop a *"complexity scorecard"*, consisting of KPIs (based on data available in the organization's information system) which could inform in a simple way about changes in complexity in order to indicate when to initiate preventive and corrective actions. One should also analyze in more detail the *organizational routines constituting the mechanism for prevention and elimination of structural ballast, classifying them, determining their role in the functioning of the mechanism* and identifying *factors influencing their persistence*.

References

1. Ashby, R.: An Introduction to Cybernetics. Univ. Paperbacks, London (1956)
2. Ashkenas, R.: Nowe szaty organizacji. In: Hasselbein, F., Goldsmith, M., Beckhard, R. (eds.) Organizacja przyszłości. Business Press, Warszawa (1998)

3. Ashkenas, R., Siegal, W., Spiegel, M.: Mastering organizational complexity: a core competence for 21st century leaders. In: Woodman, R., Pasmore, W., Shani, A.B. (eds.) Research in Organizational Change and Development, pp. 29–58. Emerald Group Publishing Limited, Bingley (2014)
4. Benkler, Y.: The unselfish gene. Harv. Bus. Rev. **89**(7/8), 77–85 (2011)
5. Crozier, M.: Przedsiębiorstwo na podsłuchu: jak uczyć się zarządzania postindustrialnego. Państwowe Wydawnictwo Ekonomiczne, Warszawa (1993)
6. Drucker, P.F.: The Practice of Management. HarperBusiness, New York (1993)
7. Ghoshal, S., Nohria, N.: Horses for courses: organizational forms for multinational corporations. Sloan Manag. Rev. **34**(2), 23–35 (1993)
8. Girod, S.J.G., Karim, S.: Restructure or reconfigure. Harv. Bus. Rev. **95**(2), 128–132 (2017)
9. Gregory, I., Rawling, S.: Profit from Time: Speed Up Business Improvement by Time Compression. Macmillan Business, London (1997)
10. Griesbach, D., Kaudela-Baum, S., Wyss, S.: Innovationspotential von Einfachkeit. ZfO **6**, 377–383 (2016)
11. Hopej, M., Kamiński, R., Tworek, K., Walecka-Jankowska, K., Zgrzywa-Ziemak, A.: Community-oriented culture and simple organizational structure. Organizacja i Kierowanie **4A**, 75–93 (2017)
12. Hopej-Kamińska, M., Zgrzywa-Ziemak, A., Hopej, M., Kamiński, R., Martan, J.: Simplicity as a feature of an organizational structure. Argum. Oecon. **1**(34), 259–276 (2015)
13. Koch, R.: Menedżer 80/20. Pracuj mniej, osiągaj więcej. MT Biznes Ltd., Warszawa (2017)
14. Krames, J.A.: Sygnowano Jack Welch: 24 lekcje najwybitniejszego CEO na świecie. Wydawnictwo Studio Emka, Warszawa (2005)
15. Łobos, K.: Organizacje – proste idee, zasady, narzędzia. Wydawnictwo Wyższej Szkoły Bankowej w Poznaniu, Poznań (2011)
16. Maeda, J.: The Laws of Simplicity. MIT Press, Cambridge (2006)
17. Malik, F.: Kieruj działaj żyj. Zawód menedżer. MT Biznes Ltd., Warszawa (2015)
18. Mintzberg, H.: The Structuring of Organizations. A Synthesis of the Research. Prentice Hall, Englewood Cliffs (1979)
19. Nowak, M.A.: Five rules for the evolution of cooperation. Science **314**(5805), 1560–1563 (2006)
20. Pszczołowski, T.: Mała encyklopedia prakseologii i teorii organizacji. Ossolineum, Wrocław (1978)
21. Schneider, J., Hoffmann, W.H.: Die einfache Organisation. ZfO **85**(6), 372–376 (2016)
22. Steinmann, H., Schreyögg, G.: Zarządzanie. Oficyna Wydawnicza Politechniki Wrocławskiej, Wrocław (2001)
23. Sztompka, P.: Zaufanie. Fundament społeczeństwa. Znak, Kraków (2007)
24. Tichy, N.M., Charan, R.: Speed, simplicity, self-confidence: an interview with Jack Welch. Harv. Bus. Rev. **67**(5), 112–120 (1989)
25. Tworek, K., Hopej, M., Martan, J.: Into organizational structure simplicity. In: Proceedings of the 38th International Conference on Information Systems Architecture and Technology, pp. 173–183 (2017)

Information Systems Reliability and Organizational Performance

Katarzyna Tworek[✉], Katarzyna Walecka-Jankowska,
and Anna Zgrzywa-Ziemak

Department of Computer Science and Management,
Wrocław University of Science and Technology, wyb. Wyspianskiego 27,
50-370 Wrocław, Poland
{katarzyna.tworek, katarzyna.walecka-jankowska,
anna.zgrzywa-ziemak}@pwr.edu.pl

Abstract. This paper refers to current research area, which is the relation between information technology (IT) and organizational performance. The main objective of this paper is to identify the role of IT reliability in shaping the organizational performance in SMEs. Proposed theoretical concept is empirically verified. The verification is based on empirical studies conducted in 400 SMEs operating in Poland in 2017.

Keywords: Information systems · System reliability
Organizational performance · SMEs

1 Introduction

The importance of information systems (IS) for organizational performance is recognized and studied. The literature covers the topic of IS value quite extensively [6, 12, 16, 21, 25, 33]. Almost all researchers believe that IS influences the organization ability to create competitive advantage. However, there are findings that do not support the notion that IS can influence organizational performance [5]. De Wet and coworkers [11] argue that not using IS itself should be considered, that IS has to be reliable to influence organizational performance.

In the article the relation between IS reliability and organizational performance is discussed and empirically verified. The verification is based on empirical studies conducted in 400 SMEs operating in Poland in 2017. This research focuses on SMEs due to the importance of this group of organizations for the economy and due to the marginalization of SMEs in current empirical studies in the area of interest. Findings support the notion about positive relation between two studied phenomena.

2 Information Systems and Organizational Performance

It is important to underline that when almost every organization operates using some kind of IS, the simple fact of using it is no longer a factor influencing the organization ability to create competitive advantage [11, 17, 30]. Moreover, the same authors

© Springer Nature Switzerland AG 2019
Z. Wilimowska et al. (Eds.): ISAT 2018, AISC 854, pp. 201–210, 2019.
https://doi.org/10.1007/978-3-319-99993-7_18

underline that it has become necessary for the organizations to integrate reliable IS into all organizational functions in order to build its potential to influence organizational performance and competitiveness [11, 17, 30]. It is not enough to use IS, it has to be reliable and the perceived level of this reliability is influencing the ability of gaining competitive advantage from using IS and may influence its ability to influence organizational performance [11, 30]. In the same time, while facing new global internet threats (like WannaCry ransomwares, massive hacker attacks, information-based terrorists attacks), maintaining the reliability, security and accuracy of the IS in organization is becoming one of the biggest challenges of IS management and it is one of the reasons why IS is perceived to be one of the elements enabling not only the ability of gaining the competitive advantage, but also the business sustainability [30].

There are three big meta-analysis concerning IT role in generating value for organization (shown in Table 1) and they all show that indeed, IT capabilities are influencing organizational performance and building the competitive advantage of the organization. However, in all those studies It is treated in a very general way – not accounting for the fact that its reliability influences organization' ability to benefit from it.

Table 1. Meta-analysis of IS influence on organization performance.

Source	Sample	Sample size	Sample use	Performance ratios	Results
Bharadwaj [4]	IT leaders from 1991–1994 IW 500 listings	56/149	IT leader versus control company of similar size and industry	Business performance measured by profit and cost (1991 to 1994)	Statistically significant correlation between IT capabilities and organizational performance
Santhanam and Hartono [29]	IT leaders from 1991–1994 IW 500 listings	46/46	IT leaders versus industry average	Business performance measured by profit and cost ratios (1991 to 1994) Sustainability of superior business performance (1995 to 1997)	Strong, statistically significant correlation between IT capabilities and organizational performance
Chae et al. [6]	IT leaders selected from 2001–2004 IW 500	561/296	IT leaders versus control companies of similar size and industry	Business performance measured by profit and cost ratios (2001 to 2004) Sustainability of superior business performance (2005 to 2007)	Statistically significant correlation between IT capabilities and organizational performance in case of some hypothesis covering the relation

Source: own work

Many authors underline that it has become necessary for the organizations to integrate IT into all organizational functions in order to build its potential to influence organizational competitive advantage, performance and competitiveness [3, 11, 17, 30]. However, since the relevance and the need for IT use in organization seems to finally be undisputable, there is a need for the analysis and evaluation of its use in organization. The concept of 3 R (reliability, resilience and robustness) emerged in the literature few years ago [18] and underline that the key factor influencing the ability to profit from using IT is its appropriate functioning in organization. The question arises – what should be considered as appropriate functioning of IT in organization? It seems that the most important features are:

- IT is able to support all tasks it was designed to support,
- employees may rely on the fact that IT will be functioning without major, unpredictable breaks, based on reliable data and based and standardized processes,
- IT is easy to use, predictable and offers support services for employees.

Therefore, it may be concluded that considering the theoretical model of IS reliability, the key factor influencing the proper functioning of IT in organization is its reliability, which concerns all major features listed above. However, it is a relatively new field of study and almost no publications are available concerning this matter. Hence, there is a definite and immediate need for formulation and verification of theories concerning this notion.

3 Information Systems Reliability

Organizations are managing a diverse portfolio of information systems (IS) applications [8] that are building the IT in organization. Therefore, the notion of IT reliability connects directly to the concept of IS reliability and IT in organization simply consists of all IS in this organization. That is why in many publications those two acronyms are used interchangeably. However, it is important to underline at the beginning that reliability of IT or all IS in organization is a different concept than reliability of software or hardware, which are notions well known and broadly discussed in literature [1]. Hence, it is crucial to not confuse it with those embedded directly in computer sciences, which do not include the IT management standpoint. And this standpoint is a crucial part for obtaining and retaining a competitive advantage from IT use in organization.

Therefore, reliability of IS in organization is understood as measurable property of IS, useful for its control and management, identifying its quality level and pointing out potential problems [34] and it is directly linked to the efficiency of IS components, especially those critical to its proper operations. Therefore, it can be said that IS reliability in organization is a notion build by factors connected to 3 different IS theories. First one is DeLone and McLean success model [13], second one is Lyytinen [20] 4 types of IS failure and third one is TAM model [9]. Therefore, in order to fully develop the notion of IS reliability it is crucial to identify factors that are constructs for each of 4 identified variables proposed in the IS reliability model (see Fig. 1). To identify all of them, the search of articles published from 2000 to 2018, with key words "IS in organization", "measurement" was conducted with EBSCO and ProQuest databases. From all available publications, those concerning lists of factors describing

IS in organization were purposefully selected. Based on those research [15, 26, 27], all factors potentially related to IS reliability in the context of above-mentioned 3 IS theories were identified and assigned to proposed 4 variables [26].

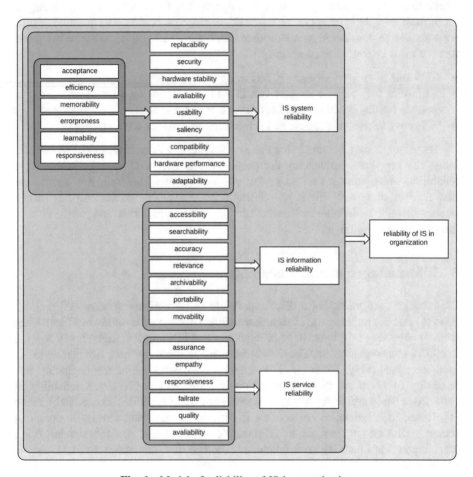

Fig. 1. Model of reliability of IS in organization

Model of IS reliability in organization has been developed, detailed description is published by Tworek [31] and it is presented on Fig. 1. The reliability of IS in organization consists of 4 factors: reliability of information included in IS in organization, reliability of support services offered for IS in the organization and reliability of system itself, which also includes the usability of this system. Each factor is built by series of items, listed on Fig. 1.

4 Information Systems Reliability and Organizational Performance

The analysis of research concerning IT use in organization offers some insights on this notion. However, the field of science concerning IT role in organization is quite chaotic. Hence, the analysis of literature should always start with some kind of organization of all available knowledge. In order to fully analyze the potential role of IT reliability in organization, the in-depth literature analysis was done. In the literature, there is a lot of IT business value models that can be useful in pinpointing the particular ways, in which IT reliability supports the organization ability to gain advantage from using IT. The most commonly known models are presented in Table 2. Presented models offer different approaches to the process of creating value for the organization from using IT, however similar conclusions may be drawn from all of them.

Table 2. The importance of reliability shown based on various IT value models.

Source	Independent variable	Dependent factors building IT value
Melville et al. [25]	Organizational performance	IT resources $(R)^*$ Complementary business resources Business process Industry characteristic Trading partner resources Country characteristic
Lucas [19]	Performance	Design of technology $(R)^*$ Information technology $(R)^*$ Other variables Appropriate use $(R)^*$
Peppard and Ward [28]	IS capabilities	Strategy Investment allocation IS competencies $(R)^*$ Processes Structure Roles Business Skills, knowledge and experience Technical skills, knowledge and experience Behaviour and attitudes $(R)^*$
Bharadwaj et al. [4]	Business value	Raw materials IT managements roles and processes $(R)^*$ IT impacts: Transformed business processes $(R)^*$ Enriched organizational intelligence Dynamic organizational structures
Marchand and Raymond [23]	IT value	Hard IT management competencies Hard information management competencies Soft competencies $(R)^*$ Information orientation $(R)^*$

$(R)^*$ - the importance of reliability for creating business value from using IT
Source: own work

First of all, almost all of them includes stages such as: "Appropriate use", "IT assets", "IT resources" in which authors underline the need for the 3R (reliability, resilience and robustness) implementation to ensure the proper functioning of the IT in organization and the fact that without users' trust in the IT actual positive influence on their task completion and job performance, creating IT value for organization is impossible. It underlines the role of IS usage reliability and IS system reliability. Second of all, stages such as "Processes", "IT impacts", "information orientation" underline the need for the reliable information input for the IT, which is the basis for reliability of the information output. Moreover, Bharadwaj [4] suggests that IT in organization is the tool, which can ensure the reliability of information and information flows in the entire organization creating standardized methods for including them in the ISs in organization and ensuring the proper exchange of this information between ISs in organization. However, all models presented in Table 2 are considering IT in general, without direct focus on its reliability as a key factor influencing the organizations ability to benefit from it.

5   Research Methodology

The survey was conducted in order to identify the relation between the level of IS reliability in organization and organizational performance. The pilot survey was conducted in 2017 among the group of 100 organization, indicating the issues concerning ambiguity of several questions. It led to the collection of random answers given as a response for those questions. They were rewritten in order to obtain the more reliable results, ensuring the informed response from the respondent. The main survey was conducted later in 2017, among small and medium enterprises (SMEs) located in Poland, using online survey service: SurveyMonkey. Only one survey was carried out in one organization. The research was anonymous. Efforts had been made to make sure that the questionnaire was filled in by employees who have a broad view of the entire organization. The statistical population (SMEs operating in Poland) is finite, but very large. 400 valid responses were collected. Since the responses were collected using properly prepared form, the online system counted only those fully and correctly filled in.

Respondents were asked to evaluate the IS in the organization based on the list of factors using the Likert scale (from very poor to very good with the middle point: fair). They were asked for the general opinion concerning reliability of system, usage, information and service, and then they were asked to evaluate each factor constructing those 4 variables. Using a Likert scale to measure IS reliability seems to be an appropriate choice. First of all, reliability of IS in organization is a subjective notion. Employees own perspective and opinion concerning aspects of IS reliability is the best source of knowledge, since their perception matters the most, that is because IS influences the organization mainly through its potential to influence every-day work of the employees. Quantitative methods are commonly used to assess the software and hardware features linked to the reliability. However, they do not give the information concerning the actual perception of this notion within the organization.

Respondents were also asked to evaluate the organizational performance. The performance measures were developed based on a literature review. A resulting 7-item scale captures the extent to which organizations achieve organizational performance, including financial and non-financial measures [24] and short-term and long-term measures [24], covering financial and market performance, quality performance and innovation performance from concept [22]. Due to the research objects - SMEs modifications were made and market share was excluded as irrelevant in case of SMEs. The evolution of the performance during the previous three years was conducted. Respondents were asked to rate the performance using the Likert scale (from well below expectations to well above expectations with the middle point: as expected). In line with literature, subjective measures of organizational performance were used [2, 22]. On the one hand, the objective performance measures (such as financial) are difficult to obtain due to confidentiality or unavailability. On the other hand, a subjective examination, although always exposed to errors, facilitates the comparison of many different organizations due to the studied aspects. Finally, there are evidence that subjective and objective performance measures are strongly correlated [7, 10, 14, 32]. Due to different industries and strategic priorities of investigated organizations, performance data need to be adjusted to evaluate each organization. For this purpose, an evaluation in comparison to the competition is usually needed. However, the object of discussed research are SMEs, which are usually significantly weaker market participants. That's why respondents were asked to answer the questions by comparison to expectations.

6   Research Results and Conclusions

There were two main variables: IS reliability and organizational performance. IS reliability consists of three dimensions theoretically distinguished. It is worth noting that Cronbach's α was 0.890 and higher for every variable, which indicates a high internal reliability of the scales and measurements. Descriptive statistics were calculated for all measured variables (Table 3).

Table 3. Descriptive statistics.

	Average	Median	Minimum	Maximum	Std. deviation
IS reliability	3,73	4,00	1,44	5,00	1,00
Information reliability	3,71	4,00	1,43	5,00	1,02
System reliability	3,74	4,00	1,39	5,00	0,98
Service reliability	3,79	4,17	1,50	5,00	1,03
Organizational performance	2,87	3,00	1,00	5,00	0,98

Source: own work

The correlation coefficients between the *IS reliability* and *organizational performance* have been calculated (Table 4). The results show that the variables are significantly correlated and the correlation is moderate.

Table 4. Correlation analysis between IS reliability and organizational performance (Spearman correlation test). (The interpretation of the r-Spearman factor: 0–0,3 means that the correlation is weak or non existant, 0,3–0,5 moderate, 0,5–0,7 strong, 0,7–1,0 very strong.)

		Organizational performance
IS reliability	Correlation coefficient	,474**
	Significance	,000
	N	371
Information reliability	Correlation coefficient	,449**
	Significance	,000
	N	383
System reliability	Correlation coefficient	,515**
	Significance	,000
	N	379
Service reliability	Correlation coefficient	,451**
	Significance	,000
	N	389

Source: own work

7   Conclusions

The analysis does not allow to conclude about the cause-and-effect relation, only about correlation. It is obvious that the organizational performance is influenced by several, different business factors. The impact of IS on organizational performance could be considered only together with these factors. However, based on results of performed study, it can be concluded that IS reliability is potentially important support for SMEs performance.

Moreover, presented research has some limitations – the statistical verification of the proposed relation is based on the sample of 400 organizations operating in Poland and further verification in different business contexts is required. Moreover, further research should concern the indication of the relation between IT reliability and organizational performance in the context of business sustainability. Presented theoretical findings and empirical research can be treated as a solid first step in that direction.

References

1. Banker, R.D., Datar, S.M., Kemerer, C.F., Zweig, D.: Software errors and software maintenance management. Inf. Technol. Manag. **3**(1), 25–41 (2002)
2. Bansal, P.: Evolving sustainably: a longitudinal study of corporate sustainable development. Strateg. Manag. J. **26**(3), 197–218 (2005)
3. Bieńkowska, A., Zabłocka-Kluczka, A.: Monitoring and improvement of managerial processes in process controlling. In: Jedlička, P. (ed.) The International Conference Hradec Economic Days 2014: Economic Development and Management of Regions. Hradec Králové, February 4th and 5th 2014: Peer-Reviewed Conference Proceedings, Pt. 4, pp. 41–50. Gaudeamus, Hradec Králové (2014)

4. Bharadwaj, A.S., Sambamurthy, V., Zmud, R.W.: IT capabilities: theoretical perspectives and empirical operationalization. In: Proceedings of the 20th International Conference on Information Systems, pp. 378–385. Association for Information Systems (1999)
5. Carr, N.G.: IT doesn't matter. EDUCAUSE Rev. **38**, 24–38 (2003)
6. Chan, Y.E.: IT value: the great divide between qualitative and quantitative and individual and organizational measures. J. Manag. Inf. Syst. **16**(4), 225–261 (2000)
7. Covin, J.G., Slevin, D.P., Schultz, R.L.: Implementing strategic missions: effective strategic, structural and tactical choices. J. Manag. Stud. **31**(4), 481–506 (1994)
8. Cummins, F.A.: Enterprise Integration: An Architecture for Enterprise Application and Systems Integration. John Wiley & Sons, Inc., New York (2002)
9. Davis, F.D.: A technology acceptance model for empirically testing new end-user information systems: theory and results. Doctoral dissertation, Massachusetts Institute of Technology (1985)
10. Dawes, J.: The relationship between subjective and objective company performance measures in market orientation research: further empirical evidence. Market. Bull. Dep. Mark. Massey Univ. **10**, 65–75 (1999)
11. De Wet, W., Koekemoer, E., Nel, J.A.: Exploring the impact of information and communication technology on employees' work and personal lives. SA J. Ind. Psychol. **42** (1), 1–11 (2016)
12. Dehning, B., Richardson, V.J.: Returns on investments in information technology: a research synthesis. J. Inf. Syst. **16**(1), 7–30 (2002)
13. Delone, W.H., McLean, E.R.: The DeLone and McLean model of information systems success: a ten-year update. J. Manag. Inf. Syst. **19**(4), 9–30 (2003)
14. Dess, G.G., Robinson, R.B.: Measuring organizational performance in the absence of objective measures: the case of the privately-held firm and conglomerate business unit. Strateg. Manag. J. **5**(3), 265–273 (1984)
15. Doherty, N.F., Champion, D., Wang, L.: An holistic approach to understanding the changing nature of organisational structure. Inf. Technol. People **23**(2), 116–135 (2010)
16. Irani, Z.: Information systems evaluation: navigating through the problem domain. Inf. Manag. **40**(1), 11–24 (2002)
17. Kohli, R., Devaraj, S.: Measuring information technology payoff: a meta-analysis of structural variables in firm-level empirical research. Inf. Syst. Res. **14**(2), 127–145 (2003)
18. Lasrado, F., Bagchi, T.: A cross-cultural evaluation of the contemporary workplace and its managerial implications. Int. J. Bus. Manag. Soc. Sci. **2**(1), 1–15 (2011)
19. Little, R.G.: Toward more robust infrastructure: observations on improving the resilience and reliability of critical systems. In: Proceedings of the 36th Annual Hawaii International Conference on System Sciences, p. 9. IEEE, January 2003
20. Lucas Jr., H.C.: The business value of information technology: a historical perspective and thoughts for future research. In: Strategic Information Technology Management, pp. 359–374. IGI Global (1993)
21. Lyytinen, K.: Different perspectives on information systems: problems and solutions. ACM Comput. Surv. (CSUR) **19**(1), 5–46 (1987)
22. Mahmood, M.A., Mann, G.J., Zwass, V.: Special issue: impacts of information technology investment on organizational performance. J. Manag. Inf. Syst. **16**(4), 3–10 (2000)
23. Maletic, M., Maletic, D., Dahlgaard, J., Dahlgaard-Park, S.M., Gomišcek, B.: Do corporate sustainability practices enhance organizational economic performance? Int. J. Qual. Serv. Sci. **7**(2/3), 184–200 (2015)
24. Marchand, M., Raymond, L.: Researching performance measurement systems: an information systems perspective. Int. J. Oper. Prod. Manag. **28**(7), 663–686 (2008)

25. Matić, I.: Measuring the effects of learning on business performances: proposed performance measurement model. J. Am. Acad. Bus. Camb. **18**(1), 278–284 (2012)
26. Melville, N., Kraemer, K., et al.: Information technology and organizational performance: an integrative model of it business value. MIS Q. **28**(2), 283–322 (2004)
27. Niu, N., Da Xu, L., Bi, Z.: Enterprise information systems architecture—analysis and evaluation. IEEE Trans. Ind. Inf. **9**(4), 2147–2154 (2013)
28. Palmius, J.: Criteria for measuring and comparing information systems. In: Proceedings of the 30th Information Systems Research Seminar in Scandinavia IRIS (2007)
29. Peppard, J., Ward, J.: Beyond strategic information systems: towards an IS capability. J. Strateg. Inf. Syst. **13**(2), 167–194 (2004)
30. Santhanam, R., Hartono, E.: Issues in linking information technology capability to firm performance. MIS Q. **27**(1), 125–153 (2003)
31. Tusubira, F., Mulira, N.: Integration of ICT in organizations: challenges and best practice recommendations based on the experience of Makerere University and other organizations. In: International ICT Conference Held at Hotel Africana, Kampala, Uganda, 5–8 September 2004
32. Tworek, K.: Model niezawodności systemów informacyjnych w organizacji. Zesz. Nauk. Organ. i Zarządzanie/Pol. Śląska **88**, 335–342 (2016)
33. Venkatraman, N., Ramanujam, V.: Measurement of business performance in strategy research: a comparison of approaches. Acad. Manag. Rev. **11**(4), 801–814 (1986)
34. Wade, M., Hulland, J.: The resource-based view and information systems research: review, extension, and suggestions for future research. MIS Q. **28**(1), 107–142 (2004)

Target Marketing Public Libraries' Vital Readers: Before

Yi-Ting Yang and Jiann-Cherng Shieh[✉]

Graduate Institute of Library and Information Studies,
National Taiwan Normal University, Taipei, Taiwan
jcshieh@ntnu.edu.tw

Abstract. Marketing is the business operation that many different companies or organizations use to promote their products or enhance their image to their customers or users. Then it derived different marketing practices and models for different situations. Target marketing is one of the most effective and popular choices. It is a kind of customer marketing, mainly using analysis of customer characteristics to segment customers, in order to achieve specific commodity marketing for the purpose. Library, especially public library, how to borrow the concept of target marketing to promote its utilization and services is a problem that library managers must actively conceives. The Pareto Principle is currently the most important and popular management rule applied to marketing, customer relationship management, services or products promotion, etc. This study will explore whether there this principle exists for patrons' borrowing in public library, thus as the basis of target marketing. And we will apply data mining technology to analyze the characteristics of the vital patrons, in order to further serve as references for setting target marketing strategy. This paper was the first research to reveal that the Pareto Principle could be found in circulation data of public libraries in Taiwan. The results can help libraries identify vital patrons and major collections, and improve the efficiency of their management and marketing in future.

Keywords: Target marketing · Pareto Principle · 80/20 rule · Public libraries
Bibliomining

1 Introduction

In recent decades, the Pareto Principle concept has become a popular guideline or tool in business and social context. The principle known as the 80/20 rule indicates that there is an unbalanced relationship between causes and results or between efforts and rewards: specifically, it maintains that 80% of rewards usually come from 20% of efforts, and the other 80% of efforts only produce 20% of the results. Therefore, if you can recognize, focus on, and control the vital 20% of efforts, you will obtain greater profits or efficiency.

Is there the Pareto Principle in library context? In recent years, a great deal of research has analyzed library data and uncovered a variety of trends, patterns, and relationships. Bibliomining, or data mining in libraries, is the application of data

© Springer Nature Switzerland AG 2019
Z. Wilimowska et al. (Eds.): ISAT 2018, AISC 854, pp. 211–221, 2019.
https://doi.org/10.1007/978-3-319-99993-7_19

mining techniques to data produced from library services [1, 2]. By applying statistic, bibliometric, or data mining tools, libraries can better understand usage patterns and rules, enabling library managers to make decisions to meet user needs based on those mining results [3, 4]. However, quality decisions must be based on quality data. Data processing is an important step in the knowledge discovery process [5, 6]. Identifying vital data and reducing the data to be analyzed can lead to huge decision-making payoffs. Similarly, identifying vital patrons and core collections would allow library managers to provide better services and popular materials to promote library utilization and user satisfaction.

Examining distributions of circulation data helps libraries understand user behaviors. Previous discussions of data mining applications in academic libraries have emphasized usage analysis [7]. There have been a number of studies that have investigated usage analysis of academic libraries [8, 9]. Renaud, Britton, Wang and Ogihara [10] analyzed data from a university library and revealed the distributions of check-out activities based on user type, academic department, LC classification, material type, life span, and so on. They also correlated the findings with student grade point averages. Goodall and Pattern [11] analyzed the usage data of electronic resources, book loans, and visits in an academic library and correlated these data points with academic achievement. However, previous research has mostly neglected to analyze circulation data in public libraries. In comparison with academic libraries, public libraries service all kinds of users. Before data analyzing or data mining, segmenting these various users and collections would conduct meaning results. At present research focusing on this issue is still very scarce.

The purpose of this study is to analyze the circulation data of one public library and to examine whether the phenomenon of Pareto principle could be found. More specifically, this study aims to discern if there are vital few patrons who borrow most of the collection, and to identify the vital patrons and their characteristics and book-borrowed distributions. If the Pareto Principle as conceptualized for this study can, in fact, be used to study patron patterns, this rule can serve as a much needed tool for public libraries to target patrons and effectively market their services and even promote their collection development.

2 Literature Review

2.1 The Pareto Principle and the 80/20 Rule

The 80/20 rule originated from the Pareto Principle, named for the Italian economist Vilfredo Pareto who identified a general imbalance in property allocation: most (80%) wealth belongs to a few (20%) people. This model of imbalance has been observed repeatedly. In the late 1940s, Juran named his notion of "the vital few and the trivial many" as Pareto Principle after the Italian economist. The 80/20 rule is an extension of the Pareto Principle developed by Richard Koch based on a theoretical view of Pareto and Juran. Koch [12] pointed out that the 80/20 rule applies in various fields. In business, 80% of a company's profits come from 20% of its customers, 80% of revenues come from 20% of the products, and 80% of sales come from 20% of the sellers.

In quality management, 80% of the problems come from 20% of the faults. In computer science, most software takes 80% of the time to run 20% of the programs. The unbalanced relationship between efforts and rewards or causes and results makes delineating the vital few very important. Several studies also have suggested the benefit of applying the 80/20 rule [13, 14]. Concentrating on the groups of customers and the specific markets that are profitable can substantially improve a company's bottom line, having insight into the vital few is an important issue.

2.2   The 80/20 Rule in Library Studies

The study of the Pareto Principle in libraries was initialized by Trueswell [15] who applied the 80/20 rule to address the relationship between collections and circulation numbers. Trueswell [15] noted that about 20% of collections bring 80% of circulation numbers [16]. Hardesty [17] traced the book acquisitions and circulations of a university for five years. He found that 30% of books accounted for 80% of circulation. In recent years, Singson and Hangsing [18] analyzed usage patterns of electronic journals academic consortia. They found the user downloads for some publishers follow the 80/20 rule. The few, core journals were downloaded the most. Some research suggested the 80/20 rule could be used to identify the core collections within libraries. Burrell [19] investigated the circulation data of university libraries and public libraries and found between 43% and 58% of circulating collections are required to account for 80% of borrowings. He developed a theoretical model of library operations to help libraries identify their core collections. Nisonger [20] examined the 80/20 rule in relation to the use of print serials, downloads from electronic databases, and journal citations, concluding that the 80/20 rule is a valid method for determining core concepts in journal collection management. However, few previous studies have explored the distributions of circulation in public libraries. This paper examines a circulation dataset from a public library in Taiwan and analyzes usage patterns to understand the distributions of patrons and circulations. Moreover, this study concerns the marketing of libraries and thus mainly exams the Pareto Principle focused on patron's perspective. That's a lot different from previous studies. Our purpose is trying to identify if there are the vital patrons in public libraries.

3   Research Methods

3.1   Data Description

The data collection for analyzing purpose includes circulation data, patron data, item data, and branch location data. Circulation dataset contains more than 18 million transactions conducted over two years, and patron dataset contains data about 460 thousands patrons.

3.2   Preprocessing Data Privacy

To preserve privacy, the data had been adopted through preprocessing before we got and processed it. Columns that may have identified someone by providing information such as patron names, addresses, or phone numbers have been deleted. Furthermore, patron corresponding data and branch related data had been translated into substituted codes by some one-way hash functions respectively [21].

3.3   Data Processing Procedure

This paper analyzes public library data to determine whether the phenomenon of the Pareto Principle manifests, and then identifies the distributions of the patrons and collections. We integrate the local-borrow transactions, inter-borrow transactions, and reservation data into one table and sum up transactions for each patron ID and then rank them in the table by the number of transaction. Then we calculate the accumulative total of patrons and the accumulative total of items, and then calculate the accumulative percentage separately. Finally, we identify the datum where the accumulative percentage of patrons and the accumulative percentage of items is 100%.

3.4   Analysis Tools and System Environment

The database system used in this study is Microsoft SQL Server 2014. And we applied Microsoft SQL Server Data Tools (SSDT) and Microsoft Excel 2016 as data mining tool to analyze data. We adopted a PC workstation with Intel Core i7-7700 CPU, 16G memory and 1T SSD to support database system operation, data processing and analysis tasks.

4   Data Analysis and Findings

Data analysis results and findings of this research are presented in the following three sections.

4.1   The Pareto Principle

We rank the patrons according to the number of items they borrow and then calculate the accumulative total of patrons, the accumulative total of items, the accumulative total percentage of patrons, and the accumulative total percentage of items. As shown in Table 1, when the accumulative percentage of patrons is 24.69%, the accumulative percentage of items is 75.31%. That the accumulative percentage of patrons and the accumulative percentage of items is 100% demonstrates that the circulation data of the public library follow the Pareto Principle approximating to 80/20 rule. In other words, the vital few patrons account for most of the borrowings. The data in which the accumulative percentage of patrons is 20% and the accumulative percentage of items is 80% are also shown in Table 1.

In summary, the percentage following the 80/20 rule for the two years examined is 75.3/24.7. These findings indicate that a few patrons would borrow the most items in

Table 1. Rank of borrowing and accumulative percentage

Borrow amount	Accumulative patrons	Accumulative items	Accumulative patrons (%)	Accumulative items (%)
3212	1	3212	0.000217507	0.017778698
3159	2	6371	0.000435013	0.035264036
3088	3	9459	0.00065252	0.052356383
3030	4	12489	0.000870027	0.069127695
2990	5	15479	0.001087533	0.085677604
2973	6	18452	0.00130504	0.102133416
≈				
56	91950	12534661	19.99973899	69.38043286
56	91951	12534717	19.9999565	69.38074283
56	91952	12534773	20.00017401	69.38105279
56	91953	12534829	20.00039151	69.38136275
≈				
44	113496	13606589	24.68613786	75.31364706
44	113497	13606633	24.68635537	75.3138906
44	113498	13606677	24.68657288	75.31413415
44	113499	13606721	24.68679038	75.31437769
≈				
35	134910	14453189	29.34382586	79.99965129
35	134911	14453224	29.34404336	79.99984502
35	134912	14453259	29.34426087	80.00003875
35	134913	14453294	29.34447838	80.00023247
≈				

the case public library. The application of 80/20 rule to identify the vital few patrons can significantly improve organizational efficiency, and the public library managers would certainly benefit from embracing this approach. In doing so, they could not only improve utilization but also give vital patrons more proper services. In the next section, we will analyze the characteristics of the top 20% active patrons and the features of the 80% collections they borrowed from.

4.2   The Top 24.7% Patrons

Statistical Analysis of Patrons. After establishing that public library data conform to the 80/20 rule, we analyze the distributions of patrons. Our two-year dataset for the Taiwan public library includes about 460 thousands patrons who borrowed at least one item. Therefore, the top 24.7% is composed of about 115 thousands patrons who borrowed at least 44 items. Applying the 80/20 rule, these patrons are active users of the library. Identifying and understanding these patrons may help the library target users and market services to them efficiently.

We analyze the characteristics of these top 24.7% of patrons. Table 2 presents the distributions of the top 24.7% of patrons' birth year and gender. In general, most of the patrons were born in the 1970s, followed by the 2000s, and the percentage of female patrons is higher than that of males. Table 3 shows the results of our analysis of the distributions of the top 24.7% of patrons' types and occupations. Most of the top patrons are general patrons, followed by families. By occupation, students borrow the most, followed by businesses and finance and then children. This is consistent with the findings in the next stage of our analysis: families and children are vital patrons and children's books are also a popular material type. For public libraries, categorizing patrons is important because it enables them to develop collections for specific patrons, and recommend specific collections to targeted patrons.

Table 2. Distribution of patrons' birth year and gender

	Female	Male	Null
2010s	1156	1126	19
2000s	12690	10544	67
1990s	7693	3713	23
1980s	10693	4254	131
1970s	24498	8749	334
1960s	12083	6748	88
1950s	3769	3424	17
1940s	873	1217	4
1930s	183	327	2
1920s	41	113	1
1910s	10	5	0
Other	82	47	3

Bibliomining Analysis of Patrons. In order to know the characteristics of these patrons, we apply clustering of data mining technics to these 24.7% patrons. They are divided into 4 groups by K-means algorithm. Input variables are ages, occupations, and gender. The results are shown as Table 4. In the meantime, we use the same clustering method to all patrons who borrowed at least one item. There is a great difference between these 2 results. Therefore, using 20/80 rule can point out the vital patrons.

4.3   The Collections Borrowed by the Vital Users

Statistical Analysis of Collections Borrowed. Finally, we analyze the features of 75.3% collections that borrowed by the top 24.7% patrons. Table 5 shows these materials' types and subject codes in the New Classification Scheme for Chinese Libraries used for general collections. Chinese books are borrowed most frequently, followed by children's books and then the attachments. In collections organized according to the New Classification Scheme for Chinese Libraries, items belonging to 800 Linguistics and Literature are borrowed most frequently, followed by 400 Applied Sciences and then 900 Arts.

Table 3. Distribution of patrons' occupation and type

	VIP patron	General patron	Volunteer	Family	Group	Staff
Null	8	1355	2	710	3	5
Agriculture	2	424	1			
Arts				1		
Business and finance	64	18366	66	52		11
Children	91	18009	1	1		
Education and training	31	4885	29	7	8	2
Entertainment		2				
Fishery		3				
Freelance		16		4		1
General Services	1	16		11		16
Government	23	5093	23	15	13	261
Healthcare		3		1		
Homemaker		39		34		
Industrial	28	7608	14	14	1	1
Military		87		1		
Others	164	31263	449	186	20	110
Pasture		1				
Religious		2				
Student	113	24905	30	10		9
Transportation				1		
Public servant		1				

In the second layer classification of items that had been borrowed by vital patrons, items belonging to 850 Various Chinese literature are borrowed most frequently, followed by 870 Western literatures and then 860 Oriental literatures. In the third layer classification, items borrowed more than 100,000 times belong to 857 Fiction, 859 Chinese Children Literature, 861 Japanese Literature, and 874 American Literature.

Bibliomining Analysis of Collections Borrowed. We apply association rules of data mining technics to the collections borrowed by 24.7% patrons. This study uses second classification number as bases to see the association patterns. Take one association rule for example, "820,690→850" has the highest confidence, 0.996. That means, in the vital patrons who ever borrowed books of classification number 820 and 690, 99.6% of them also borrowed books of classification number 850.

Table 4. Clusters of patrons.

	Variable	Value	Percentage
Cluster 1	Patron age	0–20	98.88%
	Patron gender	Female	55.02%
		Male	44.91%
	Patron occupation	Children	64.13%
		Student	35.23%
Cluster 2	Patron age	0–20	11.35%
		21–32	84.47%
		33–43	4.19%
	Patron gender	Female	68.83%
		Male	30.76%
	Patron occupation	Business and finance	9.00%
		Children	2.43%
		Education and training	2.53%
		Government	3.88%
		Industrial	3.32%
		Others	16.61%
		Student	61.94%
Cluster 3	Patron age	21–32	15.01%
		33–43	84.99%
	Patron gender	Female	73.46%
		Male	25.44%
		Null	1.10%
	Patron occupation	Business and finance	24.12%
		Education and training	6.66%
		Government	5.69%
		Industrial	8.60%
		Others	36.11%
		Student	18.21%
Cluster 4	Patron age	0–20	0.62%
		21–32	9.33%
		33–43	37.25%
		44–81	52.80%
	Patron gender	Female	66.33%
		Male	33.30%
	Patron occupation	Agriculture	0.72%
		Business and finance	27.19%
		Education and training	7.99%
		Government	8.43%
		Industrial	10.40%
		Others	41.75%
		Student	3.38%

Table 5. Distribution of material types and subject code in New Classification Scheme for Chinese Libraries

	000	100	200	300	400	500	600	700	800	900
Chinese book	43020	353972	140993	242366	1239097	568368	82593	416972	3735515	322321
Foreign reference	–	–	–	–	4	2	2	–	30	–
Indonesian book	1	5	2	–	–	1	7	14	31	1
Audio book	298	4449	686	320	613	1285	–	61	4022	595
English book	109	376	130	287	68	98	223	791	10929	272
Foreign child book	240	256	106	2233	607	5530	714	1079	87368	1109
Infant book	107	29	–	1605	3291	81781	5	24	32558	295
Child book	44093	38286	9240	407991	57602	221949	37738	152550	2744094	132242
Child reference	311	–	–	218	15	43	4	23	102	1
Child picture book	–	–	–	–	7	–	–	–	–	–
Attachment	16724	5948	3010	65333	50334	43420	5902	12690	335886	35293
Teeanger book	–	–	–	–	–	–	–	1	136	–
Government publication	–	–	–	11	116	149	2	43	24	37
Music	4	25	65	–	12	65	–	–	2926	4443
Aboriginal book	–	–	–	–	–	2	–	–	–	–
Book box	–	–	–	–	–	–	–	–	95	1
Thai book	–	–	–	–	–	1	–	–	4	–
Malay book	2	–	–	–	–	–	2	–	–	–
Reference book	31	1	14	220	69	109	13	97	95	53
Periodical	51530	737	535	10324	39142	25130	63	1377	16835	8957
Video	713	995	618	12136	5300	7021	2038	8646	7172	353248
Vietnam book	–	4	–	–	–	–	133	3	47	10
Local government literature	3	–	–	5	2	6	4	21	1	4
E-resources	–	–	–	–	–	–	–	–	4	–
Comic book	2	307	225	635	527	538	176	573	4469	449478
Elderly book	–	–	–	–	–	–	–	12	5	–
Journal	–	–	10	–	10	1	–	–	–	–
Korean book	–	–	–	3	1	–	–	–	–	1

5   Conclusion

This paper examines the circulation data of the case public library through the lens of the Pareto Principle, finding that the Pareto Principle does manifest in its circulation data. The findings indicate that, during the two years covered by our investigation, 24.7% of the patrons borrowed 75.3% of all items borrowed. Conforming to the Pareto Principle, the majority of borrowing was done by the vital few patrons. We analyze the distributions of the vital few patrons and the collections most frequently borrowed from. Among the vital few patrons, most of them are students, families, and children. Among popular collections, the most frequently borrowed items are Chinese books and children's books. The most popular collection subject is Linguistics and Literature. This study also analyzed these patrons and collections by data mining technics. The findings of clustering have implications for public libraries to understand their patrons. Libraries could apply the Pareto Principle to identify vital patrons and collections and use that information to improve the efficiency of both their management and marketing.

Future research should analyze the correlation or connection between the vital few patrons and popular collections. And on the other hand, studies explore Pareto Principle based on collections are underway. We continue to analyze the circulation data to identify popular items in public libraries. The results of such an analysis would enable library management to make more effective management and marketing decisions, which helping libraries improve utilization and patron satisfaction and develop collections that appeal to their patrons.

References

1. Nicholson, S.: The basis for bibliomining: frameworks for bringing together usage-based data mining and bibliometrics through data warehousing in digital library services. Inf. Process. Manag. **42**(3), 785–804 (2006)
2. Shieh, J.C.: Bibliomining. Mandarin Library & Information Service, Taipei (2009)
3. Xiang, Z., Hao, Z.: Personalized requirements oriented data mining and implementation for college libraries. Comput. Model. New Technol. **18**(2B), 293–300 (2014)
4. Hajek, P., Stejskal, J.: Library usage mining in the context of alternative costs: the case of the municipal library of Prague. Libr. Hi Tech **35**(4), 565–583 (2017)
5. Han, J., Kamber, M., Pei, J.: Data Mining: Concepts and Techniques. Morgan Kaufmann, Waltham (2011)
6. Bajpai, J., Metkewar, P.S.: Data quality issues and current approaches to data cleaning process in data warehousing. Glob. Res. Dev. J. Eng. **1**(10), 14–18 (2016)
7. Siguenza-Guzman, L., Saquicela, V., Avila-Ordóñez, E., Vandewalle, J., Cattrysse, D.: Literature review of data mining applications in academic libraries. J. Acad. Librariansh. **41**(4), 499–510 (2015)
8. Al-Daihani, S.M., Abrahams, A.: A text mining analysis of academic libraries' tweets. J. Acad. Librariansh. **42**(2), 135–143 (2016)
9. Wu, F., Hu, Y.H., Wang, P.R.: Developing a novel recommender network-based ranking mechanism for library book acquisition. Electron. Libr. **35**(1), 50–68 (2017)
10. Renaud, J., Britton, S., Wang, D., Ogihara, M.: Mining library and university data to understand library use patterns. Electron. Libr. **33**(3), 355–372 (2015)
11. Goodall, D., Pattern, D.: Academic library non/low use and undergraduate student achievement: a preliminary report of research in progress. Libr. Manag. **32**(3), 159–170 (2011)
12. Koch, R.: The 80/20 Principle: The Secret of Achieving More with Less. Currency Doubleday, New York (2011)
13. Kim, B.J., Singh, V., Winer, R.S.: The Pareto rule for frequently purchased packaged goods: an empirical generalization. Mark. Lett. **28**(4), 491–507 (2017)
14. Mesbahi, M.R., Rahmani, A.M., Hosseinzadeh, M.: Highly reliable architecture using the 80/20 rule in cloud computing datacenters. Future Gener. Comput. Syst. **77**, 77–86 (2017)
15. Trueswell, R.L.: Some behavioral patterns of library users: the 80/20 rule. Wilson Libr. Bull. **43**(5), 458–461 (1969)

16. Nash, J.L.: Richard Trueswell's contribution to collection evaluation and management: a review. Evid. Based Libr. Inf. Pract. **11**(3), 118–124 (2016)
17. Hardesty, L.: Use of library materials at a small liberal arts college. Libr. Res. **3**(3), 261–282 (1981)
18. Singson, M., Hangsing, P.: Implication of 80/20 rule in electronic journal usage of UGC-infonet consortia. J. Acad. Librariansh. **41**(2), 207–219 (2015)
19. Burrell, Q.L.: The 80/20 rule: library lore or statistical law? J. Doc. **41**(1), 24–39 (1985)
20. Nisonger, T.E.: The "80/20 rule" and core journals. Ser. Libr. **55**(1–2), 62–84 (2008)
21. Schneier, B.: Applied Cryptography: Protocols, Algorithms and Source Code in C, 20th Anniversary Edition. Wiley, New York (2015)

IT Reliability and the Results of Controlling

Agnieszka Bieńkowska$^{(\boxtimes)}$, Katarzyna Tworek,
and Anna Zabłocka-Kluczka

Faculty of Computer Science and Management, Wrocław University of Science
and Technology, wyb. Wyspianskiego 27, 50-370 Wrocław, Poland
{agnieszka.bienkowska,katarzyna.tworek,
anna.zablocka-kluczka}@pwr.edu.pl

Abstract. The quality of controlling outputs is often indicated as an important factor influencing results obtained due to controlling implementation. However, efficient functioning of controlling seems to be impossible without IT. The notion of IT reliability is discussed in the paper as a potential factor determining the achievement of high quality of controlling products and results of its implementation in the organization. The proposed theoretical framework is verified empirically on the sample of 557 organizations operating in Poland. The aim of the paper is to identify the relations between the IT reliability, the quality of controlling products and the results of its functioning in organization.

Keywords: IT reliability · Controlling · Quality of controlling
Results of controlling

1 Introduction

The development of information technology (IT) significantly influenced the way organizations operate and IT quickly became one of the key factors building the company's competitive advantage [12, 15, 19, 27]. Various IT value models [9, 20] describing ways in which organizations gain value from using it) can be a source of two conclusions. First of all, when almost every organization operates using some kind of IT, the simple fact of using it is no longer a factor influencing the organization ability to create competitive advantage [19]. It has to be reliable and perceived as useful in order to create a real value [20]. Second of all, IT has the ability to positively influence almost every area of organization operation [12, 28, 29]. However, many authors underline its especially useful role in generating and processing information for the purpose of organization management. Hence, IT can create value for organization due to the support of management methods, such as controlling, which is the main topic of this paper.

Controlling is one of the most often used methods of management in contemporary organization [6]. The essence of controlling is "wide understanding of management information in terms of decisions and different decision-making options, (...) management support in decision-making processes, both at organizational and strategic level" [14, p. 40] and coordination of management tasks in in the context of the objectives set. The specificity of controlling management support manifests itself in

© Springer Nature Switzerland AG 2019
Z. Wilimowska et al. (Eds.): ISAT 2018, AISC 854, pp. 222–234, 2019.
https://doi.org/10.1007/978-3-319-99993-7_20

providing information not only about the organization as a whole, but above all, about the functioning of its smaller parts - departments, their fragments – often referred to as centers of responsibility. This approach requires an extensive accounting engine that allows for the collection and processing of data flowing from many areas of the organization, which is clearly the area in need of extensive IT support. The beginning of the 21st century (the birth of information society) brought the possibility of multidimensional information aggregation according to the needs and expectations of users, and above all, the possibility of providing information in real time, which significantly influenced the knowledge resources that are the basis for decision-making. It should translate into higher competitiveness and better results obtained due to controlling implementation [8]. However, it is important to consider costs of providing controlling information using all those possibilities, their credibility and usefulness, as well as the relation of these parameters to the results obtained by the organization due to controlling implementation. The issue of the effectiveness of controlling is already the subject of an academic dispute [1, 22], as well as the relation between controlling and IT. However, the relation between IT reliability (as a feature ensuring its ability to generate value for organization) and controlling (its direct and indirect results) has not yet been explored and is the main aim of this paper.

2   Information Technology Reliability

Many authors underline that it has become necessary for the organizations to integrate IT into all organizational functions in order to build its potential to influence organizational competitive advantage, performance and competitiveness [12, 15, 27]. However, since the relevance and the need for IT use in organization seems to finally be undisputable, there is a need for the analysis and evaluation of its use in organization. The concept of 3 R (reliability, resilience and robustness) emerged in the literature few years ago [16] and underlines that the key factor influencing the ability to profit from using IT is its appropriate functioning in organization. Hence, the main element of this concept is IT reliability. The reliability of IT in organization is a measurable property of IT, useful for its control and management, identifying its quality level and pointing out potential problems [30] and it is directly linked to the efficiency of IS components, especially those critical to its proper operations. Therefore, it can be said that IT reliability in organization is a notion build by factors connected to 3 different IT theories. First one is DeLone and McLean success model [11], second one is Lyytinen [18] 4 types of IT failure and third one is TAM model [10]. The model of IT reliability in organization has been developed by one of the authors and detailed model description is published by Tworek [28]. The reliability of IT in organization consists of 4 factors: reliability of system itself (reliable IT has high availability connected with high security), which also includes the usability of this system (reliable IT is efficient, accepted by its users and easy to use), reliability of information included in IT (reliable IT has easily accessible, and the accurate information) and reliability of support services offered for IT in the organization (reliable IT has professional, responsive and available support services).

3 The Essence and Results of Controlling

Interpretative diversity and the freedom in defining controlling are enormous. For the purposes of this paper, the controlling is understood as "method of management support (…) consisting of coordinating the process of solving specific management tasks, supervising and monitoring the course of their implementation, as well as participating in the performance of these tasks, mainly in the field of planning, controlling and providing information" [2, p. 289]. It should be emphasized that controllers do not take over tasks that belong to managers (e.g. they do not set goals and do not make decisions), but only support their implementation. They prepare and provide "methods, techniques, instruments, models, interpretative schemes and information that support planning and supervising the processes of plan implementation, as well as coordinating the course of real processes" to the organization's management [7, p. 1].

Therefore, the essence of controlling is to support managers in decision-making processes, while the scope of this support and the manner of shaping the controlling solutions in an organization is always dependent on the situational conditions and the needs of the organization's management [1]. Regardless of the adopted solutions, it is expected that they will be appropriate, i.e. adjusted to the existing conditions so much that the implementation of controlling will entail improvement in the efficiency of the organization management, improvement of the efficiency of its functioning and increase of competitiveness on the market. The results that are expected after the implementation of controlling can be considered from two perspectives:

– from the perspective of controlling clients, i.e. managerial staff, managers of responsibility centers, considering direct outputs of controlling;
– from the perspective of the organization itself, considering indirect results obtained by the organization due to controlling implementation.

The managerial staff of the organization is a direct recipient of products offered by controlling, i.e. "information generated by the controlling information system, including all kinds of reports and controlling analyzes, plans or budgets" [1, pp. 220–221]. The quality of these products should translate into better accuracy of decisions made by managers. However, due to controlling implementation, the organization as a whole is expected to achieve the improvement of the management process and improvement of the business situation of the organization, in particular reduction of its operating costs, profit growth, improvement of financial liquidity, improvement of the organization's competitiveness [4, p. 282, 13, pp. 156–165]. Obtaining these results is possible indirectly, i.e. by achieving the previously listed outputs of controlling.

4 IT Reliability and the Results of Controlling

With the high dynamics of modern economic processes and the growing demand of management for a wide range of management information necessary to make key decisions, it is difficult to imagine the implementation of controlling tasks without proper support in the form of a reliable IT. "The implementation of controlling almost automatically connects with the implementation of the IT system" [21, p. 31], enabling continuous supply of desired information and its processing for the purposes of making decisions. Creating, transforming and transferring management information is nowadays impossible without appropriate IT systems [8]. Many authors explicitly formulate the view that "modern controlling cannot be effectively implemented without IT support" [25, p. 130]. In 1997, Reichmann noticed that "IT orientating towards management or executive information, respectively, gain more and more importance as regards an efficient company controlling in practice" [23, p. 47].

In the literature and in practice, there is a discussion concerning IT requirements for the use of controlling, stressing that they should first and foremost be "structured in such a way that the use of information is as effective as possible" [17, p. 19]. Moreover, the focus on the recipient is emphasized by Rötler-Anderson and Bragg [24, p. 376]. The IT supporting controlling should also enable ongoing monitoring and consolidation of information obtained from all areas of the organization's activities, while ensuring its (information) transparency and constituting a kind of "early warning system" about emerging threats. Cooperation between the controlling subsystem, the financial-accounting subsystem and other subsystems (e.g. information) corresponding to particular areas of the organization's activity is possible due to the existence of one information flow. It is important to keep it efficient and secure, which is impossible without ensuring the information reliability [28]. Because of that, data stored in subsystems will be accurate, transparent and compatible [3]. Adequate data archiving, the ability to search and share them in different sets, (according to the needs of decision makers) determine the ability of controllers to perform tasks, and thus affect the quality of controlling analyzes and benefits obtained from its implementation. Reliable, unbroken data, delivered in a timely manner and in the proper form seem to be crucial for the quality of products offered by controlling. Interestingly, the results of research show that in practice "the information demand does not coincide with their supply: decision-makers often receive information that they did not ask for. Information is often delivered too late, it is presented in the wrong form, it is repeated [14, p. 574]. Hence, information accuracy and relevance (which are a part of information reliability) seems to be especially important in this case.

Specific IT functionalities are needed for the purpose of controlling support, such as the ability to export data to a spreadsheet, a user-friendly and customized user interface, the availability of applications over the internet and intranet, the ability to implement advanced statistical analyzes, multi-dimensional planning, reporting and analysis, and secure storage of various variants of plans, budgets and models, acceptable (possibly short) processing time and data calculation, technical assistance of the supplier and finally, flexibility of models designed in the available tool [26, p. 16] and the possibility of collecting and storing data over several accounting periods [5, pp. 75–84]. All those functionalities are highly dependable on IT reliability and without it, they are unable to properly support controlling in organization.

Considering all the above, it seems that the IT requirements are mainly focused on the IT reliability and the reliability of information processed by IT. However, it is rarely pointed out that the quality of controlling products and results obtained due to its implementation are also significantly affected by IT service reliability. Moreover, practice shows that IT providers should also focus on the ease of use [10] of the offered functionalities, since controllers and managers from operational areas rarely have high competences in the field of computer science [26]. Lack of support (technical assistance and technological support) and poor usability mean that the potential of implemented IT solutions is not used in full. Hence, service reliability and usage reliability seem to be important factors, influencing controlling as well.

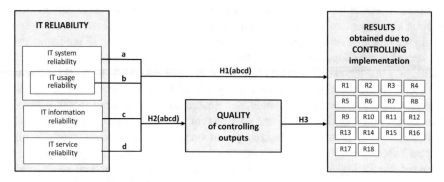

Fig. 1. The relation between IT reliability, quality of controlling outputs and results obtained due to implementing controlling. Source: own research.

In the above light, the following research hypotheses can be formulated (Fig. 1):

H1: *There is a positive relation between IT reliability (IT system reliability (a), IT usage reliability (b), IT information reliability (c), IT service reliability (d)) and results obtained due to controlling implementation.*

H2: *There is a positive relation between IT reliability (IT system reliability (a), IT usage reliability (b), IT information reliability (c), IT service reliability (d)) and quality of controlling outputs.*

In addition, it should be assumed that the better the quality of controlling products, the better the results obtained due to its implementation. In view of the above, the following hypothesis can be proposed:

H3: *There is a positive relation between the quality of controlling outputs and results obtained due to controlling implementation.*

5   Research Methodology and Results

The survey was conducted in order to verify the proposed hypotheses and identify the level of IT reliability, quality of controlling outputs and benefits obtained due to implementing controlling for every considered organization. The pilot survey was conducted in early 2018 among the group of 50 organization, indicating the issues concerning ambiguity of several questions. It led to the collection of random answers given as a response for those questions. They were rewritten in order to obtain the more reliable results, ensuring the informed response from the respondent. The main survey was conducted in March 2018, among organizations located in Poland, using online survey service: SurveyMonkey. Only one survey was carried out anonymously in one organization. Efforts had been made to make sure that the questionnaire was filled in by employees who have a broad view of the entire organization. The statistical population (organizations operating in Poland) is finite, but very large. 557 valid responses were collected, which is a very large sample for this kind of study. Sample characteristic is presented in Table 1 and clearly shows that the sample is covering organizations of all sizes and all types.

Table 1. Research sample characteristics.

Organization size	Manufacturing organizations	Service organizations	Trade organizations	Total
Micro (below 10 people)	66	31	10	106
Small (11–50 people)	48	72	20	140
Medium (51–250 people)	42	77	15	134
Large (above 250 people)	71	92	14	177
Total	227	272	59	557

Source: own work

5.1 Variables Measurement

In order to examine the relation between IT reliability and results of controlling, key variables were defined: IT system reliability (including IT usage reliability), IT information reliability, IT service reliability, which together build *IT reliability, quality of controlling outputs and results obtained due to controlling implementation.*

Respondents were asked to evaluate the IT in the organization based on the list of factors using the Likert scale (from very poor to very good with the middle point: fair). Using a Likert scale to measure *IT reliability* seems to be an appropriate choice. First of all, reliability of IT in organization is a subjective notion. Employees own perspective and opinion concerning aspects of IT reliability is the best source of knowledge, since their perception matters the most, that is because IT influences the organization mainly through its potential to influence every-day work of the employees. Quantitative methods are commonly used to assess the software and hardware features linked to the reliability. However, they do not give the information concerning the actual perception of this notion within the organization. *IT reliability* is built by 28 items and measurement scale is reliable (Cronbach's α is equal 0,945 for system reliability, 0,929 for information reliability and 0,919 for service reliability).

Respondents were asked to evaluate *the quality of controlling outputs* based on the list of 9 statements using the typical Likert scale. In particular, reference was made to elements directly affected by controlling - budgeting and information provided to managers for the purposes of decision making (including reports and analyzes). Cronbach's α was 0,916, which indicates a high coherence and reliability of the scale.

Respondents were also asked to evaluate *results obtained due to controlling implementation*. Each result (the list consists of 18 items – see Table 2) was assessed on the Likert scale (from very negative influence to very positive influence with the

Table 2. Results obtained due to controlling implementation

R1	Adaptation of the organization to changes taking place in the environment
R2	Improvement of the organization's competitiveness
R3	Increase of the chances for the long-term existence of the organization
R4	Overall increase in the efficiency of the organization's management
R5	More effective achievement of the organization's goals
R6	Shortening of the decision-making time
R7	Better flow of information inside the organization
R8	Increase of satisfaction and increase of employee morale
R9	Increase in employee involvement in achieving results
R10	Improvement of the competence of employees
R11	General improvement of the organization's financial results
R12	More effective and more rational management of the organization's resources
R13	Increase in the quality of products
R14	Improvement of the timeliness of production orders
R15	Increase in the number of changes in products/services
R16	Increase in the number of changes in business processes
R17	Increase in the number of organizational changes
R18	Increase in the number of marketing changes

Source: own work

Table 3. Descriptive statistics

	Average	Median	Min	Max	Std. deviation
IT reliability	3,73	4,00	1,44	5,00	1,00
System reliability	3,74	4,00	1,39	5,00	0,98
Information reliability	3,71	4,00	1,43	5,00	1,02
Service reliability	3,79	4,17	1,50	5,00	1,03
Quality of controlling outputs	2,60	0,90	2,56	1,00	5,00
Results obtained due to controlling implementation (based on average)*	2,31	0,98	2,02	1,00	5,00

Source: own work

middle point: no influence). The identified results related to the functioning of the organization as a whole. Descriptive statistics were calculated for all measured variables (Table 3).

5.2   Research Results

In order to verify hypotheses H1(abcd), the correlation analysis between the *IT reliability* and *results obtained due to controlling implementation* were calculated as the first part of the study (Table 4). Due to the large spread of results on variable scales, Spearman's correlation coefficients were used to assess the relation between them.

Table 4. Correlation analysis between IT reliability and results obtained due to controlling implementation.

	H1: IT reliability	H1a: IT system reliability	H1b: IT usage reliability	H1c: IT information reliability	H1d: IT service reliability
R1	r(532) = 0,348**, p < 0,001	r(532) = 0,365**, p < 0,001	r(532) = 0,299**, p < 0,001	r(532) = 0,258**, p < 0,001	r(532) = 0,336**, p < 0,001
R2	r(511) = 0,235**, p < 0,001	r(511) = 0,267**, p < 0,001	r(511) = 0,187**, p < 0,001	r(511) = 0,183**, p < 0,001	r(511) = 0,221**, p < 0,001
R3	r(498) = 0,292**, p < 0,001	r(498) = 0,326**, p < 0,001	r(498) = 0,223**, p < 0,001	r(498) = 0,198**, p < 0,001	r(498) = 0,287**, p < 0,001
R4	r(491) = 0,239**, p < 0,001	r(491) = 0,276**, p < 0,001	r(491) = 0,170**, p < 0,001	r(491) = 0,188**, p < 0,001	r(491) = 0,227**, p < 0,001
R5	r(485) = 0,172**, p < 0,001	r(485) = 0,190**, p < 0,001	r(485) = 0,158**, p < 0,001	r(485) = 0,127**, p = 0,005	r(485) = 0,155**, p = 0,001
R6	r(485) = 0,079, p = 0,083	r(485) = 0,081, p = 0,073	r(485) = 0,093*, p = 0,041	r(485) = 0,098*, p = 0,031	r(485) = 0,058, p = 0,206
R7	r(476) = 0,232**, p < 0,001	r(476) = 0,239**, p < 0,001	r(476) = 0,181**, p < 0,001	r(476) = 0,257**, p < 0,001	r(476) = 0,206**, p < 0,001
R8	r(476) = 0,220**, p < 0,001	r(476) = 0,214**, p < 0,001	r(476) = 0,210**, p < 0,001	r(476) = 0,205**, p < 0,001	r(476) = 0,201**, p < 0,001
R9	r(476) = 0,232**, p < 0,001	r(476) = 0,214**, p < 0,001	r(476) = 0,218**, p < 0,001	r(476) = 0,232**, p < 0,001	r(476) = 0,231**, p < 0,001
R10	r(474) = 0,213**, p < 0,001	r(474) = 0,213**, p < 0,001	r(474) = 0,221**, p < 0,001	r(474) = 0,179**, p < 0,001	r(474) = 0,195**, p < 0,001
R11	r(469) = 0,260**, p < 0,001	r(469) = 0,250**, p < 0,001	r(469) = 0,223**, p < 0,001	r(469) = 0,224**, p < 0,001	r(469) = 0,255**, p < 0,001
R12	r(467) = 0,232**, p < 0,001	r(467) = 0,226**, p < 0,001	r(467) = 0,212**, p < 0,001	r(467) = 0,243**, p < 0,001	r(467) = 0,187**, p < 0,001
R13	r(469) = 0,261**, p < 0,001	r(469) = 0,251**, p < 0,001	r(469) = 0,258**, p < 0,001	r(469) = 0,252**, p < 0,001	r(469) = 0,235**, p < 0,001
R14	r(466) = 0,277**, p < 0,001	r(466) = 0,263**, p < 0,001	r(466) = 0,245**, p < 0,001	r(466) = 0,288**, p < 0,001	r(466) = 0,225**, p < 0,001
R15	r(467) = 0,173**, p < 0,001	r(467) = 0,160**, p < 0,001	r(467) = 0,141**, p = 0,002	r(467) = 0,188**, p < 0,001	r(467) = 0,170**, p < 0,001
R16	r(465) = 0,258**, p < 0,001	r(465) = 0,262**, p < 0,001	r(465) = 0,231**, p < 0,001	r(465) = 0,272**, p < 0,001	r(465) = 0,206**, p < 0,001
R17	r(472) = 0,215**, p < 0,001	r(472) = 0,211**, p < 0,001	r(472) = 0,171**, p < 0,001	r(472) = 0,203**, p < 0,001	r(472) = 0,175**, p < 0,001
R18	r(474) = 0,174**, p < 0,001	r(474) = 0,142**, p < 0,001	r(474) = 0,151**, p = 0,001	r(474) = 0,180**, p < 0,001	r(474) = 0,170**, p < 0,001

Source: own work.

The results show that IT reliability is statistically significantly correlated with all benefits from implementing controlling with one only exception. Only IT usage reliability and IT information reliability are statistically significantly correlated with shortening of the decision-making time (R6) and this correlation is mild. We can also observe no statistically significant correlation between that variable and other IT reliability components.

In order to verify hypotheses H2(abcd), the correlation analysis between the IT reliability and quality of controlling outputs were calculated as the second part of the study (Table 5). The results clearly show that there is a statistically significant correlation between every aspect of IT reliability and controlling quality and it is the biggest in case of usage and information reliability.

Table 5. Relation between the IT reliability and quality of controlling outputs.

	H2: IT reliability	H2a: IT system reliability	H2b: IT usage reliability	H2c: IT information reliability	H2d: IT service reliability
Quality of controlling outputs	$r(571) = 0{,}542^{**}$, $p = 0{,}001$	$r(571) = 0{,}471^{**}$, $p < 0{,}001$	$r(571) = 0{,}621^{**}$, $p = 0{,}004$	$r(571) = 0{,}506^*$, $p = 0{,}012$	$r(571) = 0{,}329^{**}$, $p = 0{,}002$

* Correlation of r-Pearson is significant at the level of 0.05 (two-sided)
** Correlation of r-Pearson is significant at the level of 0.01 (two-sided)
Source: own work.

As a third step, in order to verify hypotheses H3, the correlation analysis between *quality of controlling outputs* and *results obtained due to controlling implementation* was performed (Table 6). The results show that controlling quality is statistically significantly correlated with only a part of them and this correlation is mild.

Table 6. Relation between the quality of controlling outputs and results obtained due to controlling implementation.

No	Quality of controlling outputs	No	Quality of controlling outputs
R1	$r(531) = -0{,}153^{**}$, $p < 0{,}001$	R2	$r(462) = 0{,}045$, $p = 0{,}314$
R5	$r(511) = 0{,}112^*$, $p = 0{,}014$	R3	$r(464) = -0{,}033$, $p = 0{,}467$
R6	$r(497) = 0{,}185^{**}$, $p < 0{,}001$	R4	$r(471) = 0{,}021$, $p = 0{,}652$
R9	$r(485) = 0{,}160^{**}$, $p < 0{,}001$	R7	$r456) = 0{,}069$, $p = 0{,}131$
R10	$r(484) = 0{,}132^{**}$, $p = 0{,}004$	R8	$r(467) = 0{,}009$, $p = 0{,}852$
R11	$r(471) = 0{,}173^{**}$, $p < 0{,}001$	R13	$r(472) = 0{,}083$, $p = 0{,}073$
R12	$r(474) = 0{,}145^{**}$, $p = 0{,}002$	R17	$r(458) = 0{,}078$, $p = 0{,}091$
R14	$r(465) = 0{,}106^*$, $p = 0{,}022$		
R15	$r(469) = 0{,}094^*$, $p = 0{,}042$		
R16	$r(470) = 0{,}150^{**}$, $p = 0{,}001$		
R18	$r(471) = 0{,}092^*$, $p = 0{,}045$		

Source: own work.

6   Discussion

In search for competitive advantage, modern organizations reach for diverse portfolio of management methods and techniques. Often, the full use of the opportunities offered by them becomes possible only with parallel application of appropriate IT solutions.

Controlling, due to the meticulousness and multidimensionality of the offered analyzes, is one of methods, which is highly sensitive to the IT influence [8].

Therefore, the obtained results are not surprising. As predicted, in the light of research findings and theoretical considerations outlined above, IT reliability is strongly positively related (or even affects) with almost all results obtained due to controlling implementation. At the same time, this means that IT unreliability causes the results received from the implementation of controlling to be assessed as small, and in extreme cases – even absent However, the lack of statistically significant correlation between shortening of the decision-making time and IT reliability, although among IT reliability components, IT usage reliability and IT information reliability are in a fact statistically significantly correlated with mentioned variable, but this correlation is mild. Decision making, which is the essence of management, always has a situational context. It seems that the more current and reliable information, the shorter the decision-making time and greater accuracy, especially in non-routine, risk-related situations, where there is no experience that managers could follow. However, the overflow of even the most accurate information extends the time of analysis and thus, decision-making. In addition, mainly people holding senior management positions spoke on behalf of the surveyed organizations. When making decisions, they are largely guided by information provided by subordinates and often by their own intuition. Therefore, it is not surprising that the IT service is not important for shortening the decision-making time.

The most interesting, because a little unexpected, results of the study concern the relation between quality of controlling outputs and results obtained due to controlling implementation. Theoretically, it could be assumed that if IT reliability is significantly correlated with the quality of controlling outputs and results obtained due to controlling implementation, then quality of controlling outputs and results obtained due to controlling implementation should be fully positively correlated as well. However, the research results did not show such dependencies. Quality of controlling output is positively related to 11 out of 18 results obtained due to controlling implementation. However, these are specific results, namely those that are clearly associated with the implementation of specific controlling solutions – i.e. solutions referring directly to the IT and reporting system as well as budgeting. There is no doubt that effective budgeting makes it possible to achieve the organization's goals (especially those relating to financial aspects), increases employees' engagement, or enables more effective and more rational management of the organization's resources. Moreover, efficient information and reporting system shortens the time of decision-making and increases the number of necessary changes in individual areas of the organization. Of course, in this context, the negative correlation between adaptation of the organization to changes taking place in the environment and quality of controlling outputs is surprising. However, this can be explained by the fact that budgets are often perceived in the organization as a solution that stiffens its functioning. Therefore, in general terms, are perceived as preventing the organization from adapting to changes taking place in the environment.

Therefore, it can be concluded that the obtained research results and the theoretical context show that IT reliability is a strong contributor that sustains controlling. Considering that it is difficult to carry out controlling tasks without IT support today, unreliable IT would be very quickly reflected in the low quality of controlling products. It can even be suggested that unreliable IT could be the basis for making a decision to abandon controlling or significantly reduce the tasks it performs, because without proper IT support, the quality of products offered by controlling would collapse.

7 Conclusions

IT solutions that support controlling are becoming crucial for its existence in modern organization. „Without the new and further developing possibilities of information system technology, such a controlling would not be possible" [23, p. 58]. This is confirmed by the results of presented study. IT reliability affects both the quality of controlling outputs and the results achieved by the organization as a whole.

However, the additional aspect, which opens up new research perspectives, should be underlined. In order for IT support to be effective, the IT should be tailored (in terms of content) to the complexity of the controlling system, should allow for the implementation of the adopted controlling concept and should change flexibly according to the circumstances. Moreover, it should first and foremost be reliable. Only then it will enable efficient delivery of controlling products that are optimal for the organization's quality.

In this context, further specific questions arise: what should be the shape of model IT solutions used for controlling in contemporary organization - taking into account, in particular, external and internal conditions of the organization functioning. The research carried out so far in this area has focused mainly on the frequency analysis of the phenomenon under consideration. However, an attempt should be made to verify model IT solutions with the use of such parameters as IT reliability, as well as the quality of management and the effectiveness of the organization as a whole. These issues should be the direction of further research.

Acknowledgements. The paper was created as a result of the research project no. 2017/01/X/HS4/01967 financed from the funds of the National Science Center.

References

1. Bieńkowska, A.: Analiza rozwiązań i wzorce controllingu w organizacji. Oficyna Wydawnicza Politechniki Wrocławskiej, Wrocław (2015)
2. Bieńkowska, A., Kral, Z., Zabłocka-Kluczka, A.: Zarządzanie kontrolingowe czy tradycyjne? In: Międzynarodowa konferencja naukowa pt. "Nowoczesne tendencje w nauce o organizacji i zarządzaniu", pp. 288–295. Oficyna Wydawnicza Politechniki Wrocławskiej, Szklarska Poręba (1998)
3. Bieńkowska, A., Kral, Z., Zabłocka-Kluczka, A.: Functional solutions of controlling - the results of the research into Lower Silesian enterprises. Econ. Organ. Enterp. 6(4), 35–57 (2009)

4. Bieńkowska, A., Kral, Z., Zabłocka-Kluczka, A.: Jakość zarządzania z uwzględnieniem controllingu. In: Kiełtyka, L., Borowiecki, R. (eds.) Przełomy w zarządzaniu: zarządzanie procesowe, Towarzystwo Naukowe Organizacji i Kierownictwa. Dom Organizatora, Toruń (2011)
5. Bieńkowska, A., Kral, Z., Zabłocka-Kluczka, A.: IT tools used in the strategic controlling process: Polish national study results. In: Šimberová, I., Kocmanová, A., Milichovský, F. (eds.) Perspectives of Business and Entrepreneurship Development in Digital Age: Economics, Management, Finance and System Engineering from the Academic and Practioners Views - Proceeding of Selected Papers, pp. 75–84. Brno University of Technology, Brno (2017)
6. Bieńkowska, A., Zgrzywa-Ziemak, A.: Współczesne metody zarządzania w przedsiębiorstwach funkcjonujących w Polsce – identyfikacja stanu istniejącego. In: Hopej, M., Kral, Z. (eds.) Współczesne metody zarządzania w teorii i praktyce. Oficyna Wydawnicza Politechniki Wrocławskiej, Wrocław (2011)
7. Błoch, H.: Controlling czyli rachunkowość zarządcza. Centrum Informacji Menedżera, Warszawa (1944)
8. Bogt, H.T., van Helden, J., van der Kolk, B.: New development: public sector controllership —reinventing the financial specialist as a countervailing power. Public Money Manag. 36(5), 379–384 (2016)
9. Chan, Y.E.: IT value: the great divide between qualitative and quantitative and individual and organizational measures. J. Manag. Inf. Syst. 16(4), 225–261 (2000)
10. Davis, F.D.: A technology acceptance model for empirically testing new end-user information systems: theory and results. Doctoral dissertation, Massachusetts Institute of Technology (1985)
11. Delone, W.H., McLean, E.R.: The DeLone and McLean model of information systems success: a ten-year update. J. Manag. Inf. Syst. 19(4), 9–30 (2003)
12. De Wet, W., Koekemoer, E., Nel, J.A.: Exploring the impact of information and communication technology on employees' work and personal lives. SA J. Ind. Psychol. 42(1), 1–11 (2016)
13. Eibisch-Stenzel, M.: Wdrożenie controllingu jako czynnik wpływający na poprawę zarządzania przedsiębiorstwem. In: Nowak, E., Nieplowicz, M. (eds.) Rachunkowość a controlling, Prace naukowe uniwersytetu Ekonomicznego we Wrocławiu, Wrocław, vol. 181, pp. 156–165 (2011)
14. Goliszewski, J.: Controlling. Koncepcje, zastosowanie, wdrożenie. Oficyna a Wolters Kluwer business, Warszawa (2015)
15. Lasrado, F., Bagchi, T.: A cross-cultural evaluation of the contemporary workplace and its managerial implications. Int. J. Bus. Manag. Soc. Sci. 2(1), 1–15 (2011)
16. Little, R.G.: Toward more robust infrastructure: observations on improving the resilience and reliability of critical systems. In: Proceedings of the 36th Annual Hawaii International Conference on System Sciences, 2003, pp. 9. IEEE (2003)
17. Litwa, P.: Podstawy informacji ekonomicznej. In: Sierpińska, M. (ed.) System raportowania wyników w controllingu operacyjnym, pp. 9–44. Vizja Press & IT, Warszawa (2007)
18. Lyytinen, K.: Different perspectives on information systems: problems and solutions. ACM Comput. Surv. (CSUR) 19(1), 5–46 (1987)
19. Mao, H., Liu, S., Zhang, J., Deng, Z.: Information technology resource, knowledge management capability, and competitive advantage: the moderating role of resource commitment. Int. J. Inf. Manag. 36(6), 1062–1074 (2016)
20. Melville, N., Kraemer, K., et al.: Information technology and organizational performance: an integrative model of IT business value. MIS Q. 28(2), 283–322 (2004)

21. Młodkowski, P., Kałużny, J.: Próba rozwiązania problemu niedopasowania systemów informacyjnych do wymagań controllingu w przedsiębiorstwie. In: Sierpińska, M., Kustra, A. (eds.) Narzędzia controllingu w przedsiębiorstwie, pp. 22–31. Vizja Press & IT, Warszawa (2007)

22. Nowosielski, K.: Quality of controlling process outputs in theory and practice. In: The International Conference Hradec Economic Days 2014: Economic Development and Management of Regions, Hradec Králové, 4–5 February 2014, pp. 83–91. Gaudeamus, Hradec Králové (2014)

23. Reichmann, T.: Controlling. Concepts of Management Control, Controllership, and Ratios. Springer, New York (1997)

24. Roehl-Anderson, J.M., Bragg, S.M.: The Controller's Function. The Work of the Managerial Accountant. Wiley, Hoboken (2005)

25. Sierpińska, M., Niedbała, B.: Controlling operacyjny w przedsiębiorstwie. Wydawnictwo naukowe PWN, Warszawa (2003)

26. Szarska, E.: Jak controllerzy oceniają systemy informatyczne BI. Controlling 12, 15–21 (2010)

27. Tusubira, F., Mulira, N.: Integration of ICT in organizations: challenges and best practice recommendations based on the experience of Makerere University and other organizations. In: International ICT Conference Held at Hotel Africana, Kampala, Uganda, 5–8 September 2004

28. Tworek, K.: Model niezawodności systemów informacyjnych w organizacji. Zeszyty Naukowe. Organizacja i Zarządzanie/Politechnika Śląska 88, 335–342 (2016)

29. Walecka-Jankowska, K., Zgrzywa-Ziemak, A.: Udział technologii informatycznych w procesach interpretacji i zapamiętywania wiedzy. In: Sposoby osiągania doskonałości organizacji w warunkach zmienności otoczenia - wyzwania teorii i praktyki. Pod red. E. Skrzypek, T.2. Lublin: Zakład Ekonomiki Jakości i Zarządzania Wiedzą. Wydział Ekonomiczny UMCS, pp. 91–99 (2006)

30. Zahedi, F.: Reliability of information systems based on the critical success factors-formulation. MIS Q. 11, 187–203 (1987)

User Experience for Small Companies – A Case Study

Miriama Dančová$^{(\boxtimes)}$ and František Babič

Department of Cybernetics and Artificial Intelligence, Faculty of Electrical Engineering and Informatics, Technical University of Kosice, Kosice, Slovakia
{miriama.dancova, frantisek.babic}@tuke.sk

Abstract. User experience (UX) refers to a person's emotions and attitudes about using a particular product, system or service. It includes a person's perceptions of various related aspects such as utility, ease of use and efficiency. In our work, we decided to test and evaluated a potential of UX use under the conditions of a small company in Slovakia. We started with an analysis of relevant business processes in the company with the aim to identify possible points for change or improvement. Based on the results, we designed a new web application to support related company's processes. During the design phase, we used suitable UX elements to ensure the most positive end-user perception with the final prototype, such as paper sketches, wireframes or functional prototypes. The final version brought a high level of customer satisfaction. He did not evaluate positively only the implemented functionalities and usability of the web application, but also the UX elements he came into contact with. The most important lessons learned are intensive communication with end-user during the whole development process; a decision about a common vocabulary to make this communication easier; an organisation of the f2f testing sessions, not only in virtual form.

Keywords: User experience · Small company · Web application

1 Introduction

Nowadays, the whole world is globalizing in every sphere of human life. Thanks to the dynamics of the trade in goods and services, there is a hyper-competition that today's businesses and businesses need to adapt. Kotler and Caslione [1] define the hyper-competition as a situation in the market at a time when technology or supplies of the companies are so new that standards and rules of mutual rivalry are still produced, thus, competitive advantages arise. However they are not sustainable. It is characterized by intense competitive action and rapid innovation, to which the other competitors must respond quickly. They should create own competitive advantages and eliminate the benefits of their rivals.

In this paper, we focused on the small companies and a situation called information mix. It means a lot of information coming from different applications that dot now work as an integrated system, such as each application collects important information but does not share it with others. Small companies often run without the suitable

© Springer Nature Switzerland AG 2019
Z. Wilimowska et al. (Eds.): ISAT 2018, AISC 854, pp. 235–244, 2019.
https://doi.org/10.1007/978-3-319-99993-7_21

support of ITC to make the work easier. In many cases, they prefer precision manual work before being effective. The entrepreneurs are confused with a lot of documents and information without the ability to deal effectively with this chaos. Disorientation could cause the risk of overlooking important information or the date on which their business is dependent. It increases the risk of error due to poor time sequence between activities, omissions or late performance. In these conditions, it is hard to deploy a new software solution. It is necessary to consider all factors, dependencies, and existing business processes to design a customized application. For this purpose, the User experience can provide a suitable framework to ensure the end-users satisfaction.

The rest of the paper is organized as follows: first, we give an overview of the User experience and Human-Centered Design with selected existing examples. Next, we describe the whole development lifecycle performed in cooperation with one small company in Slovakia. Finally, we conclude our findings and experiences.

1.1 User Experience

User experience (UX) refers to a person's emotions and attitudes about using a particular product, system or service. It includes a person's perceptions of various related aspects such as utility, ease of use and efficiency. Donald Norman brought the term user experience to wider knowledge in the mid-1990s [2]. UX is subjective, as it is the performance, feelings, and thoughts of an individual user. UX encompasses a lot of factors, some are controllable by designers and developers or some are just user preference. These factors include usability, accessibility, performance, design/ aesthetics, utility, ergonomics, overall human interaction and marketing. The feeling is a temporary phenomenon changing at a time [3].

But UX is not exactly the same thing as usability, although they are related. Usability is more about the effectiveness of a site design and how user-friendly it is. Usability is a key component of overall UX. It also encompasses "User-centered design," which is basically the same concept worded in another way.

Carrie Cousins, the chief writer at Design Shack, identified UX as a key for retail and online sales because users must be able to easily navigate the site and understand how to use it. The same claim applies to the small company because its website is the first impression to users and the future interests in goods or services depend on it.

User experience provides methods for user research, information architecture building, user-oriented design, and usability testing. The choice of specific methods depends on the company size. For example, in the small company, it is more appropriate to do a number of in-depth interviews with the target group.

1.2 Human-Centered Design

The Human-centered design (HCD) represents an adaption of the software product design to the psychological and physical needs of end-users. The HCD is not a style but a process of designing and building a product. This process uses knowledge about cognitive, physical abilities and limitations of the possible users so that all users can use the product at the highest. The process and definition of HCD are described in standard ISO 9241-210:2010 (Ergonomics of human-system interaction - Part 210:

Human-centred design for interactive systems) which replaced previous ISO 13407:1999 (Human-centred design processes for interactive systems). This standard defines following basic principles [5]:

- The design is based on the understanding of users, tasks and environments.
- The users are involved throughout design and development.
- The design addresses the whole user experience.
- The design team includes multi-disciplinary skills and perspectives.
- The process is iterative.
- The design is driven and refines by-user-centred evaluation.

During the development process, developers produce some prototypes with limited functionalities. The end-users test and evaluate these prototypes to get relevant feedback. We can use several techniques for this purpose, see Table 1. It is important to define and perform an iterative and interactive process with the target audience [4].

Table 1. Basic HCD techniques [4]

Techniques	Purpose	Stage of the design cycle
Interviews about user needs and questionnaires	Collecting data related to the user needs; evaluation of various alternatives and prototypes of the expected product	At the beginning of the design project
Interviews about business processes and questionnaires	Collecting data related to the performed activities to be supported by expected product	Early in the design cycle
Focus groups	Discussion with all related focus groups	Early in the design cycle
Observation	Collecting information about environment in which will be used the expected product	Early in the design cycle
Role playing and simulation	Evaluation of design alternatives and prototypes	Early and mid-point in the design cycle
Usability testing	Collecting quantities data related to measurable usability criteria	Final phase of the design cycle
Interviews and questionnaires	Collecting qualitative data related to user satisfaction with the product	Final phase of the design cycle

1.3 Related Work

More people are aware of what UX means and more companies are adopting UX practices to their business. We present several examples from different perspectives.

Before a small company decides to build a new ICT solution for supporting its business processes, it is necessary to analyse an issue of IT adoption. This issue motivated a research described in [6] based on reviewing and analysing current IT literature. The proposed model of effective ICT adoption is believed to provide managers, vendors, consultants, and governments with a practical synopsis of the IT adoption process in small and medium companies, which will, in turn, assist them to be successful with IT institutionalization within these businesses.

A second issue is a right place for UX in the whole development lifecycle. Existing studies concentrate mainly on early phases of user-centered design projects, while little is known about how professionals work with post-deployment end-user feedback. The paper [10] presented an empirical case study that explores the current practice of user involvement during software evolution. The authors found that user feedback contains important information for developers, helps to improve software quality and to identify missing features.

In recent years, an agile methodology has been increasingly used in software development process. A single-case case study explored the role of user experience (UX) work in agile software development [9]. The company was a large multinational telecommunication company undergoing a lean transformation process. The results showed that there were difficulties integrating UX design and software engineering work in an agile and iterative manner.

In the e-commerce domain, existing studies focused on user satisfaction and a role of UX in this case. The present [8] investigated a number of dimensions used by online customers to evaluate the e-service quality of online travel agencies, as well as the relationship of these dimensions to customer satisfaction and the repurchase intention, using both qualitative and quantitative data collection techniques. The results of multiple regression analysis showed the website functionality as the most powerful factor in predicting customer satisfaction and the repurchase intention. Similar objective motivated a work [11] presenting a set of 64 heuristics as a tool to evaluate the grade of UX achievement of the e-commerce sites. The main contribution of this work is the standardization of these recommendations for evaluation of this type of websites.

The study [12] analysed how managers of retail travel agencies perceive the antecedents and consequences of adopting e-business in their supplier relationships. The study surveyed 101 travel agents in Spain. Research findings indicate that customer pressure has a strong influence on e-communication practices, i.e. the use of the Internet is largely driven by normative pressures, and this coercive power has a detrimental impact on trust. To avoid such negative consequences, perceived reciprocity is a prerequisite for committed supplier relationships.

Finally, a user testing is one of the main UX elements. The study [13] describes a laboratory experiment examined users' reactions to a set of home pages. These reactions were captured using self-report measures and eye tracking. The authors focused on differences between two generations of online users in terms of numbers and economic impact: the baby boom (born from 1946 to 1964), and Generation Y (from 1977 to 1990). The results confirmed that both preferred pages that had images and little text. On the other hand, eye-tracking data revealed that baby boomers had significantly more fixations on the various website parts like headers or sidebars). The future of this type of testing is in innovative approaches like an ambient user interfaces [7] or collection and analyses of the user's psychophysiological measurements [14].

2 Software Development Cycle Including UX

In this section, we present crucial parts of performed software development cycle, in which we used selected UX elements to ensure the user satisfaction with the final web application.

2.1 Process Analysis

The small company operated in the field of tourism. Currently, it does not use any information system to support or optimize related business processes. It uses standard table professor to create templates for planning or organizing relevant activities and tasks. Each activity has its own spreadsheet. But, the related spreadsheets are not interconnected, i.e. all changes have to be done manually. Also, this approach lacks any notifications about upcoming deadlines or incomplete tasks. From a general perspective, it is a time-consuming and inefficient procedure.

Based on the initial communication with the customer, we identified the following bottlenecks in existing business processes: time spent on organizing work, duplicate data fills, the visibility of critical tasks and activities. As an example, we provide a part of business process model representing a trip's planning activity (Fig. 1). This model consists of five subprocesses requiring different inputs and actions.

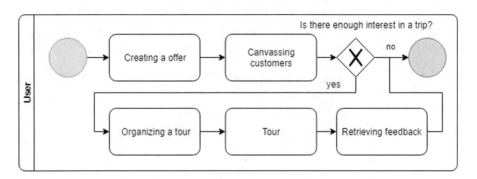

Fig. 1. Trip's planning process model.

2.2 Mock-Ups

At first, we started with a representation of a type of customer – a persona. Our customer answered the questions like who are we designing a product for? Based on the first customer feedback we created graphic sketches representing basic views of the expected software product in line with identified key user's characteristics. We focused on a simple and intuitive design helping the user to perform his daily tasks. During face to face meeting with the customer, we presented 10 mock-ups with different functionalities and all proposals were accepted (Fig. 2).

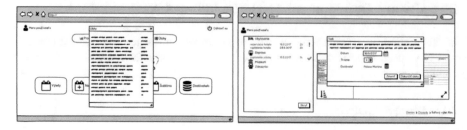

Fig. 2. Examples of created mock-ups.

2.3 Prototypes

Next phase focused on specific business processes, i.e. what inputs, what sequence, how much time, what dependency. All information was collected during face 2 face meeting with customers. Based on the second feedback and initial graphic mock-ups we started with an implementation of the first functional prototype. It offered a list of basic functionalities available within a simple user interface (Fig. 3).

Fig. 3. Prototype preview.

2.4 Testing Scenarios

We used this prototype for first customer testing. We prepared 3 test scenarios addressing various functionalities, such as:

- the user marks the "Trips" button,
- the user enters the filtering window "Rome",
- the user presses the button above the "Update".

In each scenario, we recorded the actual system response and total time needed to complete the process.

Tracking of end-user work with functional prototype brought several interesting findings. Some shortcomings have been revealed due to non-reading or inappropriate compliance with scenario steps. The customer approached the application without previous training or study manual. He expected an intuitive control. This approach has revealed some elements that are not intuitive for the end-user. Also, his behaviour uncovered typically through paths how to solve related tasks.

2.5   Simulation

We combined the knowledge from previous testing with a simulation of real process through the available prototype. The customer imitated his daily work with all spreadsheets at a certain level of simplification and explained related actions. The practical realization of the relevant processes has shown that their functioning is more complicated and complex than the problem identified in the first stage by the customer.

After completing the simulation, we discussed various alternatives to expected system behaviour. We analysed different cases using a formulation "What if …?" and we were able to specify more precisely the customer's needs and expectations.

2.6   Acceptance Testing

We organized another meeting for acceptance testing. For this purpose, we prepared 9 testing scenarios divided into 4 categories. The first one focused on user management and basic user interface. The second one dealt with the proper functioning of the trips offers creation. The third category aimed at the trips management from the first offer to the final check after a successful return. The last one represented the tests of simple analytical features to create different summary tables. This testing was associated with system validation, whether it meets the real customer requirements. The final version was accepted (Fig. 4).

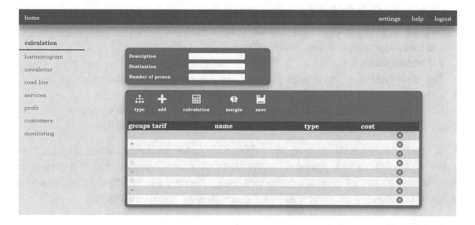

Fig. 4. User interface preview.

Figure 5 confirms that the final application appearance is very similar to the mock-up created and agreed by the customer in the early stage of the design phase.

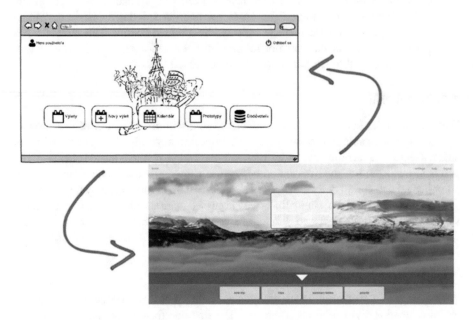

Fig. 5. Comparing the graphical design (mock-ups) with the final version.

2.7   Evaluating the Use of UX Methods

During final testing we were interested in what was a level of customer satisfaction with our chosen and applied UX methods during the whole development life cycle. For this purpose we prepared a questionnaire and it resulted in following statements.

The customer was satisfied with:

- the meetings about his needs and expectations,
- the meeting about existing business processes,
- the initial mock-ups,
- the participation in the design phase to see the prototypes evolution,
- the testing using the prepared testing scenarios.
- the iterative and interactive collaboration during the whole life cycle,
- the help of testing scenarios to perform relevant tasks or activities.

The overall satisfaction was more than 95% based on Likert scale[1] and the customer expressed his final opinion: "I consider the final web application as a very good framework that can be worked, developed, supplemented and enriched to create a very original, customized software product for our type of small company.".

---

[1] https://www.surveymonkey.com/mp/likert-scale/.

3 Conclusion

This paper described the case study on the use of appropriate UX methods in the development of a small business software solution. Based on performed business process analysis we were able to identify the most problematic points and proposed a new web application to support mainly the trips planning and organisation.

Working with the customer is often difficult because he often has no clear idea what his product should contain. He can imagine how he will use a new application in his daily work but he is not able to define the particular functionalities. The UX methods helped us to understand and identify better the customer's needs, requirements and working behaviour. At the same time, we have been able to keep customers interested in the product and its active cooperation to accomplish the mission.

The overall result is very positive and opens a possibility to deploy the developed web application into real practice. The most important lessons learned for us are communication with end-user during the whole development process; the decision about a common vocabulary to make this communication easier; to organize f2f testing sessions, not only virtual; no single method should be prescribed for a development life cycle.

Acknowledgment. The work presented in this paper was partially supported by the Cultural and Educational Grant Agency of the Ministry of Education and Academy of Science of the Slovak Republic under grant no. 005TUKE-4/2017 and the Slovak Research and Development Agency under grant no. APVV-16-0213.

References

1. Kotler, P., Caslione, J.A.: Chaotics: The Business of Managing and Marketing in the Age of Turbulence. AMACOM, New York (2009)
2. Norman, D., Miller, J., Henderson, A.: What you see, some of what's in the future, and how we go about doing it: HI at Apple Computer. In: Proceedings of Conference Companion on Human Factors in Computing Systems, Denver, Colorado, USA, p. 155 (1995)
3. Hassenzahl, M.: User experience (UX): towards an experiential perspective on product quality. In: Proceedings of the 20th International Conference of the Association Francophone d'Interaction Homme-Machine, IHM 2008, New York, NY, USA, pp. 11–15 (2008)
4. Abras, C., Maloney-Krichmar, D., Preece, J.: User-centered design. In: Bainbridge, W. (ed.) Encyclopedia of Human-Computer Interaction. Sage Publications, Thousand Oaks (2004)
5. U.S. Department of Health and Human Services: The Research-Based Web Design & Usability Guidelines, Enlarged/Expanded Edition. U.S. Government Printing Office, Washington (2006)
6. Bhobakhloo, M., Hong, T.S., Sabouri, M.S., Yulkifli, N.: Strategies for successful information technology adoption in small and medium-sized enterprises. Information 3(1), 36–67 (2012)
7. Galko, L., Poruban, J.: Tools used in ambient user interfaces. Acta Electrotechnica et Informatica 16(3), 32–40 (2016)
8. Tsang, N.K.F., Lai, M.T.H., Law, R.: Measuring e-service quality for online travel agencies. J. Travel Tour. Mark. 27, 306–323 (2010)

9. Isomursu, M., Sirotkin, A., Voltti, P., Halonen, M.: User experience design goes agile in lean transformation – a case study. In: IEEE Agile Conference (AGILE), Dallas, TX, USA, pp. 1–10 (2012)
10. Pagano, D., Bruegge, B.: User involvement in software evolution practice: a case study. In: 35th International Conference on Software Engineering (ICSE), pp. 953–962. IEEE, San Francisco (2013)
11. Bonastre, L., Granollers, T.: A set of heuristics for user experience evaluation in e-commerce websites. In: The Seventh International Conference on Advances in Computer-Human Interactions, Barcelona, Spain, pp. 27–34 (2014)
12. Andreu, L., Aldás, J., Bigné, J.E., Mattila, A.S.: An analysis of e-business adoption and its impact on relational quality in travel agency–supplier relationships. Tour. Manag. **31**(6), 777–787 (2010)
13. Djamasbi, S., Siegel, M., Skorino, J., Tullis, T.: Online viewing and aesthetic preferences of generation Y and the baby boom generation: testing user web site experience through eye tracking. Int. J. Electron. Commer. **15**, 121–158 (2014)
14. Čertický, M., Sinčák, P.: User experience optimization using psychophysiological measures. Acta Electrotechnica et Informatica **16**(3), 48–53 (2016)

Enterprise Meta-architecture for Megacorps of Unmanageably Great Size, Speed, and Technological Complexity

Matthew E. Gladden$^{(\boxtimes)}$ (ID)

Institute of Computer Science, Polish Academy of Sciences,
01-248 Warsaw, Poland
matthew.e.gladden@gmail.com

Abstract. The discipline of enterprise architecture (EA) provides valuable tools for aligning an organization's business strategy and processes, IT strategy and systems, personnel structures, and organizational culture, with the goal of enhancing organizational agility, adaptability, and efficiency. However, the centralized and exhaustively detailed approach of conventional EA is susceptible to failure when employed in organizations demonstrating exceedingly great size, speed of operation and change, and IT complexity – a combination of traits that characterizes, for example, some emerging types of "technologized" oligopolistic megacorps reflecting the Industry 4.0 paradigm. This text develops the conceptual basis for a variant form of enterprise architecture that can be used to enact improved target architectures for organizations whose characteristics would otherwise render them "unmanageable" from the perspective of conventional EA. The proposed approach of "enterprise meta-architecture" (or EMA) disengages human enterprise architects from the fine-grained details of architectural analysis, design, and implementation, which are handled by artificially intelligent systems functioning as active agents rather than passive tools. The role of the human enterprise architect becomes one of determining the types of performance improvements a target architecture should ideally generate, establishing the operating parameters for an EMA system, and monitoring and optimizing its functioning. Advances in Big Data and parametric design provide models for enterprise meta-architecture, which is distinct from other new approaches like agile and adaptive EA. Deployment of EMA systems should become feasible as ongoing advances in AI result in an increasing share of organizational agency and decision-making responsibility being shifted to artificial agents.

Keywords: Enterprise Architecture · Organizational complexity
Unmanageability · Industry 4.0 · Megacorps · Parametric design

1 Introduction

This text develops the conceptual basis for a specialized form of enterprise architecture that – unlike conventional approaches to EA – can be applied to organizations demonstrating otherwise unmanageable size, complexity, and speed of activity and

© Springer Nature Switzerland AG 2019
Z. Wilimowska et al. (Eds.): ISAT 2018, AISC 854, pp. 245–259, 2019.
https://doi.org/10.1007/978-3-319-99993-7_22

change. The fundamental feature of this variant of enterprise architecture – described here as "enterprise meta-architecture" (or EMA) – is the fact that it removes human enterprise architects by one step from the detailed work of analyzing an organization's current architecture and designing and implementing an improved target architecture. Such a model builds on existing approaches to adaptive and semi-automated EA and parametric design. Before presenting the details of the EMA model, we first consider the elements of conventional EA and the challenge posed by those organizations (including some emerging types of technologized oligopolistic megacorps reflecting the Industry 4.0 paradigm) whose size, complexity, and dynamism render the application of traditional EA unfeasible.

2 Elements of Conventional Enterprise Architecture

The goals of EA include (1) increasing an organization's capacity for managing complexity [1–3], (2) enhancing the organization's ability to resolve internal conflicts [4, 5], and (3) more effectively integrating the organization's various subsystems and constituent units, thereby providing enhanced agility that allows the organization to quickly adapt to rapidly evolving environmental conditions [5–7]. A well-designed enterprise architecture seeks to accomplish these goals by increasing the organization's degree of *alignment*. In principle, a comprehensive EA initiative strives to increase alignment between such diverse elements as an organization's business strategies, IT strategies, personnel structures, information system structures, decision-making processes, values, and organizational culture, as well as the characteristics of the external competitive ecosystem in which the organization operates [8, 9]. In practice, though, EA initiatives often focus simply on improving alignment between business and IT strategies [9].

The EA process involves analyzing an organization's current architecture, identifying its weaknesses, and formulating and implementing an improved target architecture. To facilitate this, the current architecture is captured in a detailed set of documents describing structures, processes, and systems [10] from various perspectives. Study of these documents allows the identification of areas of redundancy, inefficiency, or lack of resources that can be addressed by an improved target architecture [2]. EA frameworks for such work include TOGAF, GERAM, E2AF, and FEAF [2, 9].

3 When Conventional EA is Impossible: Technologized Megacorps and the "Unmanageable" Organization

While a considerable industry has grown up around EA – including numerous professional associations, training programs, certifying bodies, and journals – the potential benefits of EA remain debatable, the results generated by the use of different EA frameworks vary between organizations in unpredictable ways, the critical success factors for EA are unclear, and the failure rate for EA initiatives remains significant [9, 11–15]. It is not uncommon, for example, for a costly and time-consuming EA initiative to generate vast quantities of documentation that few organizational personnel will ever read or utilize [16, 17] or for a conventional EA approach to model an organization in such elaborate (and irrelevant) detail that it renders management of the organization *more* rather than *less* complex for its members [16, 18]. An especially challenging dynamic arises from the fact that the same organizational characteristics that lead an organization's decision-makers to conclude that launching an EA initiative is necessary may simultaneously make it difficult for such an EA effort to succeed.

Here we consider especially three such organizational traits that increase the perceived utility of a properly executed EA initiative while simultaneously rendering it difficult or impossible to effectively design and implement a new target architecture. These factors are extreme (1) organizational size, (2) organizational speed, (3) and organizational complexity. By itself, each of these poses a challenge for the successful execution of an EA initiative; when all three traits reach "unmanageable" levels within a single organization, conventional EA approaches can be rendered unworkable.

3.1 The Theoretical Concept of the "Megacorp"

The theoretical basis for focusing on these three characteristics in particular derives from reconsideration of the idea of the *megacorp* in light of its newly emerging "technologized" form. As conceptualized by economist Alfred Eichner in the 1960s and 1970s, the "megacorp" is not simply a "very large corporation"; rather, it represents a qualitatively distinct type of company. Namely, a megacorp is one of the leaders within an oligopolistic industry; the limited price competition that the megacorp encounters allows it to "increase the margin above costs in order to obtain more internally generated investment funds, that is, a larger corporate levy" [19, 20], which it uses to enable a state of perpetual growth [20, 21]. Because the megacorp is expected to endure permanently, sacrificing its short-term profits for investment in long-term profit growth is not risky but highly reasonable [20]. Moreover, because its shareholders (who come and go) are distanced from any involvement in the running of the company, the megacorp's professional managers are free to focus on long-term profit *growth* and the development of the company, rather than maximization of short-term profits and immediate financial gains for shareholders. Such a megacorp possesses a coherence, purpose, and even "life" [22] of its own; it is essentially an autonomous and "enduring organization with survival and growth as key objectives" [20]. However, growth and survival do not depend simply on securing enough funds; they also require a firm to successfully shape or navigate a complex array of political, social, technological, and environmental factors. Thus rather than adopting typical metrics that track a company's health solely in terms of financial performance, over time such a firm – through its managerial class – may develop a complex set of (non-financial) geopolitical, social, cultural, technological, or ecosystemic strategies and objectives – which are *its own* goals and not those of its shareholders.

One might imagine that such a megacorp is ideally suited to serve as a venue for the practice of conventional enterprise architecture. After all, it is a large and stable organization, and because it formulates strategies based on a goal of long-term multidimensional growth and development (and not maximization of short-term financial profits), its strategic objectives do not lurch from one direction to the next at a rate that makes it difficult for its personnel structures, IT systems, and other elements to keep pace.

While that is true, the higher margins and resources for investment in growth enjoyed by a megacorp allow it to methodically grow larger than would otherwise be possible for companies, pushing the limits of the organizational size manageable for human personnel. Moreover, the ongoing "technologization" (and, increasingly, "technological posthumanization") [23] of megacorps resulting from their deepening and expanding incorporation of autonomous AI, social robotics, human-robot inter-action, VR systems, brain-computer interfaces, ubiquitous computing, the Internet of Things, cyber-physical systems, and other Industry 4.0 [24, 25] phenomena both enables and drives further growth in organizational size and complexity, as artificial agency (or augmented human agency) makes it possible for a company to perform more work, more types of work, and work of greater speed and complexity than could be performed by natural biological human beings alone. Such dynamics may easily allow near-future technologized megacorps to "outgrow" the capacity of conventional EA to be employed in managing them. Below we consider such dynamics in more detail.

3.2   Organizational Size as an Obstacle for EA

Metrics like market capitalization, annual revenue, or number of employees are sometimes used in an attempt to capture a company's size in a single figure, but in reality the concept of an organization's "size" is much more complex and multidi-mensional. Here an organization's "size" can be understood as its *spatial extension;* however, this is not reducible simply to its number of physical facilities or the geo-graphical span of its operations. An organization's structures, processes, and systems not only occupy a certain amount of three-dimensional physical space; they also occupy (or create) several other overlapping types of space, including temporal, informational, cognitive, social, political, and ecosystemic space [26, 27]. A given organization possesses a unique extension in each of these spaces. Such extension generates a multifaceted workspace within which the organization can form structures and operate, but it also creates corresponding types of *distance* that tend to undermine organizational alignment.

By its very nature, a large organization's internal distances work against the pos-sibility of achieving and maintaining alignment. For example, the physical distance between employees and facilities makes it impossible for one employee to directly observe what others are doing; instead of existing within a single self-adjusting cybernetic feedback loop, an organization's activity thus becomes divided into thou-sands of disjointed operations. Similarly, it takes time for information about events occurring in one part of the organization to reach another part; in this way, spatial distance gives rise to temporal distance, making it difficult to effectively synchronize activities throughout the organization. Moreover, an employee's lack of knowledge about or causal interaction with events in other parts of the organization can lead to a lack of psychological investment; this "emotional distance" between employees can also negatively impact workplace culture, making it more difficult to create a culture that is aligned with and actively supports the other organizational elements.

To some extent, distance in informational and social space can be reduced through interpersonal contact among employees; physical and temporal distance can be compressed through technological means like email, instant messaging, social media, or videoconferencing. Indeed, the recent rise of *ad hoc* "virtual organizations" [28, 29] reflects the role that new forms of ICT can play in overcoming distance and establishing connections through organizational space, thereby allowing the creation of organizations whose employees, facilities, and informational resources are so spatially distant from one another that they would not be able to form a viable organization in the absence of such technology. On the other hand, it is possible for an organization to grow so massive in size that it is no longer simply "large" but literally becomes *unmanageably* large. Researchers have discussed such "unmanageably large" organizations in a number of contexts [30–34]. The extension of such entities within their multifaceted organizational space may become so great and the distances between their elements so vast that conventional EA approaches can no longer successfully grasp them.

3.3   Organizational Speed as an Obstacle for Conventional EA

An organization's "speed" has at least two aspects: (1) the organization's speed of regular internal operations and (2) the speed of evolution of the organization and its external ecosystem. For example, a hedge fund whose primary business activity consists of high-frequency trading may rely on its automated systems to make decisions and execute transactions within a matter of microseconds; however, the basic structure of the organization might remain little changed from year to year. An online social media company may have a high speed of internal operations *and* evolution, while an airplane manufacturer's work of designing and producing a given model of airliner might span decades [35]. A high speed of regular internal operations is relatively easy for EA frameworks to deal with. A high speed of ecosystemic evolution poses a greater challenge: by the time an organization's current architecture has been analyzed, a new architecture designed, stakeholder buy-in obtained, and necessary changes in organizational structures, processes, systems, and culture implemented, the market ecosystem will already have transformed and the organization's strategy will be out-of-date. While new strategies might be rapidly adopted, reconfiguring the organization's architecture to maintain alignment with those strategies requires time. In such circumstances, an organization's architecture may trail several steps behind its strategies.

As with the case of organizational size, it is possible for an organization's speed of internal operations or speed of organizational and ecosystemic evolution to be so great that such dynamics are no longer simply "fast" but literally become *unmanageably* fast. Organizations' struggles with "unmanageably fast" dynamics have been noted in various contexts [36, 37]. Such dynamics can easily exceed the boundaries of what the typically deliberate processes of conventional EA are capable of handling. It is precisely in a case of rapid ecosystemic change that an organization needs to develop the type of flexibility that EA promises to deliver – but it is also in such cases that the techniques of conventional EA reveal their limitations: their exhaustive, detail-oriented approach is not well-suited to quickly developing an architecture, and they do not yield

250 M. E. Gladden

architectures that can continuously and automatically update themselves to match the evolving realities of the competitive ecosystem.

3.4 Organizational Complexity as an Obstacle for Conventional EA

An organization's *complexity* can be understood in various ways. If an organization's size is reflected in the degree of extension of the organizational space that comprises overlapping component spaces of physical, temporal, informational, cognitive, social, political, and ecosystemic space, then its degree of complexity is reflected in the "convolutedness" of that space. Such complexity can be analyzed from a philosophical perspective (e.g., in terms of the Deleuzean "foldedness" of the space [38] or the topology of its underlying "possibility space" [39]), or it may even be mathematically quantified (e.g., in terms of its fractal dimension D [40]). Such complexity is manifested in phenomena like the degree of recursiveness within organizational structures, processes, and systems; the degrees of interdependency between organizational elements; and the scope and depth of specialized expert knowledge needed to successfully recognize, interpret, and manipulate various aspects of the organization's functioning.

In today's world, such convolutedness is often largely a matter of the *technological* complexity of an organization and its work. Such complexity is growing hand in hand with the emergence of the types of rich, intricate ecosystems of cyber-physical systems and organizations [25] conceptualized in the "Industry 4.0" paradigm [24] – a world in which all devices (and even human workers) are networked and become capable of directly communicating with and influencing one another, thereby exponentially increasing the topological complexity of the architectures within which they are connected.[1]

It is possible for an organization's complexity to be so great that the organization becomes *unmanageably* complex. The traits of organizations confronted by such "unmanageable complexity" have been discussed in a range of contexts [42–46]. One of the key aims of enterprise architecture is to reduce organizational complexity – or at least, to create a streamlined set of interfaces by which personnel can understand and manage their organization's remaining irreducible complexity [1–3]. However, in the case of an unmanageably complex organization, the nature and degree of complexity may be so overwhelming that enterprise architects are not able to identify, conceptually disassemble, and understand the organization's components – which is a prerequisite for formulating an improved target architecture. Figure 1 reflects the manner in which

---

[1] Drawing on the philosophical notion of human culture as a "rhizome" (i.e., an array of mutual influences that lacks a central origin or genesis and that is horizontally spreading, non-hierarchical, and maximally interconnected; possesses self-healing internal links; assimilates heterogeneous elements to form symbioses or hybrids; and grows naturally without a centrally planned architecture) developed by Deleuze and Guattari [41] and the concept of the technologized oligopolistic "megacorp" discussed earlier in this text [19, 20], the dynamics that establish such immeasurably complex interconnections between constituent elements of an organization – which, aided by decentralized networking technologies, often develop in a quasi-organic, biomimetic pattern – could be understood as contributing to the emergence of a "rhizocorp."

some emerging types of technologized oligopolistic megacorps reflecting the Industry 4.0 paradigm may combine unmanageable size, speed, and complexity.

4   Enterprise Meta-architecture: A Means of Managing the Unmanageable?

Efforts to apply the techniques of conventional EA can encounter insurmountable obstacles in organizations that are unmanageably large, fast, or complex. The question thus arises whether it might be possible to develop some variant form of EA which – while perhaps delivering substandard results for organizations of "normal" size, speed, or complexity – would nevertheless possess the advantage that it could be utilized in organizations that would be considered "unmanageable" from the perspective of conventional EA. This would appear to require reconceptualizing the relationship of organizational personnel to the EA process, the level of abstraction at which EA is "managed," and the extent to which EA activities must be automated. Ongoing developments in the fields of Big Data and parametric design suggest how this might be accomplished.

4.1   Big Data, Parametric Design, and Meta-Management

Many contemporary organizations are accumulating large amounts of data that possesses great business value, insofar as it could potentially be used to identify previously unrecognized trends, personalize product offerings for individual customers, or predict the behavior of consumers or competitors; however, the datasets are so vast in size, diverse in type, complex in structure, and rapid in their growth that they cannot be effectively managed, understood, or exploited with traditional data-analysis tools like two-dimensional spreadsheets. Moreover, such rich streams of data are often generated in real time, and an organization must process, interpret, and act upon them almost instantaneously to build a competitive advantage or maintain parity with rivals.

In recent years, a range of "Big Data" approaches (e.g., involving semi-automated data mining) have been developed to allow knowledge to be extracted from such vast datasets. In comparison to earlier data-processing approaches, many automated Big Data techniques minimize the role of the human end user in manually performing steps like data selection, cleansing, or analysis: the end user may have no direct access to individual data points but is instead presented with meaningful visualizations of particular types of entities, trends, or other phenomena uncovered within the dataset by automated algorithmic processes. Such approaches remove the user from the fine-grained detail of the dataset by one step, "elevating" the user's plane of engagement into the higher-order realm of: (1) determining the types of insights that would be useful for business purposes, if they could be obtained; (2) configuring the operating parameters for data-mining software and allowing it to autonomously extract such knowledge; and then (3) determining how the organization should act on the insights that result [47–49].

In effect, such Big Data approaches offer organizational personnel a means of effectively managing datasets that would previously have proven unmanageably large,

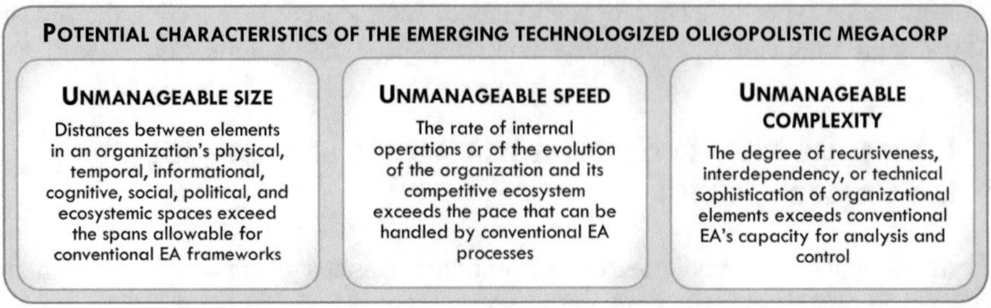

Fig. 1. Organizations of unmanageable size, speed, and complexity may face insurmountable obstacles when attempting to employ conventional EA approaches to generate alignment.

unmanageably complex, or changing in a way that is unmanageably fast – but with the constraint that such personnel are acting at one degree of remove from the data itself. Human personnel are still managing the process, but at a higher level – by establishing the broad parameters within which the automated systems will operate. In effect, automated Big Data approaches shift the role of human workers from *directly managing data* to *managing the systems that manage data;* in this way, the management of data is replaced with a higher-order "meta-management."

Similar dynamics are found in emerging approaches to parametric design and AI-facilitated form-finding used in the design of buildings: such morphogenetic techniques (e.g., based on evolutionary computing) can yield startling biomimetic designs with exceptional performance characteristics that could not have been devised by a human architect. In such an approach, the human architect serves as a "meta-designer" who (1) decides what broad criteria a building should fulfill and (2) bears legal and ethical responsibility for choosing from among the resulting designs proposed by the algorithmic system; however, the details of the design are developed by the architectural AI [38].

4.2 Distinguishing Enterprise Meta-Architecture (EMA) from Conventional EA

Drawing on Big Data and parametric design, it is possible to conceptualize a new form of EA that would be capable of developing and implementing an improved target architecture for an organization whose size, speed, and complexity place it beyond the grasp of conventional EA techniques. This proposed approach can be referred to as "enterprise meta-architecture" (or EMA),[2] insofar as the key feature distinguishing it from conventional EA is the fact that in EMA, human enterprise architects do not

---

[2] The phrase "enterprise meta-architecture" has been previously employed in other contexts, e.g., by Covvey et al. [50], who use it to describe a three-level EA incorporating the levels of "Meta-Applications," "Enterprise Middleware," and "Departmental Applications Systems," and by Ota and Gerz [51], who explain that "the development of architectures requires an enterprise (meta) architecture on how to define architectures." Similarly, Van de Wetering and Bos [52] formulate a noteworthy "meta-framework for Efficacious Adaptive EA" grounded in cybernetics and Complex Adaptive Systems theory; however, it still relies on the utilization of conventional EA frameworks by human enterprise architects.

directly design a target architecture; rather, they establish the basic goals and parameters for an automated system that generates, implements, and continuously adjusts the target architecture.[3] EMA relies on the fact that phenomena that are "unmanageable" for a human worker of a particular physiological nature and cognitive capacities may not be unmanageable for an artificially intelligent system of sufficient sophistication [23]. Differences between EMA and conventional EA are summarized in Table 1.

4.3   Toward Development of the Technological Foundations for EMA

A fully automated EMA system would require AI possessing distinct capacities for (1) analyzing an organization's structures, processes, and systems, (2) designing an improved target architecture that advances the business objectives chosen by human personnel, and (3) implementing a target architecture within the organization. Given current technological limitations, creation of a fully automated EMA system is not yet feasible. However, pieces are in place that could be employed toward its development. Beyond general semi-automated approaches to data-mining and Big Data [47–49], researchers are making progress in developing semi-automated tools for strategic analysis [54] and the gathering of data and generation of EA documentation (using tools like Nagios, Iteraplan, and SAP PI) [55–58]. Similarly, many forms of AI (including evolutionary computing) exist that could be harnessed for the automated creation of improved target architectures. Steps in that direction can be seen, for example, in algorithmic approaches to organizational design for artificial multi-agent systems [59, 60].

Perhaps the greatest obstacle to the design of a fully automated EMA system that could operate in a continuous feedback loop of architectural adjustment is the limited capacity of AI systems to implement a new target architecture within an organization. For a contemporary organization that primarily includes human workers, implementing a new architecture would require an EMA agent to successfully teach, train, monitor, and coach such workers – to persuasively communicate the rationale for actions that they may not readily accept and to negotiate with them the most contentious points of organizational change. While AI is not yet capable of effectively filling such roles, ongoing developments in the field of social robotics (especially in workplace contexts) [23] suggest that it may eventually be possible. Moreover, the need for engagement with human workers may lessen over time, as artificial agents play increasingly important and widespread roles in organizations, placing more organizational structures and dynamics under the (potential) direct influence of an automated EMA system.

4.4   Distinguishing EMA from "Adaptive" and "Agile" EA

EMA differs from recently emerging forms of "adaptive" EA [61–65] or "agile" EA [66–68] that attempt to make the process of designing and implementing a target architecture more flexible, streamlined, interactive, responsive, and outcome-oriented. Such approaches frequently rely on the use of advanced types of EA software, but as

---

[3] The determination of which organizational elements should be parameterized within the EMA system could be informed by a robust "organizational phenomenology" grounded either in the phenomenology of architecture [26] or a systems-theoretical phenomenology [53].

Table 1. A comparison of conventional anthropocentric EA suitable for use in "normal" organizations with a form of enterprise meta-architecture (EMA) that can be employed in organizations of otherwise unmanageable size, speed, and complexity.

Conventional enterprise architecture (EA)	Enterprise meta-architecture (EMA)
Human architects directly engage with the fine-grained details of architecture analysis, design, and implementation	Human architects set goals and parameters for the creation and maintenance of architectures by an automated system
Architectural software is a passive tool	Architectural software is a proactive agent
Architecture is interpreted through a handful of discrete "views" and "landscapes" comprehensible to human architects	Architecture may be captured and analyzed directly as a holistic and continuous object of endless and irreducible complexity
The range and variety of possible target architectures are limited by human experience and imagination	Processes involving, e.g., evolutionary computation may generate unexpected and counterintuitive yet effective architectures
While data-gathering may be bottom-up, architectural design and implementation is centralized and top-down	Architectural creation and implementation may be distributed among autonomous EMA agents embedded throughout an organization
An organization's actual architecture may be analyzed just once every few years	An organization's actual architecture is continuously analyzed
A new target architecture may be designed and implemented just once every few years	Adjustments to the actual architecture are ongoing and continuous
Over time, the actual architecture diverges from the nominal, normative architecture in unrecognized ways	Continuous analysis and adjustment maintain harmonization of the actual architecture with the target architecture

tools rather than actors within the architectural process: they still generally require human enterprise architects to engage with the fine-grained details of analysis, design, and implementation. EMA and agile and adaptive EA approaches may be able to mutually enhance one another: EMA that uses agile or adaptive EA as its conceptual foundation may have fewer processes to automate than EMA based on more exhaustive conventional EA approaches, while new AI-based techniques developed in pursuit of an EMA framework could further enhance the speed and flexibility of agile or adaptive EA.

4.5   The Expected Value of Enterprise Meta-Architecture

EMA could provide a means for improving alignment in organizations that have gradually grown in size, speed, or complexity to the point that it is difficult to apply conventional EA methods. Moreover, even in organizations for which conventional EA is still a practical option, the automated nature of EMA might render it less expensive and more efficient. Perhaps of greater long-term interest, though, is the potential of EMA (alongside other new management technologies) to facilitate the creation of entirely new types of architecture that could not otherwise exist or survive. For example, consider (1) a hypothetical organization whose strategies, tactics, and business processes are continuously and automatically adjusted in real time in response to the millions of interactions with customers occurring every day, or (2) a hypothetical organization whose thousands of workers all provide immediate feedback on proposed strategies as they are being developed in real time at the C-Suite and board level, with an automated system eliciting and sifting through such input to instantaneously identify and synthesize the most insightful and useful feedback. Such visions do not portray theoretical impossibilities; they simply reflect architectures that (at the moment) are unfeasible from a practical perspective. To the extent that automated EMA technologies someday allow such architectures to function, they could open the door for new organizational forms to be conceptualized and developed.[4] That would represent a qualitative shift in architecture made possible by EMA, beyond the quantitative advance of allowing conventional organizations to grow larger or more complex than is feasible today.

4.6   Potential Disadvantages of Enterprise Meta-architecture

At the same time, the use of EMA could create difficulties for organizations. Although it is meant to reduce or manage complexity, an EMA mechanism itself would be an immensely complex system subject to potential malfunctions, failures, hacking, viruses, and problems of interoperability with other systems. Moreover, if EMA allowed some large organizations to develop greater internal alignment, adaptability, and responsiveness to their markets, in the long run that might compel other organizations to adopt EMA systems for competitive reasons – despite the fact that novel and significant kinds of financial, legal, and ethical risks arise when a company delegates to automated systems higher-level functions of strategy development and implementation.

4.7   Areas for Future Research

The empirical study of EMA systems in production environments (i.e., real-world organizations) will need to wait for advances in the field of artificial intelligence that may require a considerable time to be realized. However, development of the theoretical underpinnings of EMA and its potential applications can already be pursued. As a stepping stone between theory and real-world application, simulations [39] can play a critical role; such simulations could build, for example, on existing algorithmic approaches to the development of organizations in artificial multi-agent systems [59, 60].

5   Conclusion

While conventional EA has a range of potential benefits to offer organizations, its implementation can become difficult or impossible for organizations whose size, speed, and complexity are too great for human enterprise architects to directly grasp. This challenge becomes more pronounced as organizations' spatial extension grows larger (e.g., as facilitated by Industry 4.0 technologies), the speed of organizational, technological, sociopolitical, and market change accelerates, and the technological

---

[4] The relative organizational stability of technologized oligopolistic megacorps, in particular, may provide a solid foundation for the development of such new architectural forms.

complexity of organizations and their competitive ecosystems grows ever more difficult to fathom. Through development of the types of enterprise meta-architecture approaches described in this text, it is hoped that the benefits of enterprise architecture can be more robustly enjoyed even by those organizations operating on the frontiers of unmanageability.

References

1. Højsgaard, H.: Market-driven enterprise architecture. J. Enterp. Archit. 7(1), 28–38 (2011)
2. Rohloff, M.: Framework and reference for architecture design. In: AMCIS 2008 Proceedings, Paper 118 (2008)
3. Sundberg, H.P.: Building the enterprise architecture: a bottom-up evolution? In: Wojtkowski, W., Wojtkowski, W.G., Zupancic, J., Magyar, G., Knapp, G. (eds.) Advances in Information Systems Development, pp. 287–298. Springer, Berlin (2007). https://doi.org/10.1007/978-0-387-70802-7_24
4. Fritz, R.: Corporate Tides. Berrett-Koehler, San Francisco (1996)
5. Hoogervorst, J.: Enterprise architecture: enabling integration, agility and change. Int. J. Coop. Inf. Syst. 13(3), 213–233 (2004)
6. Buckl, S., Schweda, C.M., Matthes, F.: A situated approach to enterprise architecture management. In: 2010 IEEE International Conference on Systems, Man and Cybernetics, pp. 587–592. IEEE (2010)
7. Caetano, A., Rito Silva, A., Tribolet, J.: A role-based enterprise architecture framework. In: Proceedings of the 2009 ACM Symposium on Applied Computing, pp. 253–258. ACM (2009)
8. Chan, Y.E., Reich, B.H.: IT alignment: what have we learned? J. Inf. Technol. 22, 297–315 (2007)
9. Magoulas, T., Hadzic, A., Saarikko, T., Pessi, K.: Alignment in enterprise architecture: a comparative analysis of four architectural approaches. Electron. J. Inf. Syst. Eval. 15(1), 88–101 (2012)
10. Nadler, D., Tushman, M.: Competing by Design: The Power of Organizational Architecture. Oxford University Press, Oxford (1997)
11. Niemi, E.: Enterprise Architecture Benefits: Perceptions from Literature and Practice. Tietotekniikan Tutkimusinstituutin Julkaisuja, vol. 1236-1615, p. 18 (2008)
12. Tamm, T., Seddon, P.B., Shanks, G.G., Reynolds, P.: How does enterprise architecture add value to organisations? Commun. Assoc. Inf. Syst. 28 (2011). Article 10
13. Donaldson, W.M., Blackburn, T.D., Blessner, P., Olson, B.A.: An examination of the role of enterprise architecture frameworks in enterprise transformation. J. Enterp. Transf. 5(3), 218–240 (2015)
14. Hope, T.L.: The critical success factors of enterprise architecture. Doctoral Dissertation. University of Technology Sydney, Sydney (2015)
15. Kotusev, S.: Enterprise Architecture Frameworks: The Fad of the Century. British Computer Society (2016). https://www.bcs.org/content/ConWebDoc/56347
16. Haki, M.K., Legner, C., Ahlemann, F.: Beyond EA frameworks: towards an understanding of the adoption of enterprise architecture management. In: ECIS 2012 Proceedings (2012)
17. Löhe, J., Legner, C.: Overcoming implementation challenges in enterprise architecture management: a design theory for architecture-driven IT management (ADRIMA). Inf. Syst. E-Bus. Manag. 12(1), 101–137 (2014)

18. Schekkerman, J.: How to Survive in the Jungle of Enterprise Architecture Frameworks: Creating or Choosing an Enterprise Architecture Framework. Trafford Publishing, Victoria (2004)
19. Eichner, A.S.: The Megacorp and Oligopoly: Micro Foundations of Macro Dynamics. Cambridge University Press, New York (1976)
20. Sawyer, M., Shapiro, N.: The macroeconomics of competition: stability and growth questions. In: Lavoie, M., Rochon, L.-P., Seccareccia, M. (eds.) Money and Macrodynamics: Alfred Eichner and Post-Keynesian Economics, pp. 83–95. M.E. Sharpe, Armonk (2010)
21. Eichner, A.S.: The Macrodynamics of Advanced Market Economies. M.E. Sharpe, Armonk (1987)
22. Gladden, M.E.: The artificial life-form as entrepreneur: synthetic organism-enterprises and the reconceptualization of business. In: Proceedings of the 14th International Conference on the Synthesis and Simulation of Living Systems, pp. 417–418. The MIT Press, Cambridge (2014). https://doi.org/10.7551/978-0-262-32621-6-ch067
23. Gladden, M.E.: Sapient Circuits and Digitalized Flesh: The Organization as Locus of Technological Posthumanization. Defragmenter Media, Indianapolis (2016)
24. Gorecky, D., Schmitt, M., Loskyll, M., Zühlke, D.: Human–machine-interaction in the industry 4.0 era. In: 2014 12th IEEE International Conference on Industrial Informatics (INDIN), pp. 289–294. IEEE (2014)
25. Gladden, M.E.: Strategic management instruments for cyber-physical organizations: technological posthumanization as a driver of strategic innovation. Int. J. Contemp. Manag. 16(3), 139–155 (2017)
26. Norberg-Schulz, C.: Genius Loci: Towards a Phenomenology of Architecture. Rizzoli, New York (1980)
27. Stanek, Ł.: Architecture as space, again? notes on the 'spatial turn'. SpecialeZ 4, 48–53 (2012)
28. Fairchild, A.M.: Technological Aspects of Virtual Organizations: Enabling the Intelligent Enterprise. Springer, Dordrecht (2004). https://doi.org/10.1007/978-94-017-3211-6
29. Shekhar, S.: Managing the Reality of Virtual Organizations. Springer, Berlin (2016). https://doi.org/10.1007/978-81-322-2737-3
30. Conant, M.: Railroad consolidations and the antitrust laws. Stanf. Law Rev. 14, 489–519 (1962)
31. Dikmen, I., Birgonul, M.T., Ataoglu, T.: Empirical investigation of organisational learning ability as a performance driver in construction. In: Kazi, A.S. (ed.) Knowledge Management in the Construction Industry: A Socio-Technical Perspective, pp. 166–184. Idea Group Publishing, Hershey (2005)
32. Bhidé, A.: An accident waiting to happen. Crit. Rev. 21(2–3), 211–247 (2009)
33. Mamun, M.Z., Aslam, M.: Conflicting approaches of managers and stockholders in a developing country: bangladesh perspective. Int. Corp. Responsib. Ser. 4, 317–335 (2009)
34. Samli, A.C.: From a Market Economy to a Finance Economy. Palgrave Macmillan, New York (2013)
35. Fine, C.H.: Clockspeed: Winning Industry Control in the Age of Temporary Advantage. Perseus Books, Reading (1998)
36. Lafleur, C.: The meaning of time: revisiting values and educational administration. In: Begley, P.T., Leonard, P.E. (eds.) The Values of Educational Administration, pp. 170–186. Falmer Press, London (1999)
37. Aziza, B., Fitts, J.: Drive Business Performance: Enabling a Culture of Intelligent Execution. Microsoft Executive Leadership Series, vol. 15. Wiley, Hoboken (2010)

38. Januszkiewicz, K.: O projektowaniu architektury w dobie narzędzi cyfrowych: Stan aktualny i perspektywy rozwoju. Oficyna Wydawnicza Politechniki Wrocławskiej, Wrocław (2010)
39. DeLanda, M.: Philosophy and Simulation: The Emergence of Synthetic Reason. Continuum, New York (2011)
40. Gladden, M.E.: A tool for designing and evaluating the temporal work patterns of human and artificial agents. Inf. Ekonom. 3, 61–76 (2014)
41. Deleuze, G., Guattari, F.: A Thousand Plateaus. Massumi, B. (trans). University of Minnesota Press, Minneapolis (1987)
42. Manning, M.R., Binzagr, G.F.: Methods, values, and assumptions underlying large group interventions intended to change whole systems. Int. J. Organ. Anal. 4(3), 268–284 (1996)
43. Ploetner, O., Ehret, M.: From relationships to partnerships – new forms of cooperation between buyer and seller. Ind. Mark. Manag. 35(1), 4–9 (2006)
44. Curran, C.J., Bonilla, M.: Taking OD to the bank: practical tools for nonprofit managers and consultants. J. Nonprofit Manag. 14(1), 22–28 (2010)
45. Thygesen, N.T.: The 'polycronic' effects of management by objectives – a system theoretical approach. Tamara J. Crit. Organ. Inq. 10(3), 21–32 (2012)
46. Taródy, D., Hortoványi, L.: Ambidextrous management in different growth phases. In: Strategica: International Academic Conference, Third Edition: Local versus Global, pp. 133–143. SNSPA Faculty of Management, Bucharest (2015)
47. Provost, F., Fawcett, T.: Data Science for Business. O'Reilly Media, Inc., Sebastopol (2013)
48. Dean, J.: Big Data, Data Mining, and Machine Learning: Value Creation for Business Leaders and Practitioners. Wiley, Hoboken (2014)
49. Marr, B.: Data Strategy: How to Profit from a World of Big Data, Analytics and the Internet of Things. Kogan Page Publishers, New York (2017)
50. Covvey, H.D., Stumpf, J.J.: A new architecture for enterprise information systems. In: Proceedings of the AMIA Symposium, pp. 726–730. American Medical Informatics Association (1999)
51. Ota, D., Gerz, M.: Benefits and challenges of architecture frameworks. In: 16th International Command and Control Research and Technology Symposium, Québec City (2011)
52. Van de Wetering, R., Bos, R.: A meta-framework for efficacious adaptive enterprise architectures. In: Abramowicz, W., Alt, R., Franczyk, B. (eds.) Business Information Systems Workshops: BIS 2016 International Workshops, Revised Papers, Leipzig, Germany, 6–8 July 2016, pp. 273-288. Springer, Cham (2016). https://doi.org/10.1007/978-3-319-52464-1_25
53. Ingarden, R.: O odpowiedzialności i jej podstawach ontycznych. In: Węgrzecki, A. (trans.) Książeczka o człowieku, pp. 71–169. Wydawnictwo Literackie, Kraków (1987)
54. Pedrinaci, C., Markovic, I., Hasibether, F., Domingue, J.: Strategy-driven business process analysis. In: Abramowicz, W. (ed.) Business Information Systems: BIS 2009, pp. 169–180. Springer, Heidelberg (2009). https://doi.org/10.1007/978-3-642-01190-0_15
55. Hauder, M., Matthes, F., Roth, S.: Challenges for automated enterprise architecture documentation. In: Aier, S., Ekstedt, M., Matthes, F., Proper, E., Sanz, J.L. (eds.) Trends in Enterprise Architecture Research and Practice-Driven Research on Enterprise Transformation, pp. 21–39. Springer, Heidelberg (2012). https://doi.org/10.1007/978-3-642-34163-2_2
56. Grunow, S., Matthes, F., Roth, S.: Towards automated enterprise architecture documentation: data quality aspects of SAP PI. In: Morzy, T., Härder, T., Wrembel, R. (eds.) Advances in Databases and Information Systems, pp. 103–113. Springer, Heidelberg (2013)
57. Farwick, M., Schweda, C.M., Breu, R., Hanschke, I.: A situational method for semi-automated enterprise architecture documentation. Softw. Syst. Model. 15(2), 397–426 (2016)

58. Sáenz, J.P., Cárdenas, S., Sánchez, M., Villalobos, J.: Semi-automated model-based generation of enterprise architecture deliverables. In: Abramowicz, W. (ed.) Business Information Systems: 20th International Conference, BIS 2017, Proceedings, Poznan, Poland, 28–30 June 2017, pp. 59–73. Springer, Cham (2017). https://doi.org/10.1007/978-3-319-59336-4_5

59. Horling, B., Lesser, V.: Using ODML to model and design organizations for multi-agent systems. In: Proceedings of the International Workshop on Organizations in Multi-Agent Systems (OOOP), vol. 5. AAMAS (2005)

60. Sleight, J., Durfee, E.H.: Organizational design principles and techniques for decision-theoretic agents. In: Proceedings of the 2013 International Conference on Autonomous Agents and Multi-Agent Systems, pp. 463–470. International Foundation for Autonomous Agents and Multiagent Systems (2013)

61. Ribeiro-Justo, G.R., Karran, T.: Modelling Organic Adaptable Service-Oriented Enterprise Architectures. In: Meersman, R., Tari, Z. (eds.) On the Move to Meaningful Internet Systems 2003: OTM 2003 Workshops, pp. 123–136. Springer, Heidelberg (2003)

62. Wilkinson, M.: Designing an 'adaptive' enterprise architecture. BT Technol. J. **24**(4), 81–92 (2006)

63. Yu, E., Deng, S., Sasmal, D.: Enterprise architecture for the adaptive enterprise – a vision paper. In: Aier, S., Ekstedt, M., Matthes, F., Proper, E., Sanz, J.L. (eds.) Trends in Enterprise Architecture Research and Practice-Driven Research on Enterprise Transformation, pp. 146–161. Springer, Heidelberg (2012). https://doi.org/10.1007/978-3-642-34163-2_9

64. Akhigbe, O., Amyot, D., Richards, G.: A framework for a business intelligence-enabled adaptive enterprise architecture. In: Conceptual Modeling: 33rd International Conference, ER 2014, Proceedings, Atlanta, GA, USA, 27–29 October 2014, pp. 393–406. Springer, Cham (2014). https://doi.org/10.1007/978-3-319-12206-9_33

65. Korhonen, J.J., Lapalme, J., McDavid, D., Gill, A.Q.: Adaptive enterprise architecture for the future: towards a reconceptualization of EA. In: 2016 IEEE 18th Conference on Business Informatics (CBI), vol. 1, pp. 272–281. IEEE (2016)

66. Rouhani, B.D., Shirazi, H., Nezhad, A.F., Kharazmi, S.: Presenting a framework for agile enterprise architecture. In: 1st International Conference on Information Technology, IT 2008, pp. 1–4. IEEE (2008)

67. Buckl, S., Matthes, F., Monahov, I., Roth, S., Schulz, C., Schweda, C.M.: Towards an agile design of the enterprise architecture management function. In: 2011 15th IEEE International Enterprise Distributed Object Computing Conference Workshops (EDOCW), pp. 322–329. IEEE (2011)

68. Rouhani, B.D., Nikpay, F.: Agent-oriented enterprise architecture: new approach for enterprise architecture. IJCSI Int. J. Comput. Sci. Issues **9**(6), 331–334 (2012)

Models of Production Management

A Genetic Algorithm to Solve the Hybrid Flow Shop Scheduling Problem with Subcontracting Options and Energy Cost Consideration

Sven Schulz[(✉)]

Faculty of Business and Economics, TU Dresden, Dresden, Germany
`sven.schulz@tu-dresden.de`

Abstract. This paper analyses the hybrid flow shop scheduling problem (HFSSP) with subcontracting options and time depending energy costs. While the consideration of energy costs in scheduling has increased considerably in recent years, subcontracting is rarely analysed in scheduling literature. A mathematical MILP formulation is given to define the exact problem and to calculate optimal solutions for small instances. The objective is to minimise the total production costs for internal and external manufacturing including transportation and energy costs. Since, already the general HFSSP is NP-hard the considered problem is difficult to solve to optimality. Therefore, a genetic algorithm (GA) based on a detailed matrix encoding procedure is proposed. To the best of my knowledge this is the first time that a heuristic approach is presented for the considered problem. An algorithm for intelligent swaps to make use of waiting time and a right-shifting procedure to take advantage of time depending energy costs prove to be suitable to improve the performance of the GA significantly. It can be shown that the GA finds nearly optimal solutions in a very short time.

Keywords: Hybrid flow shop scheduling · Energy awareness
Subcontracting · Genetic algorithm · Combinatorial optimization

1 Introduction

Since companies have to be become increasingly flexible and networked, temporary purchase of production capacities can be an important competitive advantage. Subcontracting is the possibility to allocate a single processing step of a job to an external manufacturer called subcontractor. Different forms of outsourcing are normally analysed on a strategic management basis. For example Hahn et al. [6] examine robust outsourcing decision making. However, short-term uncertainties such as machine failures or new orders may have a decisive influence on the subcontracting decisions. Nevertheless, at the operational level and in particular in production planning and scheduling subcontracting is rather less taken into account.

© Springer Nature Switzerland AG 2019
Z. Wilimowska et al. (Eds.): ISAT 2018, AISC 854, pp. 263–273, 2019.
https://doi.org/10.1007/978-3-319-99993-7_23

Choi and Chung [3] examine outsourcing in a single machine problem with processing time uncertainty. Lee and Sung [8] consider subcontracting in a single machine layout as well and weigh between delay and outsourcing costs. Parallel machine problems combined with strategic outsourcing decisions are analysed by Chen and Li [2] as well as Mokhtari and Abadi [10]. Qi [11] presents different models for subcontracting in a two stage flow shop problem. Li and Luan [9] describe a bi-objective flow shop problem with outsourcing possibilities, whereby different jobs are discounted at external machines while in-house machines have to be maintained. In HFSSP subcontracting is rarely considered. A mathematical formulation is introduced by Schulz [14].

Besides subcontracting this paper also considers time-depending energy costs. In the context of growing concern about environmental pollution and increasing energy costs as well as demand energy aware scheduling has received a lot of attention in recent years. Altogether more than 100 articles were published since 2010 (e.g. [13]). In addition to the possibility of reducing costs through intelligent planning by taking advantage of time-dependent price models or reducing peak power, scheduling can also be used to reduce the energy consumption. In principle, three approaches are conceivable. Firstly, the processing speed can be adapted to save energy in cost of longer processing times. Secondly, different machine states can by considered to reduce standby times and turn machines completely off. Thirdly, in the case of heterogeneous parallel machines, priority can be given to more efficient machines. An overview about different existing approaches is given by Biel and Glock [1] or Gahm et al. [5].

In Sect. 2 a MILP formulation is introduced to define the problem exactly. Afterwards in Sect. 3 a modified GA is proposed. In Sect. 4 follows a computational study analysing the performance of the heuristic compared to the optimal solution. Finally a summary as well as a short outlook are given.

2   Mathematical Model Formulation

Indices

$e \in E_i$	Set of b_i external machines
$i \in I$	Set of m stages
$j \in J$	Set of n jobs
$k \in K_i$	Set of o_i in-house machines
$l \in L_i$	Set of all $(o_i + b_i)$ machines
$t \in T$	Set of τ time intervals

Parameter

a_{il}	Machine-hour rate
c_i	Transportation charge
e_t	Real time energy price (RTP)
p_{ijl}	Processing time
v_{ik}	Energy demand
z_{ie}	Transportation time

The considered problem can be formulated as a MILP. A similar model can be found in [14]. However, some adjustments have been made to make the model more compact and to speed up the calculation of the optimal solution.

Altogether n jobs must be processed at m production stages. At each stage a number of o_i in-house machines as well as b_i subcontractors are available. The different machines are heterogeneous, which means that processing time and energy demand can vary for the same task. In the model, subcontractors are initially considered as further parallel machines. The production period under consideration is divided into τ equal time intervals where processing begins in time interval 1. The notation given above is used for the model formulation. Three different **decision variables** are used. The **binary** X_{ijtk} is equal to 1, if a job j is processed at stage i on machine l in time interval t. To assign job j to machine l at stage i the **binary** Y_{ijk} becomes 1. The value of **integer** C_{ij} corresponds to the completion time of job j at stage i.

The objective function (1) minimizes the total costs for a given production period. The first part displays the in-house costs, consisting of processing and separate energy cost, where the costs are added up over the individual time periods. The second part includes fixed external production costs independent of time as well as transport costs.

Minimize:

$$\sum_{j \in J} \sum_{i \in I} \left[\sum_{t \in T} \sum_{k \in K_i} X_{ijkt} \left(e_t \cdot v_{ik} + a_{ik}\right) + \sum_{e \in E_i} Y_{ije} \left(p_{ije} \cdot a_{ie} + z_{ie} \cdot c_i\right) \right] \quad (1)$$

Subject to:

$$\sum_{t \in T} \sum_{k \in K_i} \frac{X_{ijkt}}{p_{ijk}} + \sum_{e \in E_i} Y_{ije} = 1 \quad \forall i \in I, j \in J \quad (2)$$

$$\sum_{t \in T} X_{ijkt} = Y_{ijk} \cdot p_{ijk} \quad \forall i \in I, j \in J, k \in K_i \quad (3)$$

$$\sum_{t \in T} X_{ijet} = Y_{ije} \left(p_{ije} + z_{ie}\right) \quad \forall i \in I, j \in J, e \in E_i \quad (4)$$

$$\sum_{j \in J} X_{ijlt} \leq 1 \quad \forall i \in I, l \in L_i, t \in T \quad (5)$$

$$\sum_{l \in L_i} \sum_{t \in T | t > 1} |X_{ijlt} - X_{ijlt-1}| + X_{ijl1} + X_{ijl\tau} = 2 \quad \forall i \in I, j \in J \quad (6)$$

$$C_{ij} \leq X_{ijlt} \cdot t \quad \forall i \in I, j \in J, l \in L_i, t \in T \quad (7)$$

$$C_{ij} \leq (1 + X_{ijlt+1} - X_{ijlt})\tau + t \quad \forall i \in I, j \in J, l \in L_i, t \in T | t < \tau \quad (8)$$

$$C_{ij} \geq C_{i-1,j} + \sum_{t \in T} \sum_{k \in K_i} X_{ijkt} + \sum_{e \in E_i} Y_{ije} \left(p_{ije} + z_{ie}\right) \quad \forall i \in I | i > 1, j \in J \quad (9)$$

$$C_{ij} \geq \sum_{i^*=1}^{i} \sum_{k \in K_{i^*}} X_{i^* jkt} \cdot t + \sum_{e \in E_{i^*}} Y_{i^* je} \left(p_{i^* je} + z_{i^* e}\right) \quad \forall i \in I, j \in J, t \in T \quad (10)$$

$$C_{mj} \leq \tau \quad \forall j \in J \quad (11)$$

With constraint (2) each job is assigned to an in-house or external machine and it is ensured that a job does not skip a stage. Equations (3) and (4) assign a job to a machine for the entire processing and transport time if necessary. Condition (5) specifies that a machine can process a maximum of one job at a time.

Since non-preemption is assumed, Eq. (6) is introduced to avoid interruptions. In (7) and (8) the completion time C_{ij} is defined depending on X_{ijlt}. Based on that value constraint (9) ensures that a job cannot be started until it has been completed at the previous stage of production. The last two inequalities (10) and (11) are not necessary to define the problem, but reduce the solution space and thus accelerate the solution finding of the solver. The basic idea is to limit the range of C_{ij} which saves up to 85% solution time for some of the test instances. To solve the model IBM ILOG CPLEX 12.6 is used.

3   Genetic Algorithm

Since the problem is only solvable for small instances to optimality a heuristic approach is necessary for industrial size problems. For that reason a Genetic algorithm shall be suggested here. The basic procedure can be seen in Fig. 1. A detailed description is given below.

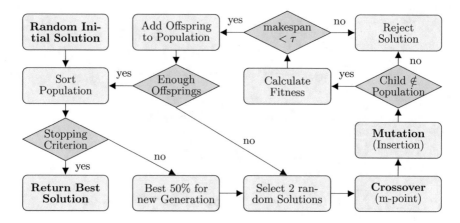

Fig. 1. General procedure of the proposed Genetic algorithm

3.1   General Procedure

Decoding and Encoding: In contrast to the classic flow shop problems, in HFSSP not only the sequence has to be encoded, but also the machine assignment influences the quality of a solution and must therefore be encrypted. It is common that only the sequence at first production stage is used to represent the solution. Afterwards the schedule is generated by list scheduling algorithms (see e.g. [12]). This procedure often proves to be advantageous, especially in the case of larger problem instances. However, many solutions are excluded with this approach. For that reason, a matrix coding procedure shall be used at this point, which contains both sequence and machine assignment in all production levels.

This procedure seems to be more suitable for problems with a small number of production stages. A similar approach is used by Dai et al. [4].

$$
\begin{array}{cccc}
\textit{Job 1} & \textit{Job 2} & \textit{Job 3} & \textit{Job 4}
\end{array}
$$
$$
\begin{array}{c}
\textit{Stage 1} \\
\textit{Stage 2}
\end{array}
\left(
\begin{array}{cccc}
1.56 & 2.42 & 1.33 & 1.67 \\
2.12 & 1.89 & 2.89 & 1.45
\end{array}
\right)
\tag{12}
$$

An example for 4 jobs and 2 production stages can be seen in (12). A decimal number is assigned to each job at each stage. The integer part assigns a machine to the Job. For example, in the first stage Job 1, 3 and 4 are processed on machine 1. The decimal places determine the sequence. The smaller this value, the earlier the job is processed. This means that in the given example the order on machine 1 at stage 1 is 3-1-4. If two jobs have the same priority the job number is used for ordering. For larger instances three decimal places are considered.

Initial Solution: To start the GA firstly a start population is needed. Since the calculation of a constructive solution takes some time and the influence within a larger population is estimated relatively low, we generate random solutions. Makespan is not an objective for the considered problem but nonetheless there is a very limited period of production (τ). The used coding procedure considers all possible solution. But with this also a lot of invalid solutions are possible. To avoid considering to many of these solutions and to accelerate the solution finding the priority values of the first stage are used for all following stages. Thus, there are $n!$ possibilities. However, the assignment to machines is completely random, which leads to $n! \cdot \prod_{i \in I}(L_i)^n$ possible initial solutions, what is the upper bound for the population size.

Crossover and Mutation: After the start population is sorted the best 50% are selected for a new population which can be seen in Fig. 1. Within the new generation, two parents are selected randomly. They produce two offspring by recombination. The recombination consists of crossover and mutation. The individuals are crossed at a different point in each production stage. These m points $\alpha_S \in \{0,..,n\}$ are generated randomly. An example can be seen in Fig. 2. The first selected parent passes the information for the first α_s jobs at stage s to child 1 and the rest to Child 2. The missing information of the offspring comes from parent 2.

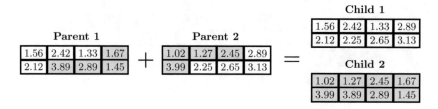

Fig. 2. m-point crossover procedure for $\alpha_s = \{3; 1\}$

Afterwards, each generated offspring is modified with a mutation probability η_m at one point. This means that a randomly selected job is either assigned to another machine or receives a different priority at one production stage.

Evaluation and Stopping Criterion: After a offspring is created, firstly it is checked if the same individual is not already part of the population. Identical solutions are rejected to maintain diversification. As can be seen in Fig. 1, the costs for new solutions, which are also used as fitness, are then calculated. Schedules that exceed the given maximum processing time τ are rejected at this point. All other solutions are included in the new population. When the new generation contains enough individuals the solutions are sorted and the procedure restarts. The algorithm is terminated if the best solution has not changed within γ generations, where γ depends on the size of the problem instance. After termination the best result is returned.

3.2   Adjustments for Improvement

Two major properties of the described problem can be identified in initial tests. Based on these findings the following two algorithms are implemented to improve the solution quality and accelerate the process.

I. Intelligent Swaps: The coding procedure enables jobs that are processed last in a stage to be scheduled very early in a subsequent stage and vice versa. This may result in long waiting and idle times. In turn it may lead to high makespan and thus to invalid solutions. We integrate intelligent swaps in the algorithm to process waiting jobs in idle times.

Simply said it is tested if a job can be scheduled before the previous job on a machine. For example in Fig. 3 it is tested if the idle time between job 3 and job 4 at stage 2 is enough to process job 2 and whether job 2 is completed at stage 1 before it should begin on stage 2. Since both requirements are met, job 2 and 4 are swapped and makespan can be reduced from 17 to 13. To swap two jobs their priority values are exchanged. Theoretically, it could also be beneficial if a job overtakes two or more jobs simultaneously, but here swaps are just tested for the direct predecessor to limit the computational effort. However, it is possible that several jobs overtake the same predecessor.

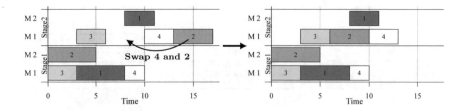

Fig. 3. Visualization of intelligent swaps

II. Right Shifting Procedure: Energy costs are not initially taken into account in the planning. Jobs are always scheduled as early as possible. With a right shifting algorithm it is tested whether the energy costs can be reduced by later processing. Therefore, beginning with the last stage the last job on a machine is stepwise shifted right until maximum completion time is reached. For each postponement the reduced costs are calculated and the best position is fixed. Subsequently the previous jobs are shifted stepwise until it finishes immediately before the next job starts. For earlier stages it must also be considered that the job must be done before processing starts at the next stage. Of course the right shifting can only be carried out for in-house production.

4 Computational Study

To analyse the performance of the introduced GA computational tests shall be examined. The results are compared with the optimal solution, which is calculated using the model in Sect. 2. For all tests an Intel Xeon 3.3 GHz CPU with 768 GB memory is used.

4.1 Test Instances

Since the problem under consideration has not yet been addressed in literature, there are no existing test instances. Thus, new examples are generated at this point. Altogether we consider 84 different instances. The various combinations of number of jobs, production stages as well as in-house and external machines are listed in Table 1. Problems with 50 and 100 jobs can be solved in around 10 respectively 30 s by the GA. Since CPLEX cannot find solutions for most of these instances, the quality of the results cannot be evaluated and therefore they are not given in detail here.

The data are randomly generated in the areas shown in Table 2. In order to be able to solve even larger instances to optimality relatively short processing times are taken into account. It is assumed that average energy costs account for approximately 10% of internal costs. In energy intensive industries like chemical industries the share can be up to 80% of production costs. (e.g. [7]) For energy

Table 1. Problem sizes

Set	Size
n (Jobs)	6, 8, 10, 12, 15, 20, 30
m (Stages)	2, 3, 4
o_i (In-house)	2, 3
b_i (External)	1, 2

Table 2. Overview of the test data

Parameter	Range
Processing time $p_{ijl}[h]$	$unif\{1, 10\}$
Energy demand $v_{ik}[kWh]$	$unif\{100, 1000\}$
Transportation time $z_{ie}[h]$	$unif\{1, 3\}$
Transportation charge $c_i[\text{€}/h]$	$unif\{30, 70\}$
Internal machine cost $a_{ik}[\text{€}/h]$	$unif\{100, 200\}$
External costs $a_{ie}[\text{€}/h]$	$\lfloor \max_{k \in K_i}(a_{ik}) \cdot 1.2 \rfloor$

prices we use Phelix spot market prices from 16th of March, 2017. External production costs are 20% higher than those of the most expensive internal machine. Added to this are the transport costs and times, which initially make external production appear significantly more expensive.

Furthermore, the maximum completion time largely influences the scheduling procedure. It should be possible that all jobs can be produced in-house and simultaneously the time horizon must not be too high to produce only in times of low electricity prices. The term 13 is used to calculate the maximum completion time for each instance. Logically, the production can be completed earlier.

$$\tau = \left\lceil \frac{\sum_{i \in I} \sum_{j \in J} \max_{k \in K_i}(p_{ijk})}{\sum_{i \in I} o_i} \right\rceil + (m-1) \cdot \max_{i \in I, j \in J, k \in K_i}(p_{ijk}) \tag{13}$$

4.2   Numerical Results

While CPLEX optimizes each of the 84 problems exactly once, we run the GA for each instance 30 times to analyse the scatter in terms of the solution quality and computing time. To run the GA firstly the **parameters** have to be set. Different computational tests were carried out for this purpose. A key finding is that because of the same sequence at all stages in the initial solutions, a high mutation rate proves to be advantageous. Furthermore a problem-dependent stopping criteria seems to have a valuable influence on the performance. Due to the limited scope of this article, we waive detailed explanations at this point. Overall the following settings are made:

- Population size: 100
- Mutation probability (η_m): 0.9
- Generations with unchanged best solution (γ): $10 + 5 \cdot n$

The results of the numerical study are shown in Table 3. The 84 test instances are grouped once according to the amount of jobs, once according to the number of stages and once according to type of machines in order to identify effects.

For the **optimal solution**, the minimum costs and the average calculation times are shown. The calculation time of the solver is limited to 1 h, which means that some instances with 20 or 30 jobs are not optimally solved. The average distance to the lower bound is shown in the Gap column. Logically, the problem is harder to solve with an increasing number of stages which can also be seen in CPU time. In contrast to the classic HFSSP, the computing time does not decrease with an increase in parallel internal machines. However, more external machines seem to make the problem easier to solve. With regard to costs, there is logically an increase with more jobs or stages. Additional machines reduce costs, with another in-house machine having a greater impact, as internal production is on average cheaper.

The average minimum costs found and the calculation time (CPU) are also given for the **GA**. Furthermore, the relative standard deviation (SD) is given

Table 3. Numerical results

	Optimal solution			Genetic algorithm				
	Costs [€]	Gap [%]	CPU [s]	Costs [€]	SD [%]	CPU [s]	SD [%]	DTO [%]
Results depending on number of jobs								
$n = 6$	8820.56	-	4.86	8853.61	0.02	0.08	17.47	0.31
$n = 8$	11806.59	-	11.52	11856.82	0.03	0.15	18.29	0.31
$n = 10$	13858.40	-	70.96	13937.65	0.07	0.27	22.94	0.44
$n = 12$	17630.41	-	108.56	17714.72	0.05	0.40	22.39	0.39
$n = 15$	19876.05	-	389.95	19984.00	0.08	0.74	25.74	0.37
$n = 20$	28787.24	0.03	1146.40	28943.58	0.08	1.53	26.25	0.34
$n = 30$	42647.98	0.11	2324.68	42905.19	0.12	4.52	30.70	0.38
Results depending on number of production stages								
$m = 2$	13513.93	-	53.75	13555.16	0.03	0.56	21.86	0.21
$m = 3$	19913.97	0.01	506.95	20033.12	0.07	1.09	22.83	0.43
$m = 4$	28040.91	0.05	1177.99	28209.82	0.09	1.64	25.49	0.45
Results depending on number of in-house machines and subcontractors								
$o_i = 2\|b_i = 1$	24105.41	0.02	626.79	24239.78	0.08	1.03	27.02	0.37
$o_i = 2\|b_i = 2$	20418.56	0.01	462.20	20491.25	0.04	1.10	21.85	0.21
$o_i = 3\|b_i = 1$	18953.95	0.03	737.03	19076.72	0.08	1.08	22.06	0.41
$o_i = 3\|b_i = 2$	18480.49	0.01	492.23	18589.72	0.06	1.18	22.66	0.46
Total	**20489.61**	**0.02**	**579.56**	**20599.37**	**0.06**	**1.10**	**23.40**	**0.36**

Notation: CPU - Computation time; SD - Standard Deviation; DTO - Distance to Optimum

for both values. Since the GA is executed 30 times for each instance and the algorithm is partly influenced by chance the results may scatter. With regard to the best objective found, the algorithm always finds similarly good solutions and there seem to be hardly any outliers. Overall, the results scatter by only 0.06%. The calculation time is quite different. Here there is an average of 23.4% deviation. The reason for these large differences in the calculation time for the same problem lies mainly in the stopping criterion. The algorithm ends if the best solution has not changed for γ generations. If a good solution is found in the beginning and can not be improved for γ generations the algorithm can be stopped very early. Vice versa it can take time if many smaller improvements are achieved.

When the results are compared, it can first be determined that the GA is significantly faster. The heuristic approach needs 1.1 s while the solver takes almost 10 min in average. Nevertheless, the results of the GA deviate only slightly from the optimal solution. The last column DTO shows the relative deviation of the best solution found for an instance from the optimum. This means that the implemented approach deviates from the optimum by only 0.36%. Especially for small problem sizes, the approach often even finds the optimal solution. These values also show that the difference remains at a similar level with an increase in the number of jobs as the termination criterion is job-dependent.

With more production stages the performance of the GA slightly declines. Possibly the number of stages should also be taken into account when selecting parameters.

5   Summary and Outlook

This paper analyses the influence of subcontracting and time depending energy prices on HFSSP. It can be shown that the temporary purchase of production capacities can be beneficial and reduce the total production costs. Especially in time of high capacity utilizations as well as high energy costs the production schedule can be improved by intelligent subcontracting.

To analyse and solve the problem a MILP formulation is given and a GA is proposed. The heuristic approach is based on a matrix encoding procedure which considers all possible schedules. To improve the performance besides basic adoptions in crossover and mutation two improvement algorithms are included. Firstly, intelligent swaps are examined to make use of waiting times. Secondly, a right shifting procedure makes use of the fluctuating energy prices. With these adoption the GA seems to be highly suitable.

In the proposed model the production time horizon is assumed to be a given value. In a future work, the makespan shall be taken into account in a multi-objective approach. Thus, depending on the situation, a production period with associated minimum costs can be selected using the Pareto front. Furthermore, subcontracted jobs could be exactly scheduled to increase the possibilities of subcontracting.

References

1. Biel, K., Glock, C.H.: Systematic literature review of decision support models for energy-efficient production planning. Comput. Ind. Eng. **101**, 243–259 (2016)
2. Chen, Z.L., Li, C.L.: Scheduling with subcontracting options. IIE Trans. **40**(12), 1171–1184 (2008)
3. Choi, B.C., Chung, K.: Min–max regret version of a scheduling problem with outsourcing decisions under processing time uncertainty. Eur. J. Oper. Res. **252**(2), 367–375 (2016)
4. Dai, M., Tang, D., Giret, A., Salido, M., Li, W.: Energy-efficient scheduling for a flexible flow shop using an improved genetic-simulated annealing algorithm. Robot. Comput. Integr. Manuf. **29**(5), 418–429 (2013)
5. Gahm, C., Denz, F., Dirr, M., Tuma, A.: Energy-efficient scheduling in manufacturing companies: a review and research framework. Eur. J. Oper. Res. **248**(3), 744–757 (2016)
6. Hahn, G.J., Sens, T., Decouttere, C., Vandaele, N.J.: A multi-criteria approach to robust outsourcing decision-making in stochastic manufacturing systems. Comput. Ind. Eng. **98**, 275–288 (2016)
7. IEA: Tracking industrial energy efficiency and CO_2 emissions. Report, International Energy Agency (2007)
8. Lee, I.S., Sung, C.S.: Single machine scheduling with outsourcing allowed. Int. J. Prod. Econ. **111**, 623–634 (2008)

9. Li, L., Luan, J., Qiu, Y.: Two-stage flowshop scheduling with outsourcing allowed. Int. J. u- e-Serv. Sci. Technol. **9**(10), 245–254 (2016)
10. Mokhtari, H., Abadi, I.N.K.: Scheduling with an outsourcing option on both manufacturer and subcontractors. Comput. Oper. Res. **40**(5), 1234–1242 (2013)
11. Qi, X.: Outsourcing and production scheduling for a two-stage flow shop. Int. J. Prod. Econ. **129**, 43–50 (2011)
12. Ruiz, R., Maroto, C.: A genetic algorithm for hybrid flowshops with sequence dependent setup times and machine eligibility. Eur. J. Oper. Res. **169**(3), 781–800 (2006)
13. Schulz, S.: A multi-criteria MILP formulation for energy aware hybrid flow shop scheduling. In: Fink, A., Fügenschuh, A., Geiger, M. (eds.) Operations Research Proceedings 2016, pp. 543–549. Springer, Cham (2018)
14. Schulz, S., Apelmeier, S., Buscher, U.: Hybrid flow shop scheduling with subcontracting options and time-depending energy costs. In: Large, R.O., Kramer, N., Radig, A.K., Schäfer, M., Sulzbach, A. (eds.) Logistikmanagement - Beiträge zur LM 2017, Stuttgart, pp. 163–170 (2017)

Calibration of the Risk Model for Hazards Related to the Technical Condition of the Railway Infrastructure

Piotr Smoczyński$^{(\boxtimes)}$ ⓘ, Adam Kadziński, Adrian Gill ⓘ,
and Anna Kobaszyńska-Twardowska

Poznan University of Technology,
pl. Marii Skłodowskiej-Curie 5, 60-965 Poznan, Poland
piotr.smoczynski@put.poznan.pl

Abstract. Risk management is an increasingly important process for business entities operating in various industries. The essential problem in effective risk management is an appropriate calibration of the chosen risk model, which makes it possible to get results that are compatible with the feelings of the persons responsible for safety issues in a a given business entity. The paper presents the method of calibration of the risk model on the example of a model intended for hazards related to the technical condition of the railway infrastructure. A set of sample diagnostic recommendations ranked according to their subjectively evaluated validity was used for this purpose. The area of feasible solutions was defined and possible values of weights of accepted criteria were determined through computer simulation based on random number generator.

Keywords: Risk management · Risk model calibration · Railway infrastructure

1 Introduction

Risk management is an increasingly important process for business entities operating in various industries. More than twenty reasons explaining this increase in interest in issues related to risk can be mentioned, including [1]:

- introduction of new legal regulations; e.g. in the European Union risk management is mandatory for entities operating in the field of medicine and finance
- the withdrawal of public authorities from the management of large state-owned enterprises, e.g. in the field of transport; private entities created in place of state-owned enterprises must manage risk on their own
- increase in the cost of insurance and its dependence on the implementation of the risk management process
- the possibility of loss of reputation as a result of the rapid spread of information about the negative issues, e.g. in social networks.

There are many alternative methods of risk management, but an essential part of each of them is to estimate the risk of identified hazards. The estimated risk is used to indicate those hazards, which in the first place should be the subject of further analysis.

© Springer Nature Switzerland AG 2019
Z. Wilimowska et al. (Eds.): ISAT 2018, AISC 854, pp. 274–283, 2019.
https://doi.org/10.1007/978-3-319-99993-7_24

In the case of a too high level of risk, these analyses aim to propose additional measures to reduce the risk, transfer it to another entity or to resign from the activity that the risk is related to.

The problem of risk estimation is the subject of many scientific papers. They indicate the inherent subjectivism of estimates [2] and warn against treating the results of the risk management process as an excuse to take difficult decisions [3]. On the other hand, documenting the rationale for the decision making using risk management makes later verification of assumptions possible. That makes it easier to understand the intentions of the decisions taken [4].

The essential problem in effective risk management is an appropriate calibration of the chosen risk model. Correctly calibrated risk model will allow to get the results that are compatible with the feelings of the persons responsible for safety issues in a given business entity. This is particularly important in the context of assuming responsibility for the implementation of the safety policy.

The aim of the work is to present the calibration method of the example risk model described in Sect. 2. This model is designed for the risk assessment of the hazards associated with the technical condition of the railway infrastructure, but the method presented in Sect. 3 can be applied to any risk model. The results obtained are summarized in Sect. 4.

2 Materials and Methods

2.1 A Set of Criteria of the Proposed Risk Model

Risk is defined as a two-dimensional combination of consequences and related uncertainty [5], understood as the possibility of hazard activation, sometimes expressed as mathematical probability [6]. Risk metrics can be defined as the value of the function R that to each of the hazards from a set Z assigns values from a subset V of the set of real numbers $\mathbb{R}$ [7]:

$$R : Z \rightarrow V \subset \mathbb{R}. \tag{1}$$

Each of the hazard is assessed in respect to m criteria $Ci(i = 1, 2, \ldots, m)$, whereas each of the criteria describes the probability and/or consequences of hazard activation.

Function R can be written in the generic form:

$$R(z_k) = f(a_i, r_i(z_k)); \quad i = 1, 2, \ldots, m; \quad k = 1, 2, \ldots, n, \tag{2}$$

where $r_i(z_k)(i = 1, 2, \ldots, m)$ denotes the risk component of hazard $z_k(k = 1, 2 \ldots, n)$ according to criterion $Ci(i = 1, 2, \ldots, m)$; a_i denotes the weight of each of the criteria.

In the further part of the paper, the risk model developed specifically for analysis domains covering linear infrastructure, e.g. railway lines, will be considered. The initial version of this model was presented in the paper [7]. There are two types of criteria that can be distinguished:

- array criteria: C1 'type of traffic' and C2 'importance for traffic'
- risk magnifiers C3 'permissible speed' and C4 'dispatcher's work experience'.

Array criteria are commonly used in the standardized risk assessment methods such as FMEA [8]. However, they are not well adaptable to describe the situation in domains, where the risk value changes approximately linearly with the change of certain features of a given domain, e.g. permissible speed or traffic flow. It was assumed that the risk components resulting from this type of criteria have the character of coefficients affecting the risk value resulting from other criteria. A similar concept was used at work [9], calling it 'risk magnification', hence the criteria of the considered nature proposed in this model will be called 'risk magnifiers'. It is also assumed that:

1. Risk magnifiers are characterized by natural numbers w ($w \in \mathbb{N}_0$), such as the number of tracks or the number of trains in the 24-h period
2. The maximum allowable value of the characteristic w is known, marked as w_{max}.

The effect of such an magnifier on the overall size/value of a risk metrics is determined by the coefficient α, which indicates the set of risk metrics values according to a given criterion:

$$\Omega_w = \left\{ 1 + \alpha \cdot \frac{0}{w_{max}}, 1 + \alpha \cdot \frac{1}{w_{max}}, \ldots, 1 + \alpha \cdot \frac{w_{max}}{w_{max}} \right\}, \tag{3}$$

where Ω_w denotes a set of risk metrics values; α denotes a coefficient of risk magnifier; w_{max} denotes the maximum admissible value of the characteristics w.

In such a situation, the risk component is given by the following equation:

$$r_w = 1 + \alpha \cdot \frac{w}{w_{max}}, \tag{4}$$

where r_w denotes the risk metrics value according to the risk magnifier; w denotes the value of the characteristics considered; α denotes a coefficient of the risk magnifier; w_{max} denotes the maximum admissible value of the characteristics w.

Due to the nature of risk magnifiers, it is assumed that their weight (Eq. 2) is always equal to 1.

The risk component quantification scheme for criterion C1 is presented in Table 1. It was assumed that the risk component according to this criterion is the lowest by freight-only traffic, and the highest in the case of a line with both types of traffic. Lines with exclusively passenger traffic have the meaning of the lines of local importance, operated by light railway vehicles.

Table 1. Risk component quantification scheme for criterion C1 'type of traffic'

j	Level $\omega_{1,j}$	Description
1	Green	Freight-only traffic
2	Yellow	Exclusively passenger traffic (light rail vehicles)
3	Red	Freight and passenger traffic

The risk component quantification scheme for criterion C2 'importance for traffic' is presented in Table 2. In the case of this criterion it is distinguished between the segments covering the tracks in stations and between stations.

Table 2. Risk component quantification scheme for criterion C2 'importance of traffic'

Segment type	j	Level $\omega_{2,j}$	Parameter	Description
Tracks at stations	1	Green	Type of track	Side tracks
	2	Yellow		Track with access to sidings
	3	Red		Main tracks
Tracks between stations	1	Green	Permissible restrictions of traffic	Traffic can be suspended
	2	Yellow		The possibility of introducing a detour or bus communication
	3	Red		The lack of the possibility to introduce a detour and bus communication

Types of tracks at stations are distinguished according to the internal work instruction of the infrastructure manager [10]:

- side tracks are used for shunting (only within the station) and parking of vehicles
- access tracks to sidings connect the railway station with the railway infrastructure of industrial plants, reloading terminals, etc.
- main tracks are tracks used for train runs (transit through the station).

The risk component according to the risk magnifier C3 'permissible speed' depends on the speed limit value (expressed in full kilometers per hour), as follows:

$$r_3 = 1 + \alpha_3 \cdot \frac{v}{160}, \tag{5}$$

where α_3 denotes the coefficient of risk magnifier C3; v denotes the permissible speed; 160 is the highest permissible speed on railway lines which are not equipped with the European Train Control System (ETCS).

The risk component according to the risk magnifier C4 'dispatcher's work experience' is given by the formula:

$$r_4 = 1 + \alpha_4 \cdot \frac{d}{50}, \tag{6}$$

where α_4 denotes the coefficient of risk magnifier C4; d denotes the work experience (in full years); 50 is the assumed maximum work experience in the position of the traffic dispatcher.

The presented criteria C1-C4 can be classified into two groups, reflecting the generally accepted definition of risk:

- criteria describing the consequences of hazard activation: C1 'type of traffic', C2 'importance for traffic' and C3 'permissible speed'
- criteria for describing the possibility of hazard activation: C3 'permissible speed' and C4 'dispatcher's work experience'.

Criterion C3 'permissible speed' has been qualified for two groups at the same time, because with the increase of speed the possibility of hazard activation increases (e.g. it is more difficult to stop the train), while higher kinetic energy translates into higher consequences of hazard activation.

2.2   Recommendations Used for Calibration of the Risk Model

Risk assessment criteria described in Sect. 2.1 are ready to use for determination of risk components. The overall risk metrics, however, can be calculated only if the weights of array criteria and coefficient of risk magnifiers are known. These values should be chosen in the way that the response of the risk model corresponds to the subjective feelings of the person in charge of the risk management process. Only under this condition the obtained results can be helpful in taking further steps.

The procedure used for determining the weights and coefficients is refereed to in this paper as 'calibration'. The input for the procedure is composed of several recommendations of diagnosticians, where the technical state of the track is described and some background information is given. The recommendation will be used to create a condition for the simulation described in Sect. 3.

Recommendation 1
Due to the contaminated ballast (Fig. 1) it is recommended to perform a local cleaning of the ballast at a length of about 35 m under track No. 2 in km 124.200 of the railway line No. 272 Kluczbork - Poznań Główny, between the stations Witaszyce and Kotlin. The line is used in mixed traffic, the maximum speed for passenger trains is 120 km/h and the line connects the city of Poznań with Ostrów Wielkopolski (approx. 72 thousand inhabitants).

Fig. 1. Contaminated ballast [11]

Recommendation 2

It is recommended to tamp the track locally due to the geometry parameters exceeding limits at the railway line No. 14 Łódź Kaliska-Tuplice in km 191.200, between stations of Krobia and Poniec. The line is mainly used by light rail vehicles of Koleje Wielkopolskie railway undertaking, providing the connection between Ostrów Wielkopolski and Leszno (approx. 65 thousand inhabitants). The maximum speed for rail buses and passenger trains is 100 km/h.

Recommendation 3

It is recommended to replace the broken sleepers in quantity of 57 elements on the railway line No. 369 Mieszków-Czempiń in km 19.109 (train stop Konarskie). Only occasional traffic of freight trains to Śrem under special conditions is run on this line.

Recommendation 4

It is recommended to replace damaged rail fastenings to the sleepers (Fig. 2) in track No. 1 at the Żerków station. The station is located on the 272 Kluczbork-Poznań Główny line and in this section is used only for freight traffic. The maximum speed of trains is 80 km/h, the traffic dispatcher at the station has a work experience of 10 years.

Fig. 2. Damaged rail fastenings [12]

Recommendation 5

It is recommended to replace the switch leading to the parking track for passenger trains at the Ostrów Wielkopolski station.

Recommendation 6

It is recommended to replace biologically damaged sleepers (Fig. 3) in track No. 5 at the Środa Wielkopolska station. The track leads to sugar plants. Dispatchers working at the station have working experience of not less than 20 years.

Fig. 3. Biologically damaged sleeper [13]

3 Results

For the adopted risk assessment criteria, the sets of risk metrics values according to the criteria C1 'type of traffic' and C2 'importance for traffic' are expressed by the following formula:

$$\Omega_i = \{1,3,5\}, \quad i = 1,2. \tag{7}$$

The set of risk metrics values according to the risk magnifier C3 'permissible speed' is given by the formula:

$$\Omega_3 = \left\{ 1, 1 + \alpha_3 \cdot \frac{1}{160}, 1 + \alpha_3 \cdot \frac{2}{160}, \ldots, 1 + \alpha_3 \cdot \frac{160}{160} \right\}, \tag{8}$$

and for the risk magnifier C4 'dispatcher's work experience' the formula is as follows:

$$\Omega_4 = \left\{ 1, 1 + \alpha_4 \cdot \frac{1}{50}, 1 + \alpha_4 \cdot \frac{2}{50}, \ldots, 1 + \alpha_4 \cdot \frac{50}{50} \right\}. \tag{9}$$

The overall risk metrics can be calculated according to a formula which generic form is given by Eq. (2). Taking into account the nature of the impact of individual criteria on the overall risk metrics, this formula can be written as:

$$
\begin{aligned}
R(z_k) &= (a_1 \cdot r_1 + a_2 \cdot r_2) \cdot \frac{a_3 \cdot r_3}{a_4 \cdot r_4} \\
R(z_k) &= (a_1 \cdot r_1 + a_2 \cdot r_2) \cdot \frac{a_3 \cdot \left(1 + \alpha_3 \cdot \frac{v}{160}\right)}{a_4 \cdot \left(1 + \alpha_4 \cdot \frac{d}{50}\right)},
\end{aligned}
\tag{10}
$$

where z_k denotes k-th hazard for which the risk metrics R is determined; a_i denotes the weight of the i-th criterion, whereby $a_3 = a_4 = 1$; α_i denotes the coefficient of the i-th risk magnifier; r_i denotes the risk metrics according to the i-th ($i = 1, 2, 3, 4$) criterion; v denotes permissible speed; d denotes the work experience of the traffic dispatcher (in full years).

The weights of array criteria (values a_1 and a_2) and risk magnifier coefficients (α_3 and α_4) should be chosen in such a way that the obtain results are in line with expectations. For this aim, the recommendations described in Sect. 2.2. will be used. Their features relevant to the risk criteria adopted in the risk model are shown in Table 3. The table contains also the 'Ranking' column, which determines the expected result of the risk assessment process. It was assumed that the lower the rank, the higher the priority in implementing the given recommendation. Therefore, the level of risk associated with the analyzed hazard should be higher.

Table 3. Ranking of calibration recommendations along with their characteristics

Recommendation	Type of traffic (C1)	Importance for traffic (C2)	Permissible speed (C3) (km/h)	Work experience (C4) (years)	Ranking
Witaszyce–Kotlin (No. 1)	Mixed	Detour possible	120	n.a.	1
Krobia–Poniec (No. 2)	Passenger	Traffic suspension possible	100	n.a.	3
Konarskie, line to Śrem (No. 3)	Freight	Traffic suspension possible	No data available	n.a.	6
Żerków (No. 4)	Freight	Main track	80	10	2
Ostrów Wlkp. (No. 5)	Passenger	Track with access to siding	No data available	no data available	5
Środa Wlkp. (No. 6)	Freight	Track with access to siding	No data available	20	4

In the example considered here, it was assumed that the highest priority (rank 1) received a recommendation related to the hazard resulting from the poor ballast condition between the Witaszyce and Kotlin stations; the lowest priority (rank 6) received a recommendation issued for a practically abandoned line to Śrem.

The determination of the weights and coefficients is performed with help of computer simulation based on generating random numbers. Four numbers are drawn in each iteration (a_1, a_2, α_3, α_4) from a given set; in this example, a set of:

$$W = \{1, 2, 3, 4\}. \tag{11}$$

Then, in accordance with the relationship (10), the risk metrics is calculated related to the individual recommendations of the diagnostics (Table 3) and the mutual dependencies between the calculated risk metrics values are checked. For the given input data, the dependencies are expressed by the following conditions:

$$R(z_1) > R(z_4) > R(z_2) > R(z_6) > R(z_5) > R(z_3). \tag{12}$$

Iterations are repeated until the expression (12) is met. If it is impossible to designate four numbers $(a_1, a_2, \alpha_3, \alpha_4)$ fulfilling this condition, the set of acceptable solutions (11) can be increased. If this method turns out to be ineffective, it may indicate the existence of unconscious criteria affecting the adopted ranking of calibration recommendations.

There are several four-element-sets $(a_1, a_2, \alpha_3, \alpha_4)$ in the example under consideration meeting the condition (12), e.g. (1, 4, 2, 1) and (1, 3, 3, 1) and any of them can be used. Finally, formula (10) can take the following form:

$$R(z_k) = (1 \cdot r_1 + 4 \cdot r_2) \cdot \frac{\left(1 + 2 \cdot \frac{v}{160}\right)}{\left(1 + 1 \cdot \frac{d}{50}\right)}, \tag{13}$$

where z_k denotes k-th hazard for which the risk metrics R is determined; a_i denotes the weight of the i-th criterion, whereby $a_3 = a_4 = 1$; α_i denotes the coefficient of the i-th risk magnifier; r_i denotes the risk metrics according to the i-th $(i = 1, 2)$ criterion; v denotes permissible speed; d denotes the work experience of the traffic dispatcher (in full years).

The results of the calculation of risk metrics according to the formula (13) for the risks associated with the calibration recommendations are presented in Table 4.

Table 4. List of risk metrics reflecting calibration recommendations

Recommendation	Ranking	Risk metrics R
Witaszyce–Kotlin (No. 1)	1	42.5
Żerków (No. 4)	2	35.0
Krobia–Poniec (No. 2)	3	15.8
Środa Wlkp. (No. 6)	4	9.3
Ostrów Wlkp. (No. 5)	5	7.0
Konarskie, line to Śrem (No. 3)	6	5.0

Based on the response of the risk model to calibration recommendations (Table 4) it is also possible, with the help of professional judgment, to define the boundaries of risk categories (acceptable, tolerable, unacceptable).

4   Summary and Conclusions

Many facts indicate that tasks related to risk management are becoming more and more important for business entities in various industries. Hence, many risk management methods have been developed and they often use original risk models. The basic problem in effective risk management is the appropriate calibration of these models. This article presents the method of calibration of an example risk model for hazards related to the condition of railway infrastructure.

The proposed calibration method can be used for any risk model. It allows to obtain risk model weights and coefficients in a simulation manner, using random numbers generator. In order to make full use of the method's possibilities, it is necessary to prepare appropriate calibration data, e.g. in the form of diagnostic recommendations.

The results of the selection of the parameters of the example risk model allow us to believe that the proposed method can make the risk model consistent with the knowledge, intuition and experience of persons responsible for safety in a given business entity.

Acknowledgements. The research work was financed with the means of statutory activities of Faculty of Machines and Transport, Poznan University of Technology, 05/52/DSPB/0280.

References

1. Sadgrove, K.: The Complete Guide to Business Risk Management (2015)
2. Rae, A., Alexander, R.: Forecasts or fortune-telling: when are expert judgements of safety risk valid? Saf. Sci. **99**, 156–165 (2017)
3. Durodié, B.: Theory informed by practice. Application informed by purpose. Why to understand and manage risk, cultural context is the key. Saf. Sci. **99**, 244–254 (2017)
4. Hokstad, P., Steiro, T.: Overall strategy for risk evaluation and priority setting of risk regulations. Reliab. Eng. Syst. Saf. **91**, 100–111 (2006)
5. Aven, T.: Risk Analysis (2015)
6. Johansen, I.L., Rausand, M.: Foundations and choice of risk metrics. Saf. Sci. **62**, 386–399 (2014)
7. Smoczyński, P., Kadziński, A.: Estimation and evaluation of risk in the railway infrastructure. In: LNNS, pp. 182–191 (2018)
8. International Electrotechnical Commission: IEC 60812:2006 Analysis techniques for system reliability – Procedure for failure mode and effects analysis (FMEA) (2006)
9. Marais, K.B., Robichaud, M.R.: Analysis of trends in aviation maintenance risk: an empirical approach. Reliab. Eng. Syst. Saf. **106**, 104–118 (2012)
10. PKP Polskie Linie Kolejowe S.A.: Instrukcja o prowadzeniu ruchu pociągów Ir-1 (R-1) (2017)
11. Urząd Transportu Kolejowego: Kontrola przejazdów na linii kolejowej Kraków Płaszów – Oświęcim. https://utk.gov.pl/pl/aktualnosci/6833,Kontrola-przejazdow-na-linii-kolejowej-Krakow-Plaszow-Oswiecim.html
12. Urząd Transportu Kolejowego: Prezes UTK stwierdził brak właściwego utrzymania infrastruktury przez PKP PLK na linii kolejowej nr 7 Warszawa Wschodnia Osobowa – Dorohusk. https://utk.gov.pl/pl/aktualnosci/6226,Prezes-UTK-stwierdzil-brak-wlasciwego-utrzymania-infrastruktury-przez-PKP-PLK-na.html
13. Urząd Transportu Kolejowego: Działania Prezesa UTK w zakresie nadzoru nad infrastrukturą wykorzystywaną w przewozie towarów niebezpiecznych - naruszenie przepisów przez PKP PLK na stacji Siemianówka. https://utk.gov.pl/pl/aktualnosci/6039,Dzialania-Prezesa-UTK-w-zakresie-nadzoru-nad-infrastruktura-wykorzystywana-w-prz.html

Selected Aspect of IT Tools Application in Process Improvement in Industrial Laundry Services

Piotr Grudowski◉ and Mateusz Muchlado(✉)◉

Quality Management and Commodity Science Department,
Gdansk University of Technology, Gdansk, Poland
mmuchlado@zie.pg.gda.pl

Abstract. The laundry services sector in Poland is growing dynamically and currently has around 13,000 enterprises, mainly micro, small and medium size. The increasing competitiveness of this industry, causes more emphasis on solutions that can in-crease the efficiency of laundry processes and thereby strengthen the market position. One of the methods that can be used for this purpose is Lean Six Sigma, aided by specialized computer software. The paper describes the process of professional washing of work clothes, analyze of its losses and according to that presents the implementation of IT solutions to support the implementation of LSS objectives. The main purpose of this article is to present the lean concept based identification of selected sources of losses in the key operational process of the company providing laundry services and to determine the possibilities of their reduction through the use of appropriate IT tools. This paper contains an analysis of the case study in which this system was applied.

Keywords: Lean Six Sigma · Laundry · IT support

1 Introduction

The laundry services sector in Poland according to data from the Central Statistical Office, there are about 13,000 enterprises that deal with washing and dyeing. A growth of approx. 2% per year is recorded for the whole category [1]. The market of washing and servicing work clothes [2] achieved in 2010 net income of 82 million PLN [3]. The above data indicate an increase in this sector in the economy and an increase in competitiveness. In order to meet the client's requirements, the companies should focus on increasing the effectiveness of their processes [4]. One of the key methodologies to improve the effectiveness of processes that can be used to improve the efficiency of processes is lean management [5]. The lean concept is a combination of the best low-volume and mass production practices, while avoiding their dis-advantages - high costs and low flexibility [6]. It mainly depends on the redesign of processes, with the application of various tools, including IT tools, so as to identify and limit the causes of wastage. Typically, overproduction, waiting, excessive transport, improper processing, excessive storage, unnecessary traffic, shortages, production errors or the creation and

© Springer Nature Switzerland AG 2019
Z. Wilimowska et al. (Eds.): ISAT 2018, AISC 854, pp. 284–292, 2019.
https://doi.org/10.1007/978-3-319-99993-7_25

design of services and goods not adapted to the customer's needs are distinguished as waste (jap. muda) within lean concept [7]. Currently, the hybrid concept of Lean Six Sigma is becoming more and more popular, combining the advantages of lean management and the Six Sigma methodology focused on the reduction of process variability [8].

The main purpose of this article is to present the lean concept based identification of selected sources of losses in the key operational process of the company providing laundry services and to determine the possibilities of their reduction through the use of appropriate IT tools. A case study of a small enterprise providing washing and cleaning of work clothing located in the Pomeranian Voivodship is presented.

2   Process of Industrial Washing and Service of Work Clothing

The process of improvement was initialized with its description, the effect of which is the block diagram (Fig. 1). The laundry process consists of four phases. The collection of clothing for washing, the washing process selected for the type of clothing, the finishing phase and the return of clothing to the customer. The operations that caused waste (muda) are marked with black background. After being picked up from the client, clothing was manually registered on the list of clothes provided for washing. The inventory was of a quantitative nature, it lacked the distinction that clothing belongs to a specific employee of the client. Then, as a result of the analysis of the laundry worker, the appropriate washing method was selected. In any case, it began with the segregation of clothing taking into account its characteristics and type of dirt, then proceeded as shown in the figure. After finishing the washing phase, the clothing goes to the optional service. Service of work wear includes its repair, i.e. patching, replacement of zippers or entire parts, e.g. trousers, sleeves. The service is followed by a quantitative verification of clothing based on previously prepared statements. The hand-made repairs made on clothing are added to this list for further settlements. If the amount of clothing is identical to the one that was accepted for washing in the first phase, clothes are packed and prepared for shipping to a customer.

After shipping clothes are handled to the customer. In the case of quantitative or qualitative shortages, corrective actions were taken. In the block diagram shown, the elements are marked with black background, which have been improved by the introduction of computer support.

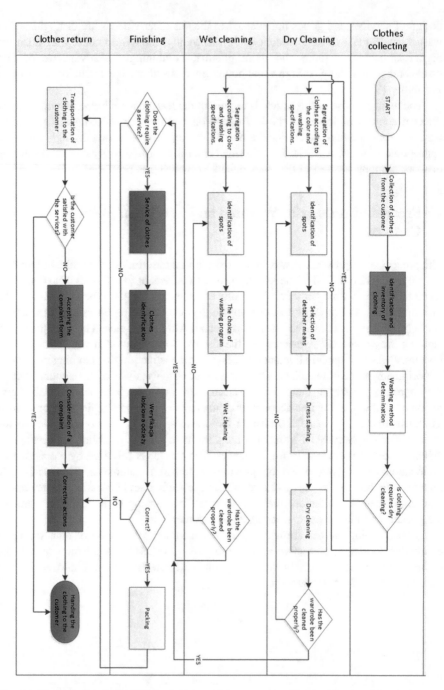

Fig. 1. Diagram of industrial work clothes washing process with identification of typical waste localization (Source: own elaboration)

3 Analysis of Waste in the Laundry Main Process

During the analysis of the process made by industry specialists, the number of wastes in the process were specified. In the garment collection phase, the manual laundry inventory control was very time-consuming. This activity involved both the service contractor and the client, due to the need for mutual confirmation. In addition, the inventory of clothing due to the lack of an electronic copy could be lost. The Fig. 2 shows how much time was required for manual inventory of clothes It shows the comparison of the times distribution before and after the improvements. Histograms were prepared based on data from 140 observations, carried out on the same client, before and after computer support. The reduction of the amount of time needed for the laundry inventory and its standardization was obtained thanks to the introduction of an electronic clothing circulation.

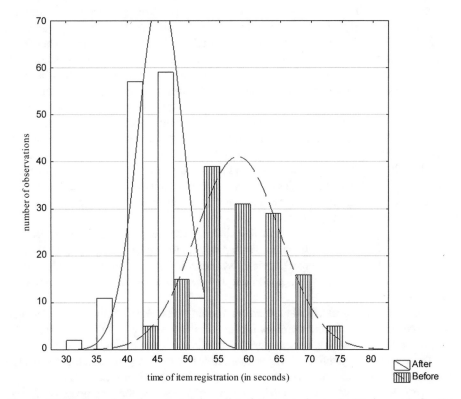

Fig. 2. Distribution of the time needed for a list of clothes to be washed before and after the introduction of IT support (Source: own elaboration)

Clothing, for the purpose of recording, has been permanently marked with bar codes along with the employee's name (Fig. 3). Barcode also includes information about type of clothes and has coded a specific, unique number.

S0000006168 4-2018 BEZRĘKAWNIK ROB.

Fig. 3. An example of a label containing a bar code used to mark work clothes (Source: own elaboration)

Specially created application allows to read barcodes and enter them automatically into the records. Thanks to this solution, the need for handwriting during delivery was eliminated, which contributed to the reduction of time needed for service. The average time of this activity has been reduced by 12 s for one piece of clothing.

The lack of electronic records also resulted in the lack of information about how long a piece of clothing is currently in the laundry and at what stage of the washing process it is. Customers have reported the need to have this type of information. With the introduction of an individual marking of work clothes and an electronic inventory, it was also possible to identify at what stage of the washing process it is located. Customers can verify whether the clothing has been picked up from them, whether it is in the laundry and whether and when it was released from the laundry. This functionality was created on the basis of assumptions taken from the Kanban Board tool. Due to the lack of personal designation of clothing, there was also no possibility of assigning a specific wardrobe to the employee. This caused problems for customers due to the need to separate clothing and assigned to a specific employee. The personal identification of clothing eliminated this problem. Clothing always re-turns to the same employee who has been assigned to it. The customer through the web browser has not only the ability to track the washing process, but also access to the history of washing. This functionality is of particular importance in the case of clothing which after a certain number of washing cycles must be recycled or specialized service procedures. For example, firefighting clothing that have to be impregnated every 20 cycles to maintain flame-resistant parameters.

Further problems were noted in the clothing service phase. Due to the lack of electronic clothing records allowing to assign clothes to a specific employee, it was also not possible to assign a repair service to a specific piece of clothing. This caused misunderstandings between the contractor and the client related to the settlement of these services.

The implemented electronic system of keeping records of clothes in the laundry also stores information about clothes repairs. Thanks to the possibility of assigning a specific service to clothing, it was possible to eliminate this problem.

The process of quantitative verification after the service was completed was also time-consuming, since there was a need to manually add support services.

Before the introduction of digital clothing control system, the employee was forced to manually check compliance with the paper census. During this operation, it took a long time to find the owner of the clothing on the list to write down the exact type of the service on his clothing. The software used to record work clothes automatically, groups the clothing depended to the one employee and only requires the closure of the individual items of clothing that he gave to wash to change the status to packed, prepared for delivery to the customer. The time losses for this activity is shown in Fig. 4.

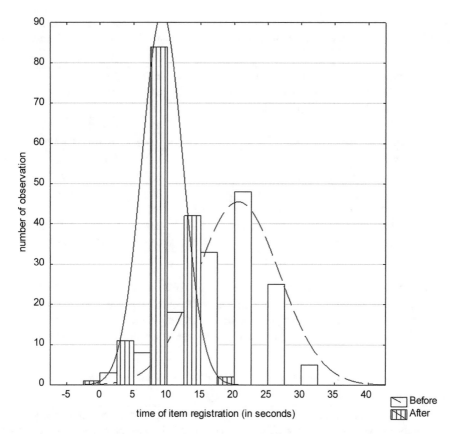

Fig. 4. Distribution of time necessary to complete clothing after the washing process and service -(before and after the introduction of IT support (Source: own elaboration)

A higher standardization was also observed for the time needed to complete clothing, which allows more accurate determination of the time needed to perform this activity, thus making the process more transparent.

Thanks to the introduction of electronic document circulation, in particular access to washing history and its online status, the number of quantitative complaints was drastically reduced. Thanks to the online preview, the customer's employees can check whether their clothes are in the laundry or have been transferred back to the workplace. Before the implementation of the computer aided system, cases of unjustifiable complaints were often recorded. Table 1 shows data regarding quantitative and qualitative complaints in the analyzed laundry. The data from year 2016, before the introduction of the electronic circulation of the laundry documentation and data from year 2017 when the system was functioning are compared. The introduction of the system took place at the beginning of January 2017. For a greater transparency, a quarterly summary was applied.

Table 1. Comparison of the number of defects in the process before and after the introduction of the IT system. (Source: Own elaboration based on data obtained from the surveyed enterprise.)

Year 2016	IQ	IIQ	IIIQ	IVQ
Number of washed clothes (in pcs)	46987	59876	55214	49506
Number of quantitative complaints (in pcs)	27	34	32	36
Number of quality complaints (in pcs)	45	49	31	48
Number of quantitative complaints (in %)	0.057%	0.057%	0.058%	0.073%
Number of quality complaints (in %)	0.096%	0.082%	0.056%	0.097%
Year 2017	IQ	IIQ	IIIQ	IVQ
Number of washed clothes (in pcs)	53625	65489	60547	56910
Number of quantitative complaints (in pcs)	18	21	14	18
Number of quality complaints (in pcs)	36	39	25	36
Number of quantitative complaints (in %)	0.034%	0.032%	0.023%	0.032%
Number of quality complaints (in %)	0.067%	0.060%	0.041%	0.063%

The electronic clothing identification system also records data on the laundry worker, who deals with a particular element of the laundry process for the wardrobe. Thanks to this provision, it was possible to control employees responsible for a given element of the process. This reduced the percentage of quantitative complaints.

Table 2. Implemented process improvements and their effects

Improvement	Effect
Introduction of an electronic circulation system for clothes delivered for washing	Reduction of the time needed for a list of clothing to be washed for an average of 12 s for each piece of clothing
	Reducing the time needed to check whether the amount of clothing after washing agrees with that given to laundry on average by 11 s for each piece of clothing
The introduction of bar codes that allow the identification of clothing	The ability to assign and return a specific piece of clothing to a specific customer employee
Introduction of Kanban board for clothing that is in the process of carrying out a laundry service	Increased process transparency for the client
Introduction of historical data on the number of washing cycles	Possibility of accurate monitoring and proper maintenance for clothing requiring disposal or specific service after a certain number of washing cycles
	A drastic reduction in the number of quantitative complaints by about 45%
Assignment and storage of information about the contractor of a laundry or service item for a specific item of clothing	Reduction the number of quality complaints by 21%

Table 3. List of digital system implementation costs

Type	Amount
Implementation costs	
Software purchase	12 000.00 PLN
POS computers	3 500.00 PLN
Bar code readers	600.00 PLN
Barcode printer	1 500.00 PLN
Total	17 600.00 PLN
Costs of system maintenance	
IT service (monthly)	500.00 PLN
VPS server rental (monthly)	320.00 PLN
Total (monthly)	820.00 PLN

Table 4. List of monthly financial benefits related to streamlining the process.

Type	Amount
Time reduction (clothes inventory before washing)	1191,00 PLN
Time reduction (clothes inventory after washing)	1092,00 PLN
Reducing the amount of compensation paid for clothing that went missing in the wash	380,00 PLN
Total	2663,00 PLN

4   Concluding Remarks

The case study presented in this article, shows how the support of the process of a professional laundry service using digital records within computer system can significantly contribute to the improvement of this process.

The Table 2 presents categories of improvements and their effects.

In conclusion, the presented improvements significantly and positively influenced the effectiveness of the process of professional washing of work clothes. On the other hand, the implementation of the computer aided record system was associated with costs for the enterprise.

Therefore, the additional analysis of costs and economic profits was conducted [9]. The Table 3 shows the costs of system implementation and its maintenance.

The Table 4 presents the savings resulting from the implemented improvements of the system, the data were also compiled monthly.

Calculations show that approximately 60 h of employee's working time were saved per month. The reduction of the time needed for the handling of clothing after washing was about 54 man-hours. The implementation of the lean based IT system resulted in financial savings of 1843,00 PLN a month. The investment costs assumed above will be refunded in less than 10 months. Taking into account the above results of analyzes, it should be recognized that the introduction of an IT system for the purpose of keeping electronic records of clothes delivered for washing fully achieved its objectives. In addition to economic advantages and the imminent return on investment, the company also gained advantages related to the effectiveness of its processes. The digitalization of identification and traceability systems, due to its availability for laundry customers, has led to increased transparency of processes, new functionalities and positive image creation of the company.

References

1. Główny Urząd Statystyczny: Miesięczna informacja o podmiotach gospodarki narodowej w rejestrze REGON luty 2018 (2018)
2. IMPEL: IMPEL – Prentacja Zarządu Q1- Prezentacja Zarządu Q1-3 2015 (2015)
3. Grupa FOREX: Monitor B nr 39, z dnia: 08-01-2010 (2010)
4. Piasecka-Głuszak, A.: Skuteczność działań lean w polskich przedsiębiorstwach – wybrane aspekty rozwiązań przy zachowaniu ducha kaizen. Badania empiryczne. Int. Bus. Glob. Econ. nr nr 33 European Union-10 Years after Enlargement, pp. 595–608 (2014)
5. Grudowski, P., Leseure, E.: LSS Plutus - Leasn SiX Sigma dla małych i średnich przedsiębiorstw. WNT, Warszawa (2013)
6. Womack, J.P., Jones, D.T., Roos, D.; The machine that changed the world: the story of lean production, pp. 1–11. World (1990)
7. Holweg, M., Bicheno, J.: Supply chain simulation – a tool for education, enhancement and endeavour. Int. J. Prod. Econ. **78**(2), 163–175 (2002)
8. Snee, R.D.: Lean six sigma – getting better all the time. Int. J. Lean Six Sigma **1**(1), 9–29 (2010)
9. Eckes, G.: The Six Sigma Revolution: How General Electric and Others Turned Process into Profits. Wiley, New York (2001)

Risk Analysis in the Appointment of the Trucks' Warranty Period Operation

Irina Makarova[1]([✉]) [iD], Anton Pashkevich[2], Polina Buyvol[1],
and Eduard Mukhametdinov[1]

[1] Kazan Federal University, Naberezhnye Chelny, Russia
kamIVM@mail.ru
[2] Cracow University of Technology, Kraków, Poland
anton.pashkevich@gmail.com

Abstract. Reliability is currently one of the factors affecting the competitiveness of vehicles. Vehicles become highly reliable due to the integrated approach to all stages of their "life cycle": design, production and operation. The most important for the manufacturer is the warranty service, because this stage is characterized by an increased probability of failure, and the manufacturer takes responsibility for their elimination. The authors propose to take into account the risks of failures of various units, assemblies and systems of the car during the appointment of the warranty period. This study used the statistics of failures received from the dealer and the service network, and from the reliability department of the engine manufacturer. The results of the study show that the use of the proposed methodology will reduce the costs of the truck manufacturer.

Keywords: Risk analysis · Warranty period · Reliability of tracks

1 Introduction

In the globalization and high competition in automotive markets, automakers create branded services systems of (BSS) in the sale regions. It is a network of dealer and service centers (DSS) authorized by the manufacturer's standards. The BSS operation quality affects the corporation's competitiveness of the, since this system works with customers. The complex system's operations an involves risks. Considering risks from the management view point, you must consider the fact that risks arise in all spheres and they cannot be completely eliminated, but they can be managed. In the corporate governance system, to automatically identify risks mechanisms should be built, evaluate and make appropriate decisions. In this regard, it is necessary to create self-regulating methods: the recognition of risks, an assessment of their severity, the choice of methods and tools, the development of a strategy and tactics for risk management.

Risk management is a strategic tool to managing a company. This means evaluating certain significant factors, calculating the likelihood of positive and negative changes, and developing and coordinating appropriate counter strategies. The term "risk" means a combination of the probability of an event and its consequences. Risk can and should be managed using methods that allow the course of risk events to predict a certain extent, correct it in a timely manner so as to minimize undesirable consequences.

© Springer Nature Switzerland AG 2019
Z. Wilimowska et al. (Eds.): ISAT 2018, AISC 854, pp. 293–302, 2019.
https://doi.org/10.1007/978-3-319-99993-7_26

When designing the BSS, as well as in the strategic planning of the DSS, this task can be solved using the intelligent systems. Intelligent control systems that have mechanisms for system's current condition monitoring, prompt and timely information's transmission and qualitative data analysis for subsequent planning and forecasting, allow to develop an effective mechanism for risk management.

The article [1] authors classified the risks at all life cycle stages of vehicles: including the development of terms of reference, design, manufacture and disposal. According to the authors, the specific risk situation causes may be the activities of the enterprise or entrepreneur, lack of information on the external environment condition that has an impact on this activity result, etc. At the same time, the authors consider the exploitation phase from the point of view of the transportation process organization, allocating only two types of risks, respectively, criminal and ecological. In our opinion, the commercial exploitation's stage of is inseparable from the stage of technical operation, which means maintaining the vehicle, as a technical object, in good condition. In this case, it is necessary to take into account the technological risks of the process of maintaining efficiency. These processes are coupled with the spare parts logistics, so it is needed to take into account the spare parts logistics processes risks, including the failed spare parts' reverse logistics. In conditions of transition to the circus economy, this becomes even more urgent.

2 Theoretical Risk Management Preconditions

2.1 Classification of Risks

Risk analysis is the analysis of the probability that significant undesirable events will occur and their adversely affect the achievement of the objectives. Risk analysis includes the risks assessment and research on methods to reduce risks or the associated adverse effects [1]. Risk assessment is the quantitative or qualitative way to determination of the magnitude (degree) of the risk. Since the vehicle belongs to the class of science-intensive and high-tech products, and its operation is associated with an efficiency loss risk, which can lead to the people death, then throughout its operation it is necessary to monitor its technical condition. Therefore, the exploitation stage includes two components: "commercial operation" (the transportation process implementation) and "technical operation" (in fact – service support, i.e. maintenance of workability). These two components are inextricably linked; as commercial exploitation is provided by high-quality technical operation (Fig. 1).

When service network development strategy creating, it is necessary to take into account the possible risks associated with the vehicles' service system functioning, which includes a large number of different processes, as the proprietor invests in the creation of such a system and he is interested in the optimal mode of its operation. At the same time, the BSS specifics, combining the functions of sales and t services provision, is a complex system, that can have risks that are typical for trade organizations on the one hand, and for industrial enterprises on the other. Moreover, the specificity of the BSS is that, unlike production, where the processes are deterministic, the service system is characterized by non-stationary demand for operations, therefore

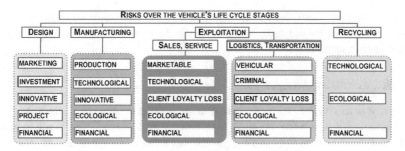

Fig. 1. Classification of automotive industry risks

the process parameters are stochastic, which causes additional risks, due to the non-stationary demand causing uneven load of positions [3–5].

It is important to understand that all risks types are related to one degree or another and the emergence of one their species can have a negative impact on the activities of related subsystems. For example, a decrease in the vehicles sales volume will negatively affect the need for spare parts and services, which in turn will result in under loading of positions, personnel and equipment.

Production risk is caused by malfunctions in the production system functioning. This is a violation in the operation of any of the DSS subsystems: a decrease in sales volumes, volumes of services, a violation in the supply chain, and so on. *Technological* risk is due to improperly chosen technology of maintenance and repair. *Investment* risk is associated with the costs of new projects, for example, with the expansion of DSN. This risk characterizes the likelihood or amount of possible investment losses in the creation, equipping and maintenance of additional work positions when underutilizing production capacity or loss of profit from loss of customers due to deficient production opportunities. *Innovation* risk is the losses probability to investing in the new services production, which, perhaps, will not find the market expected demand. The risk of *client loyalty loss* arises when the customer's loyalty is reduced (non-qualitative service). *Criminal risk* is associated with risk situations that affect the health and life of a person: when vehicle's operating, this is the so-called risk of premature failure, while providing services – the risk associated with production technology violations. *Financial risk* is associated with the specific of investing money in various projects.

Technical risk determines the production organization degree, the possibility of carrying out preventive measures (regularity preventive check of equipment, security measures), and equipment repair by the company's own capabilities. Technical risk is:

- probability of losses due to negative research results;
- probability of losses due to failure to reach the planned technical parameters during design and technological development;
- the probability of losses due to low production technological capabilities, which does not allow the application of the new developments results;
- the likelihood of loss due to side effects or problems with delayed manifestation when using new technologies and products;
- probability of losses due to failures and equipment breakdown, etc.

Technical risk is attributed to the group of internal risks, because these risks can be directly influenced by the firm, and their occurrence, as a rule, depends on the itself production activity.

2.2   Stages of Risk Analysis

Risk analysis means the analysis of the probability that certain undesirable events will occur, because of which the specified goals can not be achieved. Risk analysis includes the assessment of risks, methods for preventing them or reducing the adverse conse-quences associated with their occurrence. When analyzing risks, the following assumptions are possible [6]: losses from risk are independent of each other; loss in one activity area does not necessarily increase the loss probability in another (with the exception of force majeure circumstances); maximum possible damage should not exceed the participant's financial capabilities.

Risk analysis aims to achieve the following main objectives:

1. formation at the person making the decision of a risks holistic picture that threaten the interests of the system under consideration;
2. ranking of risks by the influence degree on the organization activities and identi-fying among them the most dangerous;
3. comparison of alternative versions of projects and technologies;
4. creation of databases and knowledge bases for expert systems to the adoption of technical and other solutions;
5. justification of risk mitigation measures [7].

The risk area is determined by the general market losses zone, within the limits of which the losses do not exceed the limit value of the established risk level. Risk analysis can be divided into two mutually complementary types: qualitative and quantitative. Qualitative analysis is carried out using the expert assessments method. Quantitative risk analysis is necessary in order to assess how the most significant risk factors can affect the performance indicators. Qualitative analysis aims to identify (identify) factors, areas and types of risks. Quantitative risk analysis should provide an opportunity to numerically determine the size of individual risks and the risk of a manufacturing firm as a whole. To perform the quantitative analysis, a risk map is drawn up, which makes it possible to identify the most significant of them. A quanti-tative analysis of the identified risks is carried out by the sensitivity analysis method. Sensitivity analysis can be carried out using specialized software packages.

3   Results and Discussion

3.1   Designing an Intelligent Risk Management System

Design and implementation of the strategy for the development of the DSN is linked to risk assessment, which should ensure the adoption of scientifically sound decisions and minimize the all categories risks. For these purposes, a comprehensive assessment is needed, for which a Balanced Scorecard (BSC) can be used. BSC was developed by

David Norton and Robert Kaplan. It was based on the research of American companies, where the concept of tableau de board, developed in the 30s in France was applied [8].

The basic principle of BSC, which largely caused the high efficiency of this control technology is – "can only be managed by what can be measured" [9]. Thus, when developing a strategy to achieve the goals, a system of measured indicators is created on the basis of which decision-makers (DM) can choose the sequence of actions necessary to achieve success, as well as assess the correctness of the decisions made in this moment. Analyzing the risks of the company within the framework of the Norton–Kaplan model, it is possible to single out: resource-threats, process-threats and result-threats [10]. Such a scheme is obtained by modernizing the Norton–Kaplan model [8] and aligning it with the Robert Dilts [11] SCORE model describing the company's transition from the available state to the desired one in the time scale. When modernization, a chain of transitions is obtained: strategic resources → business processes → interaction with partners → financial implications (Fig. 2).

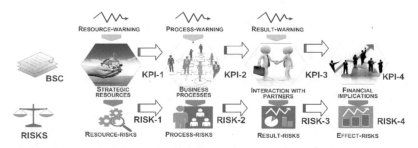

Fig. 2. Accordance of the BSC-card and risk card parameters

The standards system witch interconnected key performance indicator (KPI is associated with a system of risks, each KPI has its own risk, i.e. if a certain event occurs, the corresponding KPI may go beyond its KPI normative value.

The integrated algorithm to choose the sequence of actions for risk management consists of three main stages: risk analysis; making decisions; monitoring results (Fig. 3). Reduction of losses caused by an abnormal situation is possible due to: prevention (prophylaxis), monitoring (timely situation's recognition) and management of the critical situation (correct actions in case of its occurrence). Timely problem recognition often allows you to minimize losses or reduce them to zero.

The risk management system should ensure the implementation of three main objectives: risk identification, risk measurement, risk optimization (Fig. 4).

Risks identification is associated with the creation of a system of indicators and monitoring, which allows assessing the situation on the basis of statistical information analysis. The feedback mechanism implemented in such a system will allow estimating the probability of losses at each stage and correcting the actions depending on the changed external conditions.

Fig. 3. The scheme of choosing the approach to risk management

Fig. 4. Risk management flowchart

3.2 Accounting for Risks in Determining the Warranty Period

For companies engaged in the production of vehicles and their subsequent sale and warranty service, the definition of risk can be formulated as follows: Risk (R) is a value characterizing the probability and amount of possible losses as a result of exceeding the amount of costs for warranty repair of the transport fleet in relation to the reserve amount. The reserve amount (C_{WF}) – is the amount that is formed at the enterprise as a result of deductions of the established percentage from each sold vehicle and spent for covering warranty repairs. The reserve amount or "Enterprise's Warranty Fund" should be formed in such a way that the following conditions are met:

- the value of the enterprise's risk should be minimal, i.e. lie in a risk-free area;
- the warranty fund should be such that as little as possible of the enterprise's assets is in an illiquid state.

$$\left| \frac{R}{C_{WF}} \right\rangle min \qquad (1)$$

These two conditions contradict each other, since in order to reduce the risk, an increase in the guarantee fund is necessary, and with a reduction in the warranty fund, in turn, the risk of the enterprise increases. The following notation is introduced: R $(C_{WC} > C_{WF})$ - the enterprise's risk; C_{WF} - the enterprise's warranty fund for the warranty fleet; C_{WC} - the enterprise's warranty costs on the warranty fleet.

The company's guarantee fund is determined by the formula:

$$C_{WF} = C_\% \times C_V \times N_{WTF} \qquad (2)$$

where $C_\%$ – the percentage of deductions to the enterprise's warranty fund from the cost of the sold vehicle; C_V – average cost of a new vehicle; N_{WTF} – warranty vehicles fleet (the average number of vehicles is on warranty service). To determine the enterprise's warranty costs formula 3 is used, but it is necessary to take into account not only the cost of parts for restoration, but also the cost of eliminating the failure. Given the above, it was received:

$$C_{WF} = \sum_{i=1}^{m} \left(c_i \cdot f_i(i) + t_i^n \cdot c^n \right) \cdot N_{WTF} \tag{3}$$

where C_i - cost of the i-th type part, rub; $f_i(t)$ - function of the distribution density for mean operating time to failure of the i-th type parts; t_i^n standard labor input of elimination of failure, man-hours; $c^n = 500$ rub/man-hour - rate of norm-hour; m - number of part groups.

To company is vehicle-produces, the risk of exceeding the warranty costs relative to the reserve amount can be considered risk-free at a value within the range $R = 0 \div 0.2$. The graph of the dependence of the total costs for warranty repair of the vehicle from the run is constructed (Fig. 5) and the point corresponding to the maximum amount of deductions to the enterprise's warranty fund $C_\%$ is put. The point corresponding t'_{WTF} will be the maximum possible warranty operating time. In real conditions, the establishment of such a warranty period is impossible due to the high risk of exceeding the funds for warranty repairs in relation to the reserve amount, as well as a significant spread in the total costs of warranty repairs for vehicles operating in different conditions.

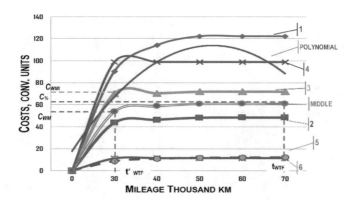

Fig. 5. Diagram of dependence of expenses for warranty repair of vehicles from operating time

C_{WF} - costs for warranty repairs; t - operating time; $C = f(t)$ - a function of the cost of warranty repair from operating time; $C_\%$ - deductions to the enterprise's warranty fund in % of the vehicle's cost; t'_{WTF} - the maximum possible warranty operating time; t_{WTF} - warranty operating time, corresponding to the risk value within the acceptable limits.

Therefore, it is necessary to determine a number of warranty costs' values C_i and build a function of the distribution density of costs for warranty repairs. As the values C_i can act data collected on vehicles operated within the same region, republic, country.

Further, it is necessary to apply a vertical line of costs corresponding to the amount of deductions to the enterprise's warranty fund $C_\%$. Thus the resulting curve area formed by lines $f(C)$ and $C_\%$ will correspond to the risk for a certain operating time and the established amount of deductions to the warranty fund.

To ensure the possibility of an operative assessment of the enterprise's risk under the warranty, a functional dependence of the risk's value on the warranty operating time and the amount of deductions to the enterprise's warranty fund in % of the vehicle's value was established.

The check was carried out using data on the failures of KAMAZ vehicles operated in the near and far abroad countries (in 28 countries). Since for each country purchases are carried out by lots for a certain work type (construction, agriculture, utilities, etc.), i. e. are operated under the analogous conditions, and the organization and quality of service are similar throughout the country, the average values for the country were used as the values for the warranty service costs.

$$P = 7,4 \times 10^{-12} \times t^2 + 0,76 \times C_\%^2 - 4,6 \times 10^{-7} \times C_\% \times t - 1,78 \times C_\% + 3,8 \times 10^{-7} \times t + 1 \quad (4)$$

where

t - operating time;

$C_\%$ - deductions to the enterprise's warranty fund in % of the vehicle's cost.

The graph of the risk surface is shown in Fig. 6a.

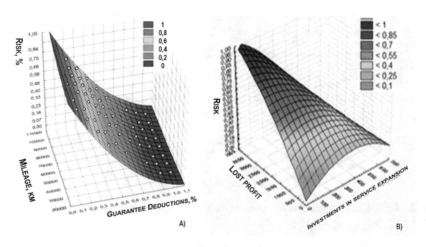

Fig. 6. Risk areas

3.3   Managerial Risks in the Corporate Service System

For DSC that sell cars and provide maintenance and repair services (M&RS), the risk can be formulated as follows: the risk (P) is the value characterizing the probability and amount of possible losses as a result of overloading or underloading of service center

[12–15]. The DSC risk which implementing M&RS will be measured by the lost profit from the customers loss due to insufficient DSN capacity and the risk of inefficiency of investments invested in the content of additional work positions. The lost profit of the enterprise from loss of clients will make Z_{check} th. rub. and is defined as the product of the number of lost car races by the average cost of one M&RS.

$$Z_{check} = L_p \times C_{MR} \tag{5}$$

where

L_p - lost profit DSC,
C_{MR} - average cost M&RS, conventional units (c.u.).

The vehicle owner's risk will be expressed in profit loss during his downtime in a queue to service and be determined based on cost and downtime in a queue:

$$Z_{queue} = t_{queue} \times C_{dt} \tag{6}$$

where t_{queue} - waiting time for the car in the service queue;
C_{dt} - the average rate of the cost of idle time of the car (only the wages of the driver are taken into account).

In Fig. 6b shows the enterprise's risk surface, which shows the balance between investing in the maintenance of redundant work positions and the lost profit from the client's going due to long waiting in the queue.

4 Conclusion

The article shows that reliability is currently one of the factors affecting the competitiveness of vehicles. Vehicles become highly reliable due to the integrated approach to all stages of their "life cycle": design, production and operation. The most important for the manufacturer is the warranty service, since this stage is characterized by an increased probability of failure, and the manufacturer takes responsibility for their elimination. The authors propose to take into account the risks of failures of various units, assemblies and systems of the car at the appointment of the warranty period. In addition, the risks that occur at all stages of the life cycle of a vehicle are considered. This study used statistical data on errors received from the dealer and the service network, as well as from the reliability department of the engine manufacturer. The results of the research show that using the proposed methodology will reduce the costs of the truck manufacturer.

References

1. Zadornova, Y.S., Zaytsev, S.A.: Risk management in the automotive industry. «Automotive industry», №2 2004. http://www.avtomash.ru/guravto/2004/20040201.htm
2. Kokcharov, I.: What is risk management? http://www.slideshare.net/igorkokcharov/what-is-project-risk-management
3. Makarova, I., Khabibullin, R., Belyaev, E., et al.: Improving the logistical processes in corporate service system. Transp. Probl. 11(1), 5–18 (2016)
4. Belyaev, E.I., Khabibullin, R.G., Makarova, I.V.: Optimization of the control system of spare parts delivery to improve the automotive engineering warranty service quality under the operating conditions abroad. In: 3rd Forum of Young Researchers: in the Framework of International Forum Education Quality, pp. 194–202 (2012)
5. Makarova, I., Khabibullin, R., Belyaev, A., et al.: Dealer-service center competitiveness increase using modern management methods. Transp. Probl. 7(2), 53–59 (2012)
6. Makarova, I., Khabibullin, R., Buyvol, P., et al.: System approach at risk management of the autoservice enterprise. Transp. Probl. 8(4), 5–16 (2013)
7. Khabibullin, R.G.: Development of methods to improve the effectiveness of the firm service trucks. Transp. Co. 8, 40–42 (2011)
8. McGivern, G., Fischer, M.D.: Reactivity and reactions to regulatory transparency in medicine, psychotherapy and counseling (PDF). Soc. Sci. Med. 74(3), 289–296 (2012)
9. Aguilar, F.: Scanning the Business Environment. Macmillan, New York (2006)
10. Committee Draft of ISO 31000 Risk management (PDF): International Organization for Standardization. 2007-06-15. Archived from the original (PDF) on 25 March 2009
11. DeLozier, J., Dilts, R.: Encyclopedia of Systemic Neuro-Linguistic Programming and NLP New Coding. NLP University Press, Santa Cruz (2000)
12. Buddhakulsomsiri, J., Siradeghyan, Y., Zakarian, A., et al.: Association rule-generation algorithm for mining automotive warranty data. Int. J. Product. Res. 44(14), 2749–2770 (2006)
13. Rai, B., Singh, N.: Forecasting warranty performance in the presence of the 'maturing data' phenomenon. Int. J. Syst. Sci. 36(7), 381–394 (2005)
14. Huang, H.-Z., Liu, Z.-J., Murthy, D.N.P.: Optimal reliability, warranty and price for new products. IIE Trans. 39, 819–827 (2007)
15. Xie, W., Liao, H., Zhu, X.: Estimation of gross profit for a new durable product considering warranty and post-warranty repairs. IIE Trans. 46, 87–105 (2014)

Rationalization of Retooling Process with Use of SMED and Simulation Tools

Joanna Kochańska$^{(\boxtimes)}$ and Anna Burduk

Faculty of Mechanical Engineering, Wroclaw University of Science and
Technology, Wybrzeze Wyspianskiego 27, 50-370 Wroclaw, Poland
{joanna.kochanska, anna.burduk}@pwr.edu.pl

Abstract. In the paper, the problem of limited flexibility of manufacturing companies is discussed. The problem consists in the necessity to manufacture in big lots with respect to long-lasting retooling of machines. This is illustrated on the example of an automotive enterprise. Implementation process of the SMED method (Single-Minute Exchange of Die) is presented. Potential improvements were verified with use of simulation models of basic and auxiliary processes, prepared in the program iGrafx Process for Six Sigma. Combination of the SMED method with simulation tools made it possible to develop a new, more effective standard of retooling. To estimate yearly profit, calculations were made with use of a spreadsheet.

Keywords: Production support processes · Retooling of machines
Single-Minute Exchange of Die (SMED) · Computer model
Process simulation · Functional relation diagram · Lean Manufacturing (LM)

1 Introduction

In spite of common knowledge about benefits coming from manufacture in small lots, many companies still manufacture their products in relatively big lots [1, 2]. The main reason is too long time required for retooling of machines. This big-lot manufacture suppresses to some extent the problem of losses related to frequent waiting for the machine to be prepared, but generates additional problems. First of all, this is demonstrated by decreased flexibility of production management, which results in a divergence between supply and demand [3, 4]. So, the problem of wastage of unused materials and unsold finished products increases. These goods present the tied-up capital and occupy the limited storage space. The analysed problem was noticed, among others, in a company being both manufacturer and distributor of components, belonging to an international automotive consortium. As a solution of the problem of manufacture in big lots due to long retooling times, the SMED method was applied.

This work was aimed at checking possibilities to improve the retooling process in the selected manufacturing company with use of the SMED method and simulation tools. Application of computer models and simulations makes it possible to analyse [5] and predict [6] courses of processes and to solve the posed tasks [7, 8], as well as to

© Springer Nature Switzerland AG 2019
Z. Wilimowska et al. (Eds.): ISAT 2018, AISC 854, pp. 303–312, 2019.
https://doi.org/10.1007/978-3-319-99993-7_27

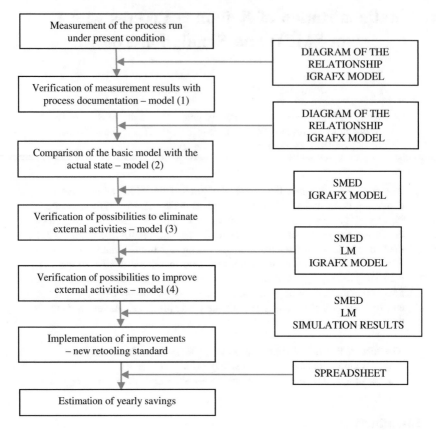

Fig. 1. Schematic presentation of implementation and verification of improvements in the retooling process.

bring to perfection [9, 10] potential rationalizations with no necessity to test them each time in practice. Figure 1 shows the plan of implementation and verification of the improvements.

Stages of the process were ordered with use of the functional relation diagram, performed measurements and observations. All simulation models were prepared in the program iGrafx Process for Six Sigma. The model iGrafx (1) is the initial model in that the process course was verified with documentation, with interruptions unconsidered. The second model iGrafx (2) reflects actual course of the process. It makes the base for testing improvements implemented by means of the SMED method and Lean Manu-facturing (LM) tools. The third model iGrafx (3) was built in order to exclude external activities. The fourth model iGrafx (4) served for checking possibilities to improve the process by rationalization of internal activities. Analysis of simulations made on the above-mentioned models made it possible to develop a new standard of retooling that was really implemented. Potential annual profits were estimated by calculations in a spreadsheet, using historical data.

2   SMED Method

The SMED method (Single-Minute Exchange of Die) is a set of techniques and tools aimed at reduction of retooling time down to less than ten minutes. The purpose is to reach such quick retoolings that manufacture in small lots becomes profitable. This will make it possible, first of all, to increase flexibility of responding to customers' demands and productivity thanks to extension of real work time at the expense of awaiting time and to minimisation of stocks [11, 12]. Even if experiences of various companies show that not all real processes can be reduced to ten minutes, application of SMED each time brings some profits.

The idea of SMED is based on proper management of retoolings depending on their type [11, 13, 14]. Distinguished are internal retoolings (requiring the machine to be stopped) and external retoolings (that can be performed during operation).

The SMED method belongs to a wide set of methods used in Lean Manufacturing (LM) [15], finding its application in various industries [16]. Its implementation is aimed not only at reduction of idle time of the machine during retooling, but also at the following:

- increase of productivity,
- elimination of rejects resulted from the necessity to manufacture a trial lot,
- improvement of the operators' safety,
- better organisation of the workspace,
- reduction of costs,
- reduction of requirements concerning skills and experiences of the employees,
- increase of flexibility of processes,
- elimination of errors,
- reduction of interoperational stocks,
- improvement of quality [11, 12, 17–19].

3   Implementation of SMED - Case Study

3.1   Problem of Long-Lasting Retoolings

It was decided to implement the SMED method in the company being a manufacturer and distributor of components, belonging to an international automotive consortium. Implementation was started from the department manufacturing ecological products. This new production was introduced as a result of global orientation towards environment protection. At first, the actions were focused on the production line that brings much larger incomes from sale of its products than any other line. However, its potential is inhibited just by too long retooling times. The retooling process on the analysed line is illustrated in Fig. 2 by a fragment of the functional relation diagram.

The analysed line is characterised by relatively variable production, since a dozen different products are manufactured during a year. The retooling process includes both physical replacement and adjustment of parts and instrumentation of the machine, but also operation of the control panel. Most of the actions require co-operation of the

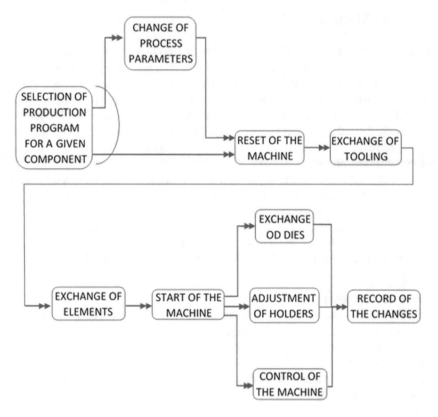

Fig. 2. Schematic presentation of retooling process – fragment of functional relation diagram.

employees. Retooling time depends on sizes of the produced elements. Average time ranges around forty minutes. It is estimated that, in average, almost four hundred of such retoolings are carried-out each year. This can be converted to over eleven three-shift work days used for retoolings only.

The acquired data inclined to verification of the obtained measurements with the process documentation. For the research, a model of the process was built in the program iGrafx Process for Six Sigma not considering the disturbing factors, i.e. the model (1). Results of the performed simulations showed that the retooling time should fluctuate about 30 min. For validation, a model of actual state was also built, i.e. the model (2) considering the occurring disturbances. The obtained results confirmed the problem of too much time-consuming retoolings of the machine. The developed model was accepted as the basis for verification of effectiveness of the suggested improvements with no necessity to implement them in practice each time.

3.2   Analysis of Retooling Actions

The SMED analysis was started from filming the retooling by foremen of each shift. The highest attention was paid to division of duties between the operators, to preparation of tools and materials, as well as to irregularities of the process. The employees were previously acquainted with basics and goal of the method. Next, analysis was carried-out in presence of the other operators involved in retooling and times were recorded for each activity. In the next stage, retooling was divided to internal and external ones with use of check lists. Identified were also wastage and risks. Results of the performed analyses are shown in Figs. 3, 4 and 5.

Fig. 3. Comparison of times of external and internal actions.

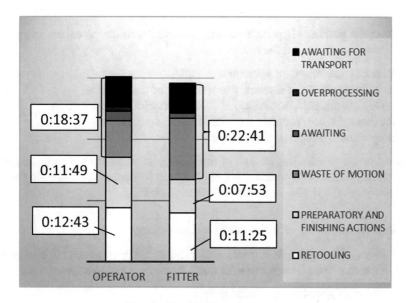

Fig. 4. Identification of wastage.

Analysis of the obtained results indicates that, at present, ca. 20 to 30% of retooling time is taken by activities potentially external. So, this time can be reduced by correct managing the groups of activities, i.e. by executing the actions potentially external in parallel with operation time of the machine.

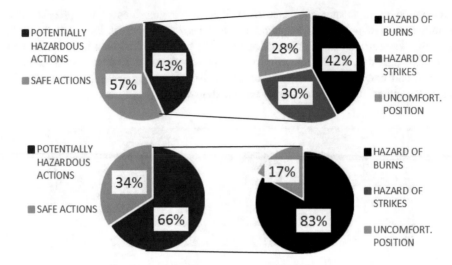

Fig. 5. (**a**) Potential risks for the operator during retooling. (**b**). Potential risks for the fitter during retooling.

Among the problems happening during observation of the processes, many irregularities were noted, related mainly to:

- unnecessary movement of operators,
- waiting for transport, components or readiness of the machine,
- relatively complicated installation of holders, requiring manual skills,
- application of non-ergonomic (or even dangerous) solutions,
- missing standardisation of the process.

Moreover, it appears that ca. half of the time spent on retooling can be dangerous for the workers. However, potential risk does not results from non-ergonomic positions during work only, but also from risks of burns or strikes.

3.3   Implementation and Verification of Improvements

After analysis of the actual state, transformation of retoolings was started, consisting in changing internal operations to external ones. Exemplary changes were to include:

- preparation of a tool truck outside the machine,
- preparation of parts for retooling – before retooling,
- removing parts from previous production – before retooling,
- feeding the machine by the logistics department at the beginning of retooling,
- preparation of parts for retooling – during manufacture of the last pieces,
- use of the truck for transporting the parts required for retooling.

In order to examine potential of the suggested improvements, a subsequent model (3) was built in the program iGrafx Process for Six Sigma. The model was based on the basic model (2) that was modified by excluding some actions outside the actual retooling. The performed simulations confirmed the possibility to reduce the actions classified as external ones. The estimated savings amounted to ca. 10 min for the average retooling time.

Next, the stage of wastage reduction in internal actions was started. Within the set of improvements, it was decided to:

- provide the work station with pneumatic drill-drivers with coloured tips corresponding to colours of bolts,
- use automatic drill-drivers,
- prepare tools, materials and safety means in a proper place on the work station,
- develop standards of manual setting of the actuator,
- develop an operation standard of the panel of automatic wheel rotation,
- install a cooling system for lamps inside the machine,
- visualise storage places for tools,
- develop a standard for wheel movement,
- visualise the ways of fitting dies,
- develop a standard for preparation and storage of dies,
- use magnets.

The suggested improvements were introduced to the computer model, free already of external actions, i.e. model (3). By designation of respective actions as inactive and by reduction of times of active parts, the model (4) was built. Results of the performed simulations showed a possibility to reduce execution time of external actions by the next 10 min.

On the grounds of the obtained results, a new plan of retooling was developed. A part of actions was planned for the machine operation time, a part before and a part after the actual retooling. A part of actions was simplified and new standards were introduced.

3.4 Result of Implementation of Improvements

As a result of implementation of the suggested changes in practice, it was possible to reduce retooling times by times of the external actions and to reduce wastage in the internal actions. This resulted in a reduction of total time of the analysed retooling, see Fig. 6.

Retooling time was reduced from 43:09 min to 23:53 min for the operator and from 41:59 min to 19:18 min for the fitter, which gives over 40% improvement. Together with the implemented improvements, a significant part of risk for the employees was eliminated. The gained working time was estimated to 120–149 h per year. With use of historical data, annual production capacity was estimated, see Fig. 7.

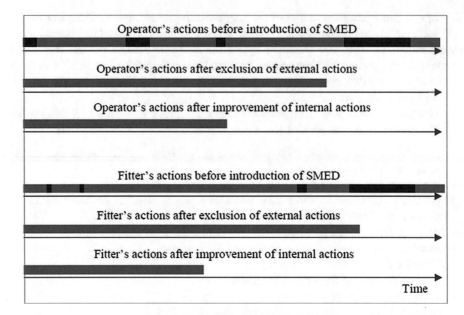

Fig. 6. Reduction of retooling times.

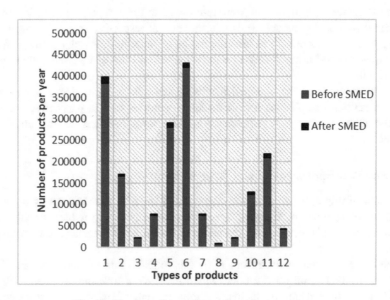

Fig. 7. Number of manufactured products per year.

Amount of annual production was calculated with use of the data concerning production and retooling times for individual products and their percentages in total production. It was estimated that reduction of the retooling time can significantly influence improvement of production. The saved time will make it possible to:

(a) increase productivity even by over 60,000 pieces, or
(b) increase the number of retoolings by even 40% to improve flexibility in responding the customers' needs.

4 Summary

The issue of time-consuming retoolings of machines was presented on the example of an automotive company. In order to improve the existing situation, the method of Single-Minute Exchange of Die (SMED) was implemented, based on simulation models prepared in the program iGrafx Process for Six Sigma. Combination of the SMED method with simulation tools made it possible to develop a new, more favourable retooling standard. The retooling process was improved mainly by execution of some actions "externally", i.e. during operation of the machine, and by elimination of the observed kinds of wastage. The introduced improvements influenced also some improvement of working conditions of the operators. Hazards at the workplace were reduced, more ergonomic solutions were applied and standardization was introduced. This resulted in over 40% reduction of retooling times. Apart from elongation of the operation time, flexibility of the company increased, which resulted in faster responses to customers' needs and improved delivery reliability index.

References

1. Liker, J.K.: The Toyota production system and art: making highly customized and creative products the Toyota way. Int. J. Prod. Res. **45**(16), 3681–3698 (2007)
2. Ohno, T.: Toyota Production System: Beyond Large-Scale Production. Productivity Press, New York (1988)
3. Cousens, A., Szwejczewski, M., Sweeney, M.: A process for managing manufacturing flexibility. Int. J. Oper. Product. Manag. **29**(4), 357–385 (2009)
4. Karwasz, A., Chabowski, P.: Productivity increase through reduced changeover time. J. Mach. Eng. **16**(2), 61–70 (2016)
5. Kłos, S., Patalas-Maliszewskal, J.: The topological impact of discrete manufacturing systems on the effectiveness of production processes. In: Rocha, Á., Correia, A., Adeli, H., Reis, L., Costanzo, S. (eds.) WorldCIST 2017. AISC, vol. 571, pp. 441–452. Springer, Cham (2017)
6. Sobaszek, Ł., Gola, A., Kozłowski, E.: Application of survival function in robust scheduling of production jobs. In: Ganzha, M., Maciaszek, M., Paprzycki, M. (eds.) Proceedings of the 2017 Federated Conference on Computer Science and Information Systems, vol. 11, pp. 575–578 (2017)

7. Kempa, W.M., Paprocka, I., Kalinowski, K., Grabowik, C., Krenczyk, D.: Study on transient queueing delay in a single-channel queueing model with setup and closedown times. In: Dregvaite, G., Damasevicius, R. (eds.) ICIST 2016. CCIS, vol. 639, pp. 464–475. Springer, Cham (2016)

8. Grzybowska, K., Kovács, G.: The modelling and design process of coordination mechanisms in the supply chain. J. Appl. Log. **24**, 25–38 (2017)

9. Górnicka, D., Markowski, M., Burduk, A.: Optimization of production organization in a packaging company by ant colony algorithm. In: Burduk, A., Mazurkiewicz, D. (eds.) ISPEM 2017. AISC, vol. 637, pp. 336–346. Springer, Cham (2018)

10. Kotowska, J., Markowski, M., Burduk, A.: Optimization of the supply of components for mass production with the use of the ant colony algorithm. In: Burduk, A., Mazurkiewicz, D. (eds.) ISPEM 2017, AISC, vol. 637, pp. 347–357. Springer, Cham (2018)

11. Moreira, A.C., Campos Silva Pais, G.: Single minute exchange of die. A case study implementation. J. Technol. Manag. Innov. **6**(1), 129–146 (2011)

12. Shingo, S.: A Revolution in Manufacturing: The SMED System. Productivity Press, New York (1985)

13. Almomani, M.A., Aladeemy, M.: A proposed approach for setup time reduction through integrating conventional SMED method with multiple criteria decision-making techniques. Comput. Ind. Eng. **66**(2), 461–469 (2013)

14. Bikram, J.S., Dinesh, K.: SMED: for quick changeovers in foundry SMEs. Int. J. Product. Perform. Manag. **59**(1), 98–116 (2009)

15. Hines, P., Holweg, M., Rich, N.: Learning to evolve: a review of contemporary lean thinking. Int. J. Oper. Product. Manag. **24**(10), 994–1011 (2004)

16. Allahverdi, A., Soroush, H.M.: The significance of reducing setup times/setup costs. Eur. J. Oper. Res. **187**(3), 978–984 (2008)

17. Cakmacki, M.: Process improvement: performance analysis of the setup time reduction-SMED in the automobile industry. Int. J. Adv. Manuf. Technol. **41**(1–2), 168–179 (2009)

18. Palanisamy, S., Siddiqui, S.: Changeover time reduction and productivity improvement by integrating conventional SMED method with implementation of MES for better production planning and control. Int. J. Innov. Res. Sci. Eng. Technol. **2**(12), 7961–7974 (2013)

19. Sherali, H.D., Goubergen, D.V., Landeghem, H.V.: A quantitative approach for scheduling activities to reduce set-up in multiple machine lines. Eur. J. Oper. Res. **187**(3), 1224–1237 (2008)

The Risk Model for Hazards Generated at Level Crossings

Anna Kobaszyńska-Twardowska$^{(\boxtimes)}$, Adam Kadziński,
Adrian Gill[iD], and Piotr Smoczyński[iD]

Poznan University of Technology,
Pl. M. Skłodowskiej-Curie 5, 60-965 Poznań, Poland
anna.kobaszynska-twardowska@put.poznan.pl

Abstract. The current operating conditions as well as the development of road and railway transport cause a significant number of dangerous events to be generated on the interfaces of the road and railway systems. This particularly applies to level crossings, as their large quantity (in Poland there are about 14 000 of them) additionally affects the scale of the problem. Such a state requires taking formal actions in the field of risk management in these analyses domains. One of such activities is the development and application of an appropriate risk model for estimating and evaluating the risk of hazards. The analysis of the literature on this issue indicates that in other countries outside Poland, many such models have been developed and applied. This work presents a risk model adequate to the functioning conditions of level crossings in Poland. For its development, generic risk model based on many criteria and integrating the resulting components of the overall risk metrics was used. As part of the developed model, an original set of criteria for estimating the risk of hazards was established and described in detail. The processes of preparing the components of the overall risk metrics that result from the established risk analysis criteria were discussed. Among others, experts' knowledge, National Reference Value and National Investigation Body reports were used. The final result of the work is the presentation of a formal risk model which meets the requirements for risk estimation and evaluation in the entities of the European Union railway system. The user interface of a program which is a computer implementation of the risk model algorithm was presented. According to the authors of the paper, the developed model is consistent with the knowledge, intuition and experience of people responsible for safety management at level crossings.

Keywords: Risk · Risk model · Risk estimation · Level crossings

1 Introduction

With the current rapid increase in the number of road vehicles (over 28 million as of 2017) and the large number of adverse events with their participation (almost 33 thousand events with injured people or fatalities in 2017), the issues of safety engineering "on the roads" acquire an important meaning. The critical elements of the railway infrastructure are the places where road and railway transport meet – level

© Springer Nature Switzerland AG 2019
Z. Wilimowska et al. (Eds.): ISAT 2018, AISC 854, pp. 313–322, 2019.
https://doi.org/10.1007/978-3-319-99993-7_28

crossings. The level of safety at level crossings is influenced i.a. by nonconformities in the operation of individual road system components, such as [1]:

- incorrect behavior of road users
- technical failures
- bad meteorological conditions
- bad natural conditions
- acts of vandalism and terrorism.

The main problems affecting the safety level at level crossings are [1]:

- railway and road infrastructure that is not adapted to the safety standards
- low safety culture
- implemented safety management system.

Railway infrastructure managers (over 93% of railway infrastructure in Poland is managed by PKP PLK S.A.) are obliged to obtain a "safety authorization" confirming the fact of developing a safety management system (SMS). It includes organizational structures, planning, responsibility, rules of conduct, processes, procedures and resources needed to develop, implement, monitor and maintain the safety policy declared by the organization [2]. Proper management of railway infrastructure (especially of level crossings and viaducts) significantly affects the safety of railway and road transport.

In Poland, there are 14 000 level crossings on 19 000 km of railway lines. In these places, adverse events occur and cause losses especially for road users. According to [3], costs incurred by entities operating in the Polish railway system, due to adverse events at level crossings, amount to approx. PLN 5 million per year. The query of available scientific resources showed that in Poland a risk model dedicated to level crossings, which could be used to estimate and evaluate the risk of hazards identified in these domains, was not developed.

In other countries, outside from Poland, many models (often referred to as approaches) related to risk management at level crossings have been developed. A review of 23 different models (approaches) used in 13 countries around the world was carried out. Using the division proposed in the study [4], it is possible to make the following breakdown of these models:

- parametric models (P) using the description of basic properties of the crossing to identify elements of the safety system, such as road and railway traffic density, number of tracks, maximum permissible speed, etc. Such models are used, among others, in Japan, Russia, Sweden and Estonia
- qualitative models (J) based on the concept of risk of hazards and taking into account the different impact of individual components (related to the probability and consequences of adverse events) on the final risk value. The result obtained when using such models is often the "goodness" of the level crossing, i.e. a single value defining the method of selecting the elements of the safety system. Developed i.a. in Australia and Great Britain
- statistical models (S) which take into account the results of statistical analysis of data from the past. This approach is represented by the United States and Canada.

This paper presents the risk model of hazards generated at level crossings.

2 Materials and Methods

In this paper, the risk of hazard is understood as the combination of the level of possibility (probability) of hazard activation in form of an adverse event and/or the level of its effects or consequences [5]. Following types of events are considered adverse (undesirable) within the railway system [6]:

- accident – unintentional sudden event or sequence of such events involving railway vehicles, causing negative consequences for human health, property or the environment; include: collisions, derailments, level crossing events, events involving persons caused by railway vehicles in motion and fires of railway vehicles
- serious accident – an event that appears in the form of collision, derailment or other negative consequences, in which at least one person is killed or five people are seriously injured or losses caused by destruction of the railway vehicle, railway infrastructure or the environment were estimated at least EUR 2 million
- operational difficulties – events that do not meet the criteria for serious accidents, accidents and incidents
- railway incident – any event other than a serious accident and accident, but related to the movement of trains and affecting their safety.

To elaborate the risk model of hazards generated at level crossings, the generic risk model developed and presented in the paper [7] was used:

1. Identification of a finite set of identified hazards in a specific domain:

$$Z = \{z_1, z_2, \ldots, z_n\} \tag{1}$$

2. Determination of risk metrics as a value of function:

$$R : Z \to V \subset \mathbf{R} \tag{2}$$

3. Determination of the number of risk analysis criteria K_i ($i = 1, 2, \ldots, m$). Determination of the risk component according to criteria that will take values from the formula:

$$\begin{aligned} r_i : X_i^R \to \Omega_i^R; \; r_i(z_k) = \omega_{i,j}, \; \omega_{i,j} \in \Omega_i^R, \\ i = 1, 2, \ldots, m; \; j = 1, 2, \ldots, s_i; \; k = 1, 2, \ldots, nZ. \end{aligned} \tag{3}$$

where:
m – the number of risk analysis criteria,
n – the number of identified hazards,
$r_i(z_k)$ – the risk component of the hazard z_k according to the i-th risk analysis criterion,
s_i – the position of the highest risk level within the i-th risk analysis criterion,
z_k ($k = 1,2,\ldots,n$) – k-th hazard from the set of identified hazards,
X_i^R ($i = 1, 2, \ldots, m$) – sets of values of hazard characteristics used during risk analysis according to subsequent criteria,

Ω_i^R ($i = 1, 2, \ldots, m$) – sets of risk level metrics used in the risk analysis according to subsequent criteria,

ω_{ij} ($j = 1, 2, \ldots, s_i$) – risk level metrics indicated during risk analysis according to the i-th criteria.

Overall risk R of hazard z_k ($k = 1, 2, \ldots, n$) (after determining levels of all risk components):

$$R(z_k) = f_1(r_1(z_k), r_2(z_k), \ldots, r_m(z_k)), \quad k = 1, 2, \ldots, n \tag{4}$$

4. Determining the weights of the analysis criteria:

$$A = \{a_1, a_2, \ldots, a_m\} \tag{5}$$

where m is the number of risk analysis criteria, a_i ($i = 1, 2, \ldots, m$) is the weight of the i-th risk analysis criterion.

The weights of a set of risk analysis criteria can be included in risk model using f_2 function as follows:

$$R(z_k) = f_2(a_i, r_i(z_k)), \quad i = 1, 2, \ldots, m, \quad k = 1, 2, \ldots, n \tag{6}$$

Function f_2 in the Eq. (6) may be determined in the form of a mathematical function, tabularly, with a graph, verbally or in some other way.

3   Results

The risk estimation is preceded by the choice of the risk model and risk metrics models for hazards identified at level crossings. On the basis of the generic risk model presented in the previous section, for the analyses domain of level crossings (denoted as PK from Polish "przejazd kolejowy"–"level crossing"), the set of hazards takes the form [8]:

$$Z_{PK} = \{z_1, z_2, \ldots, z_6\}. \tag{7}$$

Risk model of each of the hazards from set Z_{PK} is a function of the components r_i (z_k) ($i = 1, 2, \ldots, m, k = 1, 2, \ldots, n$). Decisions are made on the basis of an analysis according to six criteria K_i ($i = 1, 2, \ldots, 6$) and values a_i ($i = 1, 2, \ldots, 6$) determining the weights of these risk analysis criteria forming the set:

$$A = \{a_1, a_2, \ldots, a_6\}. \tag{8}$$

Six criteria with the following names and meanings were adopted in the risk model for level crossings:

K_1 – **criterion of safety level for railway level crossing users ($CST_{3.1}$).** Risk component metrics $r_1(z_k)$ according to this criterion is indicated depending on the size of the safety level indicator ($CST_{3.1}$):

- low, when CST3.1 $\leq$ 138
- medium, when 138 < CST3.1 < 277
- high, when $CST_{3.1} \geq$ 277.

The safety level indicator for level crossing users is expressed as follows:

$$CST_{3.1} = \frac{L_{Z-ZN}}{L_{Pkm}},\qquad(9)$$

where $CST_{3.1}$ is the safety level indicator for level crossings users, L_{Z-ZN} is the number of fatalities and serious injuries in adverse events at level crossings during the year, L_{Pkm} is the number of train-kilometres expressed in millions.

K_2 – **criterion of coverage range of losses at level crossings.** The criterion takes into account the type of material losses that may be the result of the activation of the hazard. Losses concern the following sub-areas (parts of the analyses domain of a level crossing): railway and road infrastructure (sub-area 1), railway and road vehicles (sub-area 2), natural environment (sub-area 3), people (sub-area 4). Risk component metrics $r_2(z_k)$ according to this criterion is indicated by the rule:

- low, when losses occurred only in one of the sub-areas
- medium, when losses occurred in two or three sub-areas
- high, when losses occurred in all four sub-areas.

K_3 – **criterion of material losses as a result of adverse events at level crossings.** Risk component metrics $r_3(z_k)$ according to this criterion depends on the amount of material losses (S_{TC} – total cost), which arose after the activation of the hazard expressed in PLN:

- low, when STC $\leq$ 50 million
- medium, when 50 million $<$ STC $<$ 61 million
- high, when $S_{TC} \geq$ 61 million.

K_4 – **criterion of losses resulting from the timespan of traffic interruption at the level crossing.** Risk component metrics $r_4(z_k)$ according to this criterion depends on the timespan of interruption in the traffic at the level crossing and its access roads according to the following rule:

- low, when the interruption in road and railway traffic is shorter or equal to 1 h,
- medium, when the interruption in railway or road traffic is longer than one hour,
- high, when the interruption in railway and road traffic is longer than one hour.

K_5 – **criterion of the history of hazard activation.** The criterion is based on the assertion that if the hazard has once activated, it can activate once again. Risk component metrics $r_5(z_k)$ is indicated depending on the number of hazard activations in the 5 years preceding the analysis:

- low, when the hazard has not been activated even once
- medium, when the hazard was activated no more than 4 times
- high, when the hazard has been activated more than 4 times.

K_6 – **criterion of possibility of hazard activation.** The criterion depends on the structure, configuration and location of the level crossing. Risk component metrics $r_6(z_k)$ of this criterion is determined based on four elements ($K_{6.1}$, $K_{6.2}$, $K_{6.3}$ and $K_{6.4}$) that characterize the level crossing.

Element $K_{6.1}$ – *category of level crossings*. This element of the risk component $r_6(z_k)$ indicates the possibility of hazard activation depending on the level crossing category (after taking into account the elements of the safety system applied to it):

- low, if the level crossing belongs to category A or B (safety devices across the road)
- medium, when the level crossing belongs to category C (traffic lights)
- high, when the level crossing belongs to category D (warning signs).

Element $K_{6.2}$ – *number of tracks at the level crossing*. This element of the risk component $r_6(z_k)$ allows to make the activation of the hazard conditional on the length of the level crossing (affecting the timespan when the road vehicles are leaving the so-called danger zone) according to the following rule:

- low, for level crossings with three or more tracks
- medium, for two-track level crossings
- high, for single-track level crossings.

Element $K_{6.3}$ – *density of road and railway traffic*. This element of the risk component $r_6(z_k)$ allows to make the activation of the hazard conditional on the product of road and railway traffic density (IR):

- low, when IR $\leq$ 60 000
- medium, when 60 000 < IR < 150 000
- high, when IR $\geq$ 150 000.

Element $K_{6.4}$ – *organization of road traffic outside of the level crossing*. The fourth element of the risk component $r_6(z_k)$ makes the possibility of hazard activation conditional on the location of the level crossing according to the following rule:

- low, when the access road to the level crossing gives no right of way for road vehicles,
- medium, when a road intersection with traffic lights is located before or after the level crossing at a distance of less than 200 m,
- high, when level crossings are located on the same way < 200 m from each other.

Risk component metrics $r_6(z_k)$ is determined according to the rule:

- low, when a maximum of one element of criterion K6 was assessed as medium,
- medium, when a maximum of two elements of K6 were assessed as medium,
- high, when more than one element of criterion K_6 was assessed as *high*.

Risk component metrics $r_i(z_k)$ for each of the six criteria of the risk model assume levels from the set:

$$\Omega = \{\text{low, medium, high}\}. \tag{10}$$

Elements of the set Ω (Eq. 4) of risk components metrics are assigned to a set of risk metrics values. Thus, the result of risk estimation for each hazard from the set Z_{PK}

(Eq. 1), according to the criterion K_i ($i = 1, 2, \ldots, 6$), is the level of the risk component $r_i(z_k)$ from the set of risk metrics values.

The function allowing to estimate the total risk metrics (taking into account the results of risk estimation according to six criteria and the weights of risk analysis criteria) takes the form:

$$R_{\mathrm{PK}}(z_k) = \sum_{i=1}^{6} a_i \cdot r_i(z_k); \quad i = 1, 2, \ldots, 6; \quad k = 1, 2, \ldots, n. \tag{11}$$

After Estimating the Risk, Its Evaluation Should be Made. Procedures for evaluation of the risk of hazards identified at level crossings consist in checking (by evaluating, by comparing) to which category (class) of risk (acceptable, tolerable, unacceptable) the estimated risk belongs.

Based on the mathematical risk model its computer implementation was generated. The view of the user interface of this program is shown in Fig. 1.

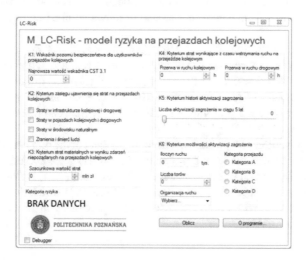

Fig. 1. The computer implementation of the risk model at level crossings – the view of the user interface in the phase before data entry

4   Discussion

Some of the criteria used in the risk model require further short comments/justifications.

Comments on criterion K_1. The limit values of $CST_{3.1}$ were determined based on the National Reference Value (NRV). The NRV shall be obtained in accordance with the procedure laid down in the decision of the European Commission [10]. This procedure applies to the adoption of a Common Safety Method for determining whether the safety requirements referred to in [11] are met. The maximum NRV for Poland is

277 [12]. On this basis, 277 was accepted as the limit for determining the criterion metrics as high. For example, in 2015, the level of NRV amounted to 93% for the value of 252,937, which indicates the need to take action in these areas.

Comments on criterion K_3. Limit values of the risk component metrics were determined based on data from the Polish Supreme Audit Office reports [13] according to which adverse events on the railway network in Poland generate (on average) costs: in the natural environment – PLN 300 000, in railway vehicles and infrastructure – PLN 19 million, costs of delays – PLN 500 000, costs related to fatalities and serious injuries multiplied by the Value of Preventing a Casualty (VPC) – PLN 166 million. It was assumed that the VPC for fatality equals to 341 000€, and for a seriously injured person – 46 500€ [13]. Events at level crossings account for about one third of all events on the railway network [14], that is why one third of all costs (about 56 million) increased by the costs of destruction of road vehicles (about 200 vehicles per year with an average value of 20 000) were adopted as a limit for determining the risk component.

Comments on criterion K_4. Analyzing accident reports [15] and press articles [16–18] it was noticed that some events at level crossings do not cause the traffic to be interrupted or the interruption timespan is short. For this reason, the risk components of this criterion are adopted in relation to one hour of road and/or railway traffic interruption at the level crossing.

Comments on criterion K_5. The number of hazard activations was adopted according to the risk analysis method proposed by the Federal Emergency Management Agency (FEMA) and described in the paper [19]. For the needs of the model, a 5-year period was assumed (FEMA assumed a 10-year period) due to the obligation of determining at maximum 5-year intervals the product of road and railway traffic density on level crossings.

Comments on criterion K_6. Limit values of the product of the number of road vehicles and trains going through the level crossing during the day were adopted in accordance with the Polish Regulation of the Minister of Infrastructure and Development [20].

Comments on element $K_{6.4}$ of the criterion K_6. In cases in which the organization of railway and road traffic does not correspond to the previously mentioned options, the element should be omitted.

5   Summary and Conclusions

Risk management of hazards generated in the analyses domains related to systems and elements of the railway and road system allows to take into account errors in the operation of these systems and/or the inability of their elements.

An important element of risk management procedures is the risk model. It is a function of components resulting from separate decisions made on the basis of analyses carried out according to many criteria. Each of the risk analysis criteria must be such that the component resulting from the decision under this criterion belongs to the group of components expressing the possibility of activating the hazard or the amount of potential losses arising after its activation in an adverse event. In addition to the levels of risk components, risk models should include measures of the weights of the analysis criteria.

Meeting social expectations related to the functioning of level crossings as elements of railway and road infrastructure depends largely on the effectiveness of risk management processes carried out in these areas of human activity. Based on the risk management requirements of the European Union's railway and road system, this paper proposes a novel approach to risk estimation and evaluation, implemented in the risk model (M_LC-Risk) of hazards generated at level crossings.

References

1. Jamroz, K., Szymanek, A.: Zintegrowane zarządzanie ryzykiem w systemie bezpieczeństwa ruchu drogowego. Międzynarodowe seminarium Gambit (2010). http://docplayer.pl/ 1887341-Zarzadzanie-ryzykiem-w-systemie-bezpieczenstwa.html
2. Dyrektywa Parlamentu Europejskiego i Rady 2004/49/WE z dnia 29 kwietnia 2004 r. w sprawie bezpieczeństwa kolei wspólnotowych oraz zmieniająca dyrektywę Rady 95/18/WE w sprawie przyznawania licencji przedsiębiorstwom kolejowym, oraz dyrektywę 2001/14/WE w sprawie alokacji zdolności przepustowej infrastruktury kolejowej i pobierania opłat za użytkowanie infrastruktury kolejowej oraz certyfikację w zakresie bezpieczeństwa (Dz. U. UE. L. 164/44 z 2004 r. ze zm.)
3. Cieślakowski, S.: Wybrane zagadnienia z zakresu bezpieczeństwa transportu. LogForum **5** (3), 1–7 (2009)
4. Little, D.A.: Use of Risk Models and Risk Assessments for Level Crossings by Other Railways. Final Report to Rail Safety and Standards Board, March 2007. s3.amazonaws.com
5. Kadziński, A., Gill, A.: Integracja pojęć (podrozdział 7.3.2). In: praca zbiorowa Krystek, R. (ed.) Zintegrowany system bezpieczeństwa transportu. Uwarunkowania rozwoju integracji systemów bezpieczeństwa transportu, tom 2, pp. 285–288. Politechnika Gdańska, WKŁ, Warszawa (2009)
6. Ustawa o transporcie kolejowym, Dz. U. z 2003 r., nr 86, poz. 789 Ustawa z dnia 28 marca 2003
7. Kadziński, A.: Studium wybranych aspektów niezawodności systemów oraz obiektów pojazdów szynowych. Wyd. Politechniki Poznańskiej, seria Rozprawy, nr 511, Poznań 2013
8. Kobaszyńska-Twardowska, A.: Zarządzanie ryzykiem zagrożeń na przejazdach kolejowo, Rozprawa doktorska. Politechnika Poznańska, Poznań (2017)
9. Kadziński, A., Kobaszyńska-Twardowska, A., Gill, A.: The concept of method and models for risk management of hazards generated at railway crossings. In: materiały International Conference Transport Means, pp. 297–302 (2016)
10. Decyzja Komisji z dnia 25 stycznia 2012 r. w sprawie technicznej specyfikacji interoperacyjności w zakresie podsystemów "Sterowanie" transeuropejskiego systemu kolei. Dziennik Urzędowy Unii Europejskiej L51/1
11. Bezpieczeństwo ruchu kolejowego w Polsce: Informacja o wynikach kontroli w 2012 r., Najwyższa Izba Kontroli, Warszawa (2013). www.nik.gov.pl. Accessed May 2014
12. Będkowski-Kozioł, M.: Status i zadania Europejskiej Agencji Kolejowej – stan obecny i perspektywy w świetle projektu IV pakietu kolejowego UE. Internetowy Kwartalnik Antymonopolowy i Regulacyjny **7**(3), 36–56 (2014)
13. NIK, Bezpieczeństwo ruchu kolejowego w Polsce, Informacja o wynikach kontroli (2012). www.nik.gov.pl. Accessed May 2014
14. Raport UTK: Ocena funkcjonowania rynku transportu kolejowego i stanu bezpieczeństwa kolejowego w 2012. www.utk.gov.pl/pl/analizy-i-monitoring/oceny-roczne/2012/3515. Accessed May 2014

15. Materiały PKP PLK S.A. Oddział Poznań – niepublikowane
16. www.dzienniklodzki.pl/artykul/887666
17. www.forum.gazeta.pl
18. www.ratownictwo.opole.pl
19. Grocki, R.: Przedsięwzięcia organizacyjne Wojewódzkiego Inspektoratu Obrony Cywilnej we Wrocławiu w zakresie bezpieczeństwa transportu i eksploatacji niebezpiecznych substancji chemicznych. In: Analiza ryzyka w transporcie i przemyśle, Młyńczak, M. (ed.) Oficyna Wyd. Politechniki Wrocławskiej, Wrocław, pp. 127–131 (1997)
20. Rozporządzenie Ministra Infrastruktury i Rozwoju Ministra z dnia 20 października 2015 r. poz. 1744, w sprawie warunków technicznych, jakim powinny odpowiadać skrzyżowania linii kolejowych oraz bocznic kolejowych z drogami i ich usytuowanie z dnia 20 października 2015

Financial Valuation of Production Diversification in the Steel Industry Using Real Option Theory

Bogdan Rębiasz, Bartłomiej Gaweł[⊠], and Iwona Skalna

AGH University of Science and Technology, Kraków, Poland
bgawel@zarz.agh.edu.pl

Abstract. Steel industry is subject to significant volatility in its output prices and market demands for different ranges of products. Therefore, it is common practice to invest in various assets, which gives the opportunity to diversify production and generate valuable switch options. This article investigates the incremental benefit of product switch options in steel plant projects. The options are valued using Monte Carlo simulation and modeling the prices of and demand for steel products as Geometric Brownian Motion (*GBM*). Our results show that the product switch option can generate a significant increase in the net present value (*NPV*) of metallurgical projects.

Keywords: Real options · Switch options · Stochastic processes
Investment decision · Monte Carlo simulation

1 Introduction

In steel industry where utilization of assets is very close to its capacity and fixed costs play significant role, the volatility of prices or product demand are main risk factors. To reduce the impact of prices and demand unpredictability, steel companies choose output product switching strategy [11].

The switch option is important because in last decades versatility of steel products has greatly increased. This causes that distinct sectors of economy demand for different steel products; and therefore, variation of product prices, even though correlated, vary [12]. However, it is difficult to valuate the output product switching strategy by using traditional methodologies such as discounted cash flow [1, 2, 5, 7, 10, 11, 13, 14]. Instead, it can be valuated as a real option. In this paper, we demonstrate how to value the switching strategy in a steel company, in the presence of many tangible investment alternatives.

The valuation of switch options starts with a cash flow simulation, in which various product mixes are tested in diverse scenarios describing product demand and prices, taking into consideration capital expenditures that may be incurred in adapting the production process. The behaviour of prices and demands is modelled as Geometric Brownian Motion.

The rest of the paper is organized as follows. In Sect. 2, a review of the bibliography on real-option valuation of projects is made. Section 3 describes the model and

© Springer Nature Switzerland AG 2019
Z. Wilimowska et al. (Eds.): ISAT 2018, AISC 854, pp. 323–332, 2019.
https://doi.org/10.1007/978-3-319-99993-7_29

methodology of valuing a product switch option. A case study is presented in Sect. 4. Section 5 contains some concluding remarks. The proposed here approach extends our results presented in [13]. Readers interested in more detailed description of the case study are encouraged to read that article.

2 Evaluation of Switch Option Value

The origin of real options is based on the idea that real assets (projects) could be evaluated in a similar way as financial options. Real Options, in contrary to *NPV*, allow a project to be modified by subsequent decisions once it has been undertaken. There are many types of options. For example, in mineral industry the most research focus on defer and abandon options. In this article we analyse the case of a metallurgical plant that model its managerial flexibility as a switch option.

Among the research carried out on switch options, it is worth mentioning the work of Bastian-Pinto et al. [1], where the option to switch final product (sugar or ethanol) in an ethanol production plant is valued. The authors calculate the value of the option that maximizes revenue at each future period taking into account variability of prices and demand. Prices are modelled using correlated mean reverting processes and options are valuated using simulation. Based on an analysis of this research, it can be stated that the option of switching products has a similar effect in the metallurgical industry. However, the specificity of the metallurgical industry must be taken into account. Ozorio *et al.* [11] value the output switch option in a hypothetical integrated steel plant composed of a blast furnace and a hot laminator. They assume the variability of prices and constant demand. The variability of prices is modelled using Geometric Brownian Motion and alternatively using the mean reverting process. The authors state that there is no consensus as to which stochastic process is more appropriate. They support the thesis of [6] that the definition of the process depends as much on statistical as on theoretical considerations. Results show that such the considered option can generate a significant increase in the *NPV* of an integrated steel plant. Other articles that are worth mentioning [2–4] concern the metallurgical industry. In particular, in [2] the problem of buying electricity through long term contracts or alternatively owning a generation unit and selling its electricity in the spot market in an aluminium processing plant was considered.

3 Model and Methodology

This section introduces the model and methodology for evaluating the product switch option. Firstly, the modelling of uncertainty using correlated Geometric Brownian Motion (GBM) is described. Next, model for estimating the value of product switch option is presented.

3.1 Modeling Uncertainty by Correlated Fuzzy Geometric Brownian Motion (GBM)

Geometric Brownian Motion is a special case of Brownian motion or Wiener process. Variable q follows *GBM* if it satisfies the following diffusion equation [1, 9, 15]:

$$dq_t = mq_t dt + \sigma q_t dw_t, \tag{1}$$

where $dw_t = \varepsilon_t \sqrt{dt}$ is the standard increment of Wiener process, and μ and σ are, respectively, the drift parameter and the standard deviation parameter. Parameter ε_t has identical independent standard normal distribution.

Let Δ be the given time interval between two observations q_t and q_{t-1}. Based on Eq. (1), we can write the formula for prediction q_t [9]

$$q_t = q_{t-1} \exp\left[\left(\mu - \frac{\sigma^2}{2}\right)\Delta + \varepsilon_t \sigma\sqrt{\Delta}\right]. \tag{2}$$

When estimating *NPV*, many correlated uncertain variables are usually encountered. Suppose that ε_{it}; $i = 1, 2, \ldots, I$, are correlated and uncertain. In order to take into account the correlation between them, they should be replaced with the set of correlated variables η_{it}; $i = 1, 2, \ldots, I$. Correlated values η_{it} can be derived based on values ε_{it} by using, e.g., Cholesky decomposition of the correlation matrix (cf. [16]). The equations for correlated uncertain values can thus be written as follows.

$$q_{ti} = q_{ti-1} \exp\left[\left(\mu_i - \frac{\sigma_i^2}{2}\right)\Delta + \eta_{ti}\sigma_i\sqrt{\Delta}\right] \tag{3}$$

Equation (3) can be used to estimate static *NPV*. When estimating the product switch option value, a Monte Carlo simulation must be carried out under the assumption that the uncertain variables involved follow a risk-neutral *GBM*. In this case, the formula

$$q_{ti} = q_{ti-1} \exp\left[\left(\mu_i - \pi - \frac{\sigma_i^2}{2}\right)\Delta + \eta_{ti}\sigma_i\sqrt{\Delta}\right] \tag{4}$$

should be used instead of the formula (3) [1, 10]. Risk-premium (π) estimation can be done as described in Hull [8].

3.2 Definition of Problem

This section presents the model for valuating product switch options in a hypothetical production setup (see Fig. 1). We analyze the effectiveness of the following two projects: construction of a new organic-coated (*OC*) sheet plant (*project 1*), and construction of a new welded tube (*WT*) plant (*project 2*).

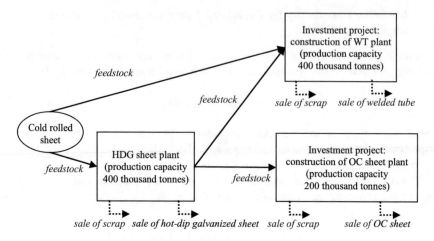

Fig. 1. Analyzed production setup.

OC sheet plant can produce *OC* sheets made from hot-dip galvanized *(HDG)* sheets with greater added value. *WT* plant can produce welded tubes and hollow sections *(WT–HS)*. The feedstock for *WT* plant is 50% *HDG* sheets and 50% cold-rolled *(CR)* sheets. For the analyzed production setup, *CR* sheets are the basic feedstock. They are partly converted into *HDG* sheets and partly into *WT–HS*. *HDG* sheets are partly sold and partly converted into *OC* sheets and *WT–HS*. The *OC* sheets and *WT–HS* are all sold. Steel scrap, i.e., a waste in the production of *HDG* sheets, *WT–HS* and *OC* sheets, is sold.

We start our analysis from the basic case, which consists of the standard valuation of a cash flow from which a static *NPV* (Eq. (5)) of the projected construction of a new *OC* sheet plant, *WT* plant separately and jointly *OC* sheet plant and *WT* plant is obtained. The annual cash flows, used to calculate the static *NPV*, obtained by investing in each variant, can be estimated from Eqs. (6)–(22):

$$NPV_s = \frac{\sum_{t=0}^{T} ICF_{st}}{(1 + r_{ris})^t} - I \qquad (5)$$

$$ICF_{st} = (ZN_{st} + DA_s + ZKO_{st}) - (ZN_{bt} + DAb + ZKO_{bt})\ (t = 0, 1, \ldots, T-1) \qquad (6)$$

$$ICF_{st} = (ZN_{st} + DA_s + ZKO_{st} + RV_{st}) - (ZN_{bt} + DAb + ZKO_{bt} + RV_{bt})\ (t = T) \qquad (7)$$

Sales of the HDG sheet in the variant when neither project is implemented:

$$SFh_t = ACh_t \times MSh\ (t = 0, 1, \ldots, T) \qquad (8)$$

$$SRh_{(1)t} = \min(SFh_t;\ CAPh)\ (t = 0, 1, \ldots, T) \qquad (9)$$

Sales of the *HDG* and *OC* sheets in the variant when construction of a new *OC* sheet plant is realized:

$$SFo_t = ACo_t \times MSo \quad (t = 0, 1, \ldots, T) \tag{10}$$

$$SRo_t = \min(SFo_t; CAPo) \quad (t = 0, 1, \ldots, T) \tag{11}$$

$$SRh_{(2)t} = \min(SFh_{bt}; (CAPh - Mo \times SRo_t)) \quad (t = 0, 1, \ldots, T) \tag{12}$$

Sales of HDG sheet and WT-HS in the variant when project construction of a new WT plant is realized:

$$SFo_t = ACo_t \times MSo \quad (t = 0, 1, \ldots, T) \tag{13}$$

$$SRo_t = \min(SFo_t; CAPo) \quad (t = 0, 1, \ldots, T) \tag{14}$$

$$SFw_t = ACw_t \times MSw \quad (t = 0, 1, \ldots, T) \tag{15}$$

$$SRw_t = \min(SFw_t; CAPw) \quad (t = 0, 1, \ldots, T) \tag{16}$$

$$SRh_{(3)t} = \min\left(SFh_{bt}; (CAPh - Mw \times SRw_t)\right) \quad (t = 0, 1, \ldots, T) \tag{17}$$

Sales of the *HDG* sheet, *OC* sheet and *WT-HS* in the variant when both projects are implemented:

$$SFo_t = ACo_t \times MSo \quad (t = 0, 1, \ldots, T) \tag{18}$$

$$SRo_t = \min(SFo_t; CAPo) \quad (t = 0, 1, \ldots, T) \tag{19}$$

$$SFw_t = ACw_t \times MSw \quad (t = 0, 1, \ldots, T) \tag{20}$$

$$SRw_t = \min(SFw_t; CAPw) \quad (t = 0, 1, \ldots, T) \tag{21}$$

$$SRh_{(4)t} = \min(SFh_{bt}; (CAPh - Mo \times SRo_t - Mw \times SRw_t)) \quad (t = 0, 1, \ldots, T) \tag{22}$$

where:

ICF_{st} – cash flow in year t for the project analyzed portfolio of projects,
r_{ris} – weighted average cost of capital,
I – capital expenditure on the project,
ZN_{st} – net profit in year t in the scenario where portfolio of projects is implemented,
ZN_{bt} – net profit in year t in the scenario where portfolio of projects is not implemented,
K_{st} – total cost in year t in the scenario where the project is implemented,
K_{bt} – total cost in year t in the scenario where the project is not implemented,
$CAPh$ – installed capacity of the *HDG* sheet plant,
$CAPo$ – installed capacity of the *OC* sheet plant,
$CAPw$ – installed capacity of the *WT* plant,

SFh_t – sales forecast for HDG sheets in year t,

SFo_t – sales forecast for OC sheets in year t,

SFw_t – sales forecast for WT in year t,

MSh – market share for HDG sheets,

MSo – market share for OC sheets,

MSw – market share for WT,

ACh_t – forecasted apparent consumption of HDG sheets in year t,

ACo_t – forecasted apparent consumption of OC sheets in year t,

ACw_t – forecasted apparent consumption of WT in year t,

SRo_t – sale of OC sheets in year t in the scenario where the project is implemented,

$SRh_{(1)t}$ – sale of HDG sheets in year t where the project is not implemented,

$SRh_{(2)t}$ – sale of HDG sheets in year t where the project OC is implemented,

$SRh_{(3)t}$ – sale of HDG sheets in year t where the project WT-HS is implemented,

$SRh_{(4)t}$ – sale of HDG sheets in year t in the scenario where the projects OC and WT-HS is implemented,

Mc – per unit consumption of CR sheet when producing HDG sheet,

Mw – per unit consumption of CR sheet when producing WT,

Mh – per unit consumption of HDG sheet when producing OC sheet,

$OPCh$ – other (with the exception of the cost of CR sheet) annual variable production costs per ton for HDG sheet,

$OPCo$ – other (with the exception of the cost of CR sheet) annual variable production costs per ton for OC sheet,

$OPCw$ – other (with the exception of the cost of CR sheet and HGD sheet) annual variable production costs per ton for WT,

DAh – annual amortization for HDG plant,

DAo – annual amortization for OC plant,

DAw – annual amortization for WT plant,

ZKO_{st} – change in net working capital in year t in the scenario where the project is implemented,

ZKO_{bt} – change in net working capital in year t in the scenario where the project is not implemented,

RV_{st} – residual value in year t in the scenario where the project is implemented,

RV_{bt} – residual value in year t in the scenario where the project is not implemented.

The static NPV is calculated N times using a Monte Carlo simulation according to Eq. (5). The result gives an estimate of the probability distribution of the static NPV. The GBM stochastic process defined by Eq. (3) is used to define the underlying uncertainty. The uncertainty of demand for HDG sheets, OC sheets, WT-HS and the uncertainty of prices for CR sheets, HDG sheets, OC sheets, WT-HS and scrap is taken into account. We also take into account the correlations between the prices of the following products: steel scrap, CR sheets, HDG sheets, WT-HS and OC sheets; and the correlation between the apparent consumption of HDG sheets and OC sheets and WT-HS.

Next, the switch option is valued. The analysis of the static NPV of the project does not take into account the managerial flexibility of being able to switch the output product. In some periods, the production of HDG sheet may be a more interesting and profitable alternative to the company than the production of OC sheet and WT–HS.

In such cases the product switch option is realized. When switch option is realized, the sales of *HDG* sheet, *OC* sheets and *WT–HS* are optimized to obtain the maximum cash flow. In the optimization process constraints (23–28) are considered:

$$SRh_{(5)t} \leq ACh_t \times MSh \ (t = 0, 1, \ldots, T) \tag{23}$$

$$SRh_{(5)t} + Mo \times SRo_t + Mw \times SRw_t \leq CAPh \ (t = 0, 1, \ldots, T) \tag{24}$$

$$SRo_t \leq ACo_t \times MSo \ (t = 0, 1, \ldots, T) \tag{25}$$

$$SRo_t \leq CAPo \ (t = 0, 1, \ldots, T) \tag{26}$$

$$SRw_t \leq ACw_t \times MSw \ (t = 0, 1, \ldots, T) \tag{27}$$

$$SRw_t \leq CAPw \ (t = 0, 1, \ldots, T) \tag{28}$$

To ensure that sale of products is switched to the most profitable structure, the values $SRh_{(5)t}$, SR_{ot}, SR_{wt} which give the highest cash flow value are sought. Here $SRh_{(5)t}$ define sale of *HDG* sheets realized in year t in the scenario where the project *OC* and *WT-HS* is implemented and switch option is realized.

The values of product switch options can be obtained by simulating the incremental cash flows defined for the level of sales of *OC* sheets, *WT–HS* and *HDG* sheets discussed above in relation to the cash flow defined according to the conditions assumed in the calculation of the static *NPV*. The following equations are used to compute the value of the switch option:

$$OPT_0 = \sum_{t=1}^{T} \frac{ICF_t^*}{\left(1 + r_f\right)^t}, \tag{29}$$

where

OPT_0 – value of the switch option at time 0,
r_f – risk free rate,
ICF_t^* – incremental cash flow related to the product switch option in year t,

Nevertheless, both the simulations and discount rate must assume risk neutrality, since when valuing options, the level of risk will change when these options are exercised. Thus, we must use the risk-free rate for discounting the incremental cash flows when the option is exercised, but these must be simulated using a risk-neutral expectation. Therefore, the *GBM* stochastic process defined by Eq. (4) is used here to model the underlying uncertainty.

4 Computational Results

The quantities used in the calculations, i.e.,

- drift parameter (μ), volatility parameter (σ), premiums (π) for the price and apparent consumption for analyzed products,
- correlation matrix of prices of analyzed products,
- correlation matrix of apparent consumption of analyzed products,
  were defined based on historical data (1996–2016).

Figures 2 and 3 show, respectively, the probability distributions of the static *NPV* for each combination of alternatives and the probability distribution of the value of a product switch option for each combination of alternatives. Probability distributions and switch option distributions are estimated with kernel density.

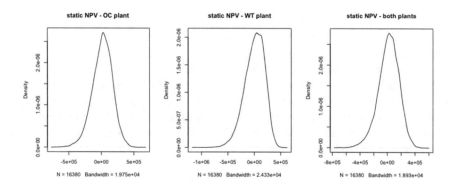

Fig. 2. Static *NPV* probability for each investment combinations.

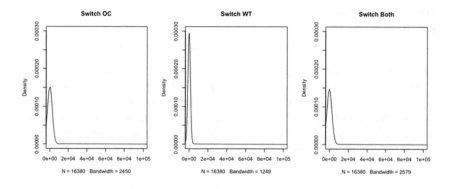

Fig. 3. Switch option distributions for each investment combination.

Table 1 summarizes the values found for the projects when a product switch option is considered. As can be seen, the value of a product switch option represents over 30% of the static *NPV*. Portfolios of real options typically interact such that the value of the whole (in our case 5 196 thousand USD) differs from the sum of the separate parts (in our case 5 718 thousands USD). This problem is called as deviations from value additivity.

Table 1. Value of product switch option and static *NPV* for analyzed projects

Description	WT plant	OC sheet plant	Both plants
Static NPV of investments, thousands USD	4 290	13 034	16 254
Extended NPV of investments, thousands USD	5 814	17 228	21 550
Product switch option value, thousands USD	1 523	4 194	5 196
Increase in comparison to static NPV (%)	35.5%	32.2%	32.0%

5 Conclusions

In this paper we have estimated the projected cash flows, which are dependent on fluctuations in demand and prices. We have also calculated the value of the option available to metallurgical industry of delivering the product that maximizes revenue in each future period. Using a *GBM* approach, we have concluded that this option significantly increases the *NPV* of an analyzed investment and considerably enhances the value created by metallurgical projects for shareholders. Mentioned above the problem of deviations from value additivity has not been adequately covered in the real options literature, whereas it is of high relevance both for academia and in practical applications. Modeling the interdependencies of options in the process of defining real options portfolios will be the subject of our further research.

Acknowledgements. Wydanie publikacji finansowane przez Akademię Górniczo-Hutniczą im. Stanisława Staszica w Krakowie (dotacja podmiotowa na utrzymanie potencjału badawczego).

References

1. Bastian-Pinto, C., Brandão, T., Hahn, W.J.: Flexibility as a source of value in the production of alternative fuels: the ethanol case. Energy Econ. **31**(3), 411–422 (2009)
2. Bastian-Pinto, C., Eduardo, L., Brandão, T., Raphael, R., Ozorio, L.M.: Financial valuation of operational flexibilities in the aluminum industry using real option theory. In: Proceedings of the 17th Annual International Conference Real Options: Theory Meets Practice, Tokyo, pp. 1–23 (2013)
3. Byko, M.: TMS plenary symposium: energy reduction in the aluminum industry. JOM J. Miner. Metals Mater. Soc. **54**(5), 39–40 (2002)
4. Das, S., Long, W., Hayden, H., Green, J., Hunt, W.: Energy implications of the changing world of aluminum metal supply. JOM J. Miner. Metals Mater. Soc. **56**(8), 14–17 (2004)

5. Dimitrakopoulos, R., Sabour, S.A.A.: Evaluating mine plans under uncertainty: can the real options make a difference? Resour. Policy **32**(3), 116–125 (2007)
6. Dixit, A.K., Pindyck, R.S.: Investment Under Uncertainty. Princeton University Press, Princeton (1994)
7. Gligoric, Z., Kricak, L., Beljic, C., Lutovac, S., Milojevic, J.: Evaluation of underground zinc mine investment based on fuzzy-interval grey system theory and geometric Brownian motion. J. Appl. Math. **2014**, 1–12 (2014)
8. Hull, J.C.: Options, Futures and Other Derivatives Securities, 6th edn. Prentice Hall, Englewood Cliffs (2006)
9. Marathe, R., Ryan, S.M.: On the validity of the geometric brownian motion assumption. Eng. Econ. **50**(2), 1–40 (2005)
10. Muharam, F.M.: Assessing risk for strategy formulation in steel industry through real options analysis. J. Global Strateg. Manag. **5**(2), 5–15 (2011)
11. Ozorio, L.M., Bastian-Pinto, C., Baidya, T.K.N., Brandão, T.: Investment decision in integrated steel plants under uncertainty. Int. Rev. Financ. Anal. **27**(3), 55–64 (2013)
12. Rebiasz, B.: Polish steel consumption, 1974–2008. Resour. Policy **31**(1), 37–49 (2006)
13. Rębiasz, B., Gaweł, B., Skalna, I.: Valuing managerial flexibility: an application of real-option theory to steel industry investments. Oper. Res. Decis. **27**, 91–110 (2017)
14. Samis, M., Davis, G.A.: Using Monte Carlo simulation with DCF and real options risk pricing techniques to analyse a mine financing proposal. Int. J. Financ. Eng. Risk Manag. **1**(3), 264–281 (2014)
15. Wattanarat, V., Phimphavong, P., Matsumaru, M.: Demand and price forecasting models for strategic and planning decisions in a supply chain. Proc. Schl. ITE Tokai Univ. **3**(2), 37–42 (2010)
16. Yang, T.I.: Simulation based estimation for correlated cost elements. Int. J. Project Manag. **23**(4), 275–282 (2005)

Some Aspects of Project Management

Behavioural Aspects of Communication in the Management Process of IT Projects

Kazimierz Frączkowski$^{(\boxtimes)}$, Barbara Gładysz, Hanna Mazur,
and Ewa Prałat

Faculty of Computer Science and Management,
Wroclaw University of Science and Technology, Wrocław, Poland
{kazimierz.fraczkowski, barbara.gladysz, hanna.mazur,
ewa.pralat}@pwr.edu.pl

Abstract. The paper presents the results of research on the significance of communication in project teams, especially its behavioral aspects influencing the success or failure of an Information Technology Project (IT project). The holistic approach to the issue of communication, apart from the commonly indicated success and failure factors of such projects, also encompasses such issues as team work culture, its size, the achieved confidence level and mutual kindness. The work presents the current state of knowledge and the results of own research on impact factors in communication quality in IT projects. The authors own work comprised the formulation of research hypotheses related to selected factors influencing IT projects in various economy sectors and next the preparation and conduction of surveys (among project managers, leaders and team members both on the side of service providers and customers) and also proving the presented hypotheses.

Keywords: Communication · Confidence trust · IT project management
Project success · Project team

1 Introduction

Inappropriate communication has led to serious problems or even the complete failure of numerous ventures. One of the earliest examples was described in the Bible "Come, let us build ourselves a city, with a tower that reaches to the heavens, so that we may make a name for ourselves; otherwise we will be scattered over the face of the whole earth" [1]. According to the Old Testament the Tower of Babel was built by all the people of the Earth – describing this endeavor in contemporary terms in project categories, one can propose the hypothesis that despite the availability of probably the best resources and technologies of those days, communication became the reason for the failure of the whole project. The sentence "Come, let us go down and confuse their language so they will not understand each other" is the first signal of the impact communication may have on the final result of any venture. Regardless of numerous interpretations of the above quotation, one thing does not raise any doubt, namely that communication is a decisive factor in the progress of civilization and project implementation.

© Springer Nature Switzerland AG 2019
Z. Wilimowska et al. (Eds.): ISAT 2018, AISC 854, pp. 335–347, 2019.
https://doi.org/10.1007/978-3-319-99993-7_30

Communication (from Latin *communicatio* – exchange, connectedness, conversation) is a very wide notion and one can find numerous definitions of this word in literature. According to Dance [2], who presented ninety five definitions of the word communication, the notion is the result of the generalization of mental operations.

Communication management is a very complex process and only a holistic approach to this issue allows to take into consideration the factors which influence the efficiency of IT projects. Numerous authors indicate the significance of communication, yet there is no extensive research enabling the identification of communication components, such as, among others, kindness and trust. These elements constitute an added value in the process of building the intellectual capital of teams allowing them to successfully implement IT projects [3].

Due to various interpretations of the notion of success, the definition adopted for the needs of the conducted research encompasses basic success attributes (factors), such as the achievement of the adopted goal in the planned timeframe line, within the allocated budget, in the planned scope and preserving appropriate quality parameters [4, 5]. Project success can be evaluated subjectively form the perspective of various stakeholders (e.g. contractors, recipients and clients) or measurable project attributes [6].

2   Communication and Its Impact on Project Success

Communication can be understood as the process of exchanging information between the members of any project team [7] or, more extensively, between the project team and project stakeholders [8]. Good communication is frequently indicated as a critical success factor in publications presenting literature surveys on project management [9–11]. Authors emphasize the significance of the careful planning and effective management of communication in achieving project success [8, 12, 13].

The significance of appropriate communication between a project team and its stakeholders was confirmed, among others, in the research which positively verified the dependence between project success and the quality of communication management with stakeholders [14]. Other authors [15], in their research related to project management according to the PMBOK methodology, also indicate that team communication is a key factor in the achievement of project goals.

Solutions such as emails and face to face meetings (F2F) lead to the development of more satisfactory relations and facilitate the flow of knowledge [16]. From the perspective of communication analysis in traditional and virtual project teams, very interesting results were obtained in the research conducted on 150 specialists working in IT departments. It showed that poor communication is a risk factor for projects done by both traditional and virtual project teams. Additionally it was stated that an insufficient glow of knowledge has a significantly bigger influence on project failure in distributed (virtual) teams and such increased risk should be taken into consideration in risk management plans [17]. On the other hand Leśniak [18] discusses problems occurring in the cooperation of virtual and multicultural project teams operating according to the Scrum methodology. The author emphasizes the need to understand differences related to culture and communication methods among team members coming from very different environments. The issues connected with the influence of cultural differences and the lack of direct communication on communication quality in virtual teams are also discussed by Levin and Rad [19], while Blenke [20] presents results confirming the significant influence of face to face meetings on the success of projects done by virtual teams.

Some publications discuss also trust between parties involved in projects [21, 22]. Trust is understood, among others, as: openness and honesty in conflict situations, assistance in taking the most important decisions, readiness to offer support [23], and also as confidence in the competence, integration and kindness of the people involved in a given group, open communication and knowledge sharing [24]. Trust was indicated as one of the most important factors whose significance starts at the first stage of project when the underlying assumptions, expectations and requirements are established. It is listed as the third most important factor influencing teamwork in both single- and multicultural teams [25], and also a necessary condition for the efficient cooperation in a virtual team [26]. The dependence between trust shared by tam members and their communication quality was confirmed in [27]. Mutual trust in a project team and the feeling of confidence that the other team members do their tasks well are also listed as three critical success factors in relational partnerships. These research results indicate that there is an essential dependence between the level of trust and the quality of communication, and between the level of trust and partner risk management [28]. Trust is also considered in the context of its influence on project success thanks to its contribution to the development of well operating teams – it is an essential factor in the minimization of project failure risk [29]. The empirical proof of the role of communication in building the atmosphere of trust and the will to share knowledge in IT projects is presented by Park and Lee [23]. On the basis of the analysis of data obtained from 135 project teams, the authors indicate the existence of a connection between communication frequency and trust level among team members and their will to share knowledge. Moreover, it was ascertained that there was a dependence between better project results and the trust between project owners and contractors [22]. The researchers also developed a model showing the influence of trust o communication, and hence also on project results [30].

Another issue discussed in the literature on project management is kindness. It is considered one of the elements which act as an incentive to contribute to mutual work [31] and, as leader's characteristic it can enhance their impact on others [32]. Kindness is also indicated as an important factor in shaping teamwork climate in culturally differentiated teams [25]. Trust is usually enhanced thanks to F2F contacts as they create conditions for the development of good communication. Communication itself is the process of creating and delivering messages and later their explicit interpretation, which has a significant influence on the risk level connected with the accuracy of arrangements, e.g. stakeholders' expectations and requirements. Hence the research on the influence of trust and kindness on communication and conditions supporting their development is justified.

The authors of numerous packages focus on the analysis dependencies between such factors such as budget, project size, deadlines and methodology, however, there are no quantitative results related to the significance of such aspects as trust and kindness for correct communication and, in consequence, for project success. The attention of the research team was also attracted project failure risk factors, such as lack of trust and kindness in project teams. These factors had not been identified and their influence on project success had not been assessed in earlier surveys [33]. There are not any references to such results in the Standish Group Reports, e.g. in [34] encompassing the research conducted in 2014.

Numerous literature sources refer to communication in teams involved projects pertaining to areas other than IT [21, 28, 30]. It should be noted that IT projects have their specificity – frequently these are unique and innovative projects, increasingly frequently large, complex and costly and, thus requiring a good knowledge of state of the art teleinformatic technologies and solutions, programming languages and tools, and also an extensive knowledge of this field. IT projects affected by a high level of uncertainty due to long implementation time, dependence on external environment, changing legal regulations and political conditions on which project managers have very limited influence. Frequently they require very good contact with the future users of their systems, knowledge sharing, gaining users' confidence and building a climate of kindness.

IT projects are a special type of ventures. Most authors explain their unsatisfactory level of success with their "specificity". One of such special characteristics, frequently undermined in IT projects, is the necessity to cooperate on an equal basis with all stakeholders in a project team, and the way leading to this is building trust and developing kindness in a continuously improved communication process. The present paper investigates the influence of communication on such attributes as communication channels and frequency, building kindness and trust, and the size of a project team on the success or failure of IT projects.

3 Research Methodology and Results

The project "Research on Factors Critical for the Success or Failure of Polish IT Projects 2015–2017" was conducted thanks to a grant awarded by the Polish Information Processing Society (pol. Polskie Towarzystwo Informatyczne – PTI) whose goal was the improvement of project management on the basis of the experience gained in IT projects in Poland. The formulated research hypotheses were verified by: development an appropriate questionnaire; recruitment of respondents; evaluation of obtained results; statistical analysis of the dependencies between selected factors influencing communication in a project team and project success; interpretation of obtained results forming a basis for the rejection or confirmation of the formulated hypotheses.

The research was conducted in the years 2016–2017 using a questionnaire posted on an open access internet platform www.interankiety.pl/ankieta.

Analysis encompassed responses given by 125 respondents, each of which offered a complete description of one IT project by answering 60 questions in open or closed categories, with one or a few answers, simple (one question) or matrix questions, obligatory or optional ones and, thus providing 496 information units on project attributes. The respondents performed various roles in the project: 84% of them were contractors (50% – project managers, 34% - team members), 23% represented the client, (13% – managers, 10% – team members). The responses were related to IT projects in various sectors – the largest number from state administration (17%), healthcare (15%), finance and social insurance (14%), education and scientific research (8%). The most commonly used project management methodology was Scrum (41%) and Prince2 (30%).

The rate of respondents who considered their project successful was 28%, while 43% classified them as project influenced by some problems (however, generally considered quite successful), 10% classified their projects as not successful, 3% as rather unsuccessful, and 16% did not have any opinion on whether the project was successful or not. In the group of 35 definitely successful projects 60% of projects were done by teams of 10 people (17% – 5-person teams, 43% in 6–10-person teams), while 23% in 11–25-person teams. In the group of projects considered by respondents as rather successful 50% of the projects were done by teams of not more than 10 people (22% – in 5-person teams, 28% in 6–10-person teams), 31% in teams of 11–25 people. In response to the question about the reason for problems during the project 57% of respondents indicated unclear requirements, 48% communication, 30% working style, 30% a lack of experience in project implementation, 26% changes in the project team, and 14% localization. The results of survey formed a basis for the analysis of the collected material with regards to the verification of the formulated research hypotheses.

4 Research Hypotheses and Their Verification

The research goal was the identification of communication components influencing project success. It was assumed that the quality of communication may be influenced, among others, by the size of project team, used communication channel, communication frequency, fluctuations in the team and psychological factors essential for team structure, such as trust and kindness. Some of these factors, mainly trust and kindness, are shaped in the maturing process of the project team. In teams in which there were no problems with communication, as many as many as 84% of projects were successful and only 16% failed.

Unclear requirements are one of the most important factors contributing to failures indicated by 58% of respondents. A vital element not only during the pre-project analysis, but also at the requirement formulation stage and the whole project work, is planning and creating favourable conditions for communication which is shaped in the maturing process of the project team.

4.1 Factors Causing Problems in Realization of IT Projects

The comparison of the project fractions in which particular factors hindered project work showed that there were two homogeneous classes of these factors (Table 1).

A two-population fraction test was used. The first class (of factors which are most frequently indicated by respondents) encompasses: unclear requirements and communication which occurred in about 50% of the projects. The other class are such factors as: style of work, lack of experience of team members, high rotation among contractors and the localization of team members. They occurred in 13.6% to 29.6% of projects.

Table 1. Homogeneous classes of factors causing problems in project work.

Factor	Subset for alpha = 0.05	
	1	2
Unclear requirements	0.568	
Communication	0.480	
Style of work		0.296
Lack of experience in project implementation		0.296
Changes in team composition (high rotation)		0.264
Localization		0.136

Fractions for groups in homogeneous subsets are displayed.
Sample size = 125. Sign. level 0.05.

Due to the fact that communication and unclear requirements are the most frequent problems occurring in project implementation, further research focused on examining whether they were factors significantly influencing project success/failure. The work presents only these hypotheses which indicated the factors significantly influencing communication effectiveness and IT project success/failure.

The following research hypotheses have been verified:

H1. Project success depends on clearly formulated requirements.
H2. Project success depends on the quality of team communication.

The verification of the so formulated hypotheses was conducted for the projects which were either clearly successful or unsuccessful. Hypotheses were verified with independence test χ^2. There was no basis to reject hypotheses H1 ($\chi^2 = 4.607$, sign. level = 0.036), H2 ($\chi^2 = 4.253$, sign. level p = 0.046). Hence, project success/failure depends on clearly formulated requirements and on problems occurring in communication.

Unclear requirements are the reason for the failure of 66.7% projects. However, in the group of successful projects unclear requirements were a problem in only 31.4% projects. In 58.3% of failed projects the indicated reason for problems in project implementation were communication problems. Simultaneously, in only 25.7% of successful projects, communication was mentioned as one of the problems occurring during project work. In projects in which the frequency of meetings with clients was sufficient, as many as 87.5% of projects were successful. Therefore, aside unclear requirements, communication is an essential factor IT project success/failure depends on.

4.2 Efficiency of Communication Channels in IT Project

In the analysis of the frequency of using particular communication channels in project management, five homogenous subsets of communication channels were distinguished, they are presented in Table 2.

The most frequently used communication channels were e-mail and direct contacts. They were used in at least 80% of projects. Phone calls were another commonly used

Table 2. Homogenous subsets of communication channels distinguished with regards to the frequency of use.

Communication channel	Subset for alpha = 0.05				
	1	2	3	4	5
E-mail	0.872				
F2F meetings	0.800	0.800			
Phone calls		0.728			
Video calls			0.512		
Internet communicators			0.504		
Official instructions/memos				0.248	
Text messages				0.160	
Social media networks					0.048

Fractions for groups in homogeneous subsets are displayed.
Sample size = 125. Sign. level 0.05.

channel (72.8% of projects). In about 50% of projects contractors communicated using video calls and various internet communicators. Official instructions (memos) and text messages were used in about 20% of projects. Communication over social media networks was the least frequently used one, in only 5% of projects.

The analysis encompassed also the assessment of the efficiency of particular communication channels in IT projects. The following research hypothesis was verified:

H3. The efficiency of particular communication channels is varied.

The hypothesis was verified with variance analysis. There was no basis to reject hypothesis H3 (F = 6.6748, sign. level p = 0). The efficiency of particular channels was varied. The conducted analysis seems to confirm preferences indicated in Table 2, however, it is also important which channel plays the main role in the process of efficient communication being the basis for project success.

Yet another hypothesis was verified:

H4. F2F contacts are characterized by higher communication efficiency.

In this case Tukey's HDS test was used. As a result, five homogenous subsets were distinguished as presented in Table 3. The highest efficiency – 4.6 (on a scale of 1–5) characterized F2F contacts, while the lowest (about 2.3) was attributed by the respondents to social media networks and text messages.

In their evaluation of communication channel efficiency, most frequently the respondents indicated F2F contacts: 74% stated that undoubtedly it was the most efficient channel, while 18% assessed it as probably the most efficient one, which altogether gives 92% positive answers. The second most efficient communication channel was the phone call (27% and 50% responses, respectively). The other communication channels, such as text messages, email, official instructions, social media networks, communicators and chats received fewer votes.

Table 3. Efficiency of communication channels in IT projects.

Tukey HSD	Subset for alpha = 0.05				
Communication channel	1	2	3	4	5
F2F meetings	4.592				
E-mail		3.920			
Phone calls		3.888			
Video calls		3.688	3.688		
Internet communicators			3.384		
Official instructions/memos				2.952	
Text messages					2.376
Social media networks					2.264
Sig. level	1	0.712	0.366	1	0.993

Means for groups in homogeneous subsets are displayed. Sample
Size = 125.

4.3   Communication versus the principles of project management

In project management methodologies [31, 34, 36], the issue of communication as the
main principle (foundation) is not raised. Our research indicates that the formulated
hypothesis was correct:

H5. Project success depends on the frequency of meetings with a client.

The hypothesis was verified with Fisher's exact test. There was no basis to reject
hypothesis H5 (F = 35.388, sign. level = 0). Therefore, project success/failure depends
on the frequency of meetings with the client.

In IT projects one should pay special attention to the frequency of contacts in the
project team in which the user plays a very important role. It is confirmed by the fact
that, as many as 87.5% of projects in which the frequency of contacts with the client
was sufficient were successful.

4.4   Trust and kindness in a project team

The investigation of the trust and kindness aspects formed the basis for the verification
of the following research hypotheses.

**H6. IT projects in which project team members show kindness to one another,
less frequently experience communication problems.**

The hypothesis was verified with Fisher's exact test. There was no basis to reject
hypothesis H6 (F = 12,789, sign. level = 0.004). The projects in which team members
show kindness to one another, communication problems occurred in only 18% of the
projects, and in 82% no such problems occurred.

H7. IT projects in which team members trust one another, communication problems are less frequent.

The hypothesis was verified with Fisher's exact test. There was no basis to reject hypothesis H7 (F = 7.122, sign. level = 0.072). The projects in which team members trust one another, communication problems occurred in only 12% of the projects, and in 88% no communication problems were indicated.

H8. IT projects in which team members show kindness to one another have better chances of success.

To verify hypothesis H8, the significance of the correlation coefficient of mutual kindness in project team and project success was investigated. The Spearman correlation coefficient is a positive value (r = 0.362) and is significantly different from zero (one-tailed sign. level p = 0). Therefore, the higher the kindness level among project team members, the higher the chances of project success.

H9. IT projects in which team members trust one another have better chances of success.

To verify hypothesis H9, the significance of the correlation coefficient of mutual trust in project team and project success was investigated. The Spearman correlation coefficient is a positive value (r = 0.383) and is significantly different from zero (one-tailed sign. level = 0). Therefore, the higher the trust level among project team members, the higher the chances of project success.

4.5   Project team size versus project success

For the purpose of this analysis, four categories of project team size depending on the number of team members: 1–10, 11–50, 51–100, more than 100 people, were adopted.
The following research hypothesis was verified:

H10. There is a dependence between project team size and project success – it is easier to achieve success in smaller teams.

To verify hypothesis H10, the significance of the correlation coefficient between project team size and project success was investigated. The Spearman correlation coefficient is a negative value (r = −0.187) and is significantly different from zero (one-tailed sign. level p = 0.037). Therefore, the bigger the project team, and hence also the IT project, the higher the probability of project failure. Only 19% of the projects with more than 50 people were successful (definitely successful), whereas there were 30% of such projects in smaller teams.

5   Discussion and Results

The earlier investigation of Polish IT projects, conducted between April and June 2010 and called pmresearch.pl [33] showed that 21% of projects were successful. The results of the present research indicate that in the last few years there has been some improvement in the successful project rate and now 28% of projects are successful. The results are concurrent with the Standish Group Reports [8, 13, 34].

The collected research material allowed to verify the formulated research hypotheses and to form the following theses:

T1. Project success depends on clearly formulated requirements.
T2. Project success depends on the quality of team communication.
T3. The efficiency of particular communication channels is varied.
T4. F2F contacts show the highest communication efficiency.
T5. Project success depends on the frequency of meetings with a client.
T6. In IT projects in which team members show kindness to one another, communication problems occur less frequently.
T7. In IT projects in which team members trust one another, communication problems occur less frequently.
T8. IT projects in which team members show kindness to one another, have better chances of success.
T9. IT projects in which team members trust one another, have better chances of success.
T10. There is a dependence between project team size and project success – it is easier to achieve success in smaller teams.

To sum up the conducted research, the following conclusions can be formulated.

1. Communication depends on project team size and in smaller team (not more than 10 people) more projects are successful. Also smaller teams find it easier to develop their mode of cooperation based on F2F meetings during which they can discuss not only project related issues, but also other topics (personal, family related and world-view), which allows to create bonds and build trust being the foundation for an increase in the involvement in project work. The research also shows that in small project (not more than 10 people) which achieved success only 19% of respondents complained about communication problems.
2. In teams in which the frequency of contacts is high, it is easier to build trust, thus the frequency of meetings translates into project success. Possibly, the popularity of agile methodologies (based on small teams and frequent meetings with customers or owners), among others, results in an opportunity to be involved in direct conversations allowing to build close emotional ties.
3. Kindness and trust have influence on communication quality – in projects in which team members trust one another and are kind the chances of success are higher. The obtained results showed that it is possible to minimize project failure risk, which corresponds with the works by Park and Lee [23], and Brewer and Strahorn [29].
4. Despite the promotion and development of new communication channels, IT project stakeholders prefer F2F communication. Indirect contact is characterized by non-verbal information deficit (gestures, facial expressions, tone of voice, appearance). In the conducted research as many as 74% of respondents indicated F2F contacts as the most efficient communication channel. In the subsequent positions there was email and video communicators (29% each) and phone calls (27%), which unequivocally indicates communication deficits resulting from indirect communication.

It is hard to define the size of losses resulting from the lack of trust and kindness in project teams. According to the PMI report [35] in the USA as many as 55% of project managers indicate communication as the most significant factor in project management. For every billion of USD spent on projects, as much as USD 75 million is at risk as a result of inefficient communication. The proportions in Poland may be similar. Therefore, the process of developing trust and kindness between project stakeholders should be more emphasized in project management methodologies. Soft management skills must form an integral part of technical management processes. Knowing other team members better at various levels (professional and social) is an essential element contributing to project success.

Continuous efforts to improve communication and the awareness of the necessity to create such a need is as important as the principles of management methodologies [36] and should not be replaced by concentration only on new technologies, techniques, communication methods and management tools.

Future work will address the answer to the question: does the success/failure of an IT project depend on measurable indicators of the communication quality of project stakeholders.

Acknowledgements. This work was partially supported by the Polish Information Processing Society as a part of the "Research on success factors and failures of IT projects in Poland" and partially supported by the Polish Ministry of Science and Higher Education.

References

1. Biblia Tysiąclecia. Wydawnictwo Pallotitnum, Poznań (2000)
2. Dance, F.E.X.: The "concept" of communication. J. Commun. **20**(2), 201–210 (1970)
3. Grudzewski, W.M., Hejduk, I.K., Sankowska, A., Wańtuchowicz, M.: Zaufanie w zarządzaniu pracownikami wiedzy. E-mentor **5**(27) (2008). www.e-mentor.edu.pl/artykul/index/numer/27/id/598. Accessed 15 Mar 2018
4. Frączkowski, K.: Sukces i porażka projektów IT z perspektywy zarządzania zespołem projektowym. In: Frączkowski, K., et al. (eds.) Zarządzanie zespołem projektowym, pp. 148–176. Wydawnictwo Texter, Warszawa (2016)
5. Mazur, Z., Mazur, H.: Ocena jakości systemu informatycznego. Logistyka. **6**(CD), 2461–2470 (2011)
6. Gładysz, B., Frączkowski, K.: Wielowymiarowa analiza czynników sukcesu projektów IT. Zeszyty Naukowe Uniwersytetu Ekonomicznego w Katowicach **248**, 80–89 (2015)
7. Pinto, M.B., Pinto, J.K.: Project team communication and cross-functional cooperation in new program development. J. Prod. Innov. Manag. **7**(3), 200–212 (1990)
8. Butt, A., Naaranoja, M., Savolainen, J.: Project change stakeholder communication. Int. J. Project Manag. **34**(8), 1579–1595 (2016)
9. Fortune, J., White, D.: Framing of project critical success factors by a systems model. Int. J. Project Manag. **24**(21), 53–65 (2006)
10. Brocke, H., Uebernickel, F., Brenner, W.: Success factors in IT-projects to provide customer value propositions. In: 20th Australasian Conference on Information Systems, Melbourne (2009)
11. Nasir, M.H.N., Sahibuddin, S.: Critical success factors for software projects: a comparative study. Sci. Res. Essays **6**(10), 2174–2186 (2011)

12. Lachiewicz, S., Wojsa, A.: Czynniki sukcesu w procesie realizacji projektów inwesty-cyjnych w sektorze nieruchomości komercyjnych Zeszyty Naukowe Politechniki Łódzkiej. Organ. Zarz. **57**, 55–64 (2014)
13. Kandfer-Winter, K., Nadskakuła, O.: Komunikacja w zarządzaniu projektami. CeDeWu, Warszawa (2016)
14. Naqvi, I.H., Aziz, S., Rehman, K.: The impact of stakeholder communication on project outcome. Afr. J. Bus. Manag. **5**(14), 5824–5832 (2011)
15. Kennedy, D.M., Sommer, S.A., Nguyen, P.A.: Optimizing multi-team system behaviors: Insights from modeling team communication. Eur. J. Oper. Res. **258**, 264–278 (2017)
16. Joshi, K., Sarker, S., Sarker, S.: Knowledge transfer within information systems development teams: examining the role of knowledge source attributes. Decis. Support Syst. **43**, 322–335 (2007)
17. Reed, A.H., Knight, L.V.: Effect of a virtual project team environment on communication-related project risk. Int. J. Project Manag. **28**, 422–427 (2010)
18. Leśniak, T.: Zagadnienie komunikacji w zespołach wirtualnych i różnokulturowych pracujących w różnych strefach czasowych według metodyki Scrum. In: Werewka, J. (ed.) Wybrane zagadnienia zarządzania projektami w przedsiębiorstwach informaty-cznych. AGH, Kraków (2013)
19. Levin, G., Rad, P.F.: Requirements for effective project communications: differences and similarities in virtual and traditional project environments. In: PMI® Global Congress. Project Management Institute, Anaheim (2004)
20. Blenke, L.R.: The role of face-to-face interactions in the success of virtual project teams. Doctoral Dissertation, Missouri University Of Science And Technology (2013)
21. Chan, A.P.C., Chan, D.W.M., Chiang, Y.H., Tang, B.S., Chan, E.H.W., Ho, K.S.K.: Exploring critical success factors for partnering in construction projects. J. Constr. Eng. Manag. **130**(2), 188–198 (2004)
22. Pinto, J.K., Slevin, D.P., English, B.: Trust in projects: an empirical assessment of owner/contractor relationships. Int. J. Project Manag. **6**(27), 638–648 (2009)
23. Park, J.G., Lee, J.: Knowledge sharing in information systems development projects: explicating the role of dependence and trust. Int. J. Project Manag. **32**, 153–165 (2014)
24. Xue, Y., Bradley, J., Liang, H.: Team climate, empowering leadership, and knowledge sharing. J. Knowl. Manag. **15**(2), 300–303 (2011)
25. Krawczyk-Bryłka, B.: Ocena klimatu pracy w zespole wielokulturowym. Marketing i Rynek **5**(CD), 1084–1090 (2014)
26. Krawczyk-Bryłka, B.: Budowanie zespołu wirtualnego – zasady i wyzwania. Studia Informatica Pomerania nr 2/2016 (2016)
27. Diallo, A., Thuillier, D.: The success in international development projects, trust and communication: an African perspective. Int. J. Project Manag. **23**, 237–252 (2004)
28. Doloi, H.: Relational partnership: the importance of communication, trust and confidence and joint risk management in achieving project success. Constr. Manag. Econ. **27**, 1099–1109 (2009)
29. Brewer, G., Strahorn, S.: Trust and the project management body of knowledge. Eng. Constr. Archit. Manag. **19**(3), 286–305 (2012)
30. Cheung, S., Yiu, T., Lam, M.: Interweaving trust and communication with project performance. J. Constr. Eng. Manag. **139**(8), 941–950 (2013)
31. Trocki, M., Grucza, B., Ogonek, K.: Zarządzanie projektami. PWE, Warszawa (2003)
32. Covey, S.R.: Zasady skutecznego przywództwa. Wydawnictwo Rebis, Poznań (2004)
33. Frączkowski, K., Dabiński, A., Grzesiek, M.: Raport z polskiego badania projektów IT 2010. Wrocław   (2011).   http://pmresearch.pl/wp-content/downloads/raport_pmresearchpl.pdf. Accessed 18 Mar 2018

34. The Standish Group Report: CHAOS (2014). https://www.projectsmart.co.uk/white-papers/chaos-report.pdf. Accessed 18 Mar. 2018
35. Project Management Institute: The high cost of low performance: the essential role of communications (2013). http://pmi.org/-/media/pmi/documents/public/pdf/learning/thought-leadership/pulse/the-essential-role-of-communications.pdf. Accessed 18 Mar 2018
36. Managing Successful Projects with PRINCE2. Axelos (2017)

Best Practices in Structuring Data Science Projects

Jedrzej Rybicki[(✉)]

Forschungszentrum Juelich, 52425 Juelich, Germany
j.rybicki@fz-juelich.de

Abstract. The goal of Data Science projects is to extract knowledge and insights from collected data. The focus is put on the novelty and usability of the obtained insights. However, the impact of a project can be seriously reduced if the results are not communicated well. In this paper, we describe a means of managing and describing the outcomes of the Data Science projects in such a way that they optimally convey the insights gained. We focus on the main artifact of the non-verbal communication, namely project structure. In particular, we surveyed three sources of information on how to structure projects: common management methodologies, community best practices, and data sharing platforms. The survey resulted in a list of recommendations on how to build the project artifacts to make them clear, intuitive, and logical. We also provide hints on tools that can be helpful for managing such structures in an efficient manner. The paper is intended to motivate and support an informed decision on how to structure a Data Science project to facilitate better communication of the outcomes.

Keywords: Data Science · Project management
Management methodologies · Tools

1 Introduction

An intuitive understanding of Data Science is to view it as a way of obtaining novel insights from collected data. It applies methods from fields of computer science, software engineering, applied statistics, and data management among the others. Data Science projects comprise of different phases: collecting the data, cleaning them, analyzing, and drawing conclusions to name just a few. An important part of this process is the communication of the findings to support research hypothesis or make business decision. In a broader sense, communication is an effort of making the project and its results understandable and thus reproducible, and transparent. In this light, the mixture of data description and programs used for data cleansing and analyzing, often accompanied by some write-up of the methodology and findings is one of the most important artifacts of communication. Yet, to our experience a lot of projects are poorly structured,

© Springer Nature Switzerland AG 2019
Z. Wilimowska et al. (Eds.): ISAT 2018, AISC 854, pp. 348–357, 2019.
https://doi.org/10.1007/978-3-319-99993-7_31

mix programs, data, and libraries and provide no hints on how to redo the analysis. Thus, the subject of this paper is how to prepare, and structure such an artifact to efficiently communicate results of a data science project. We examine best practices to form the directory structures, organize them, and provide hints on how and where store both the data and programs used.

Although, the research findings gained in projects are usually consider to be their main outputs, neglecting best practices regarding project structure can hinder their understanding and reduce the overall impact of the project. When the structure is defined, intuitive, and understandable the recipient of the artifact (i.e. another researcher) can focus her effort on the important parts: project findings, algorithms used, data integration questions, etc., rather than spending time on grasping the structure and dependencies within the project. This bears some similarities to the best practices of software engineering. Using intuitive naming conventions and sticking to established programming practices make the immersion in an existing software project much easier. Successful Data Science are deployed on production infrastructure to put in practice their findings. This process of productization can also be conducted in a less painstaking manner if the project is well structured and understandable. In fact this is our main source of motivation for the work. We are affiliated with one of the larges data centers in Europe and, thus, often confronted with a request of putting results of a project on our infrastructure.

In this paper, we compile best practices from three main sources of information. We noticed that there are differences between approaches in industry and academia. Therefore, we look in both of this worlds. Firstly, we examine the most popular methodologies for data science and data mining projects. They are already superpositions of the best practices from industry and, thus potentially incorporate important insights on project structures. Second source of information are the best practices from research communities. These can be more or less formalized and often differ in certain aspects from the more industry-driven methodologies. Lastly, we analyze the structure of artifacts published on popular data sharing platforms like Kaggle or myExperiment. Here both worlds of data science: academia and industry meet and thus the places can also provide valuable hints on the aforementioned subject. The goal of this paper is not to produce a ready-to-follow instruction but rather a list of important aspects of the subject and suggestion on solution. The reader can then compare her current solution, decide on her own what makes sense for her and what suggestions should be implemented. We also collated a brief overview of generic tools that can help to build shareable and understandable artifacts.

Data Science resides at the border between data management and software engineering. We try not to go too far in any of this fields as we believe that their respective best practices should be applied where applicable. For the best practices in research data management we refer the reader to an excellent overview by DataOne [5]. Best practices in software development depend strongly on the language used, in case of Python a good overview is provided by Reitz and Schlusser [23]. To differentiate ourselves from data and software management, we (arbitrary) assume that a Data Science project starts when data are moved into the analytic environment. The environment comprises of existing software and, if required, new project-specific software is developed. After the project ends, both software but especially the data can be moved back into their original worlds, i.e., data management solution or code repository. How such a movement is achieved, what external code is used and how it is configured should be part of the communication artifact, i.e., data science project. Still a sensible and structured way of conducting data science project can help in publishing the results, for instance, by automatically provide some of the metadata required by data repository.

The rest of the paper is structured as follows. In Sects. 2, 3, and 4, we summarize best practices originating from Data Science methodologies, research communities, and data sharing platforms respectively. We discuss the common aspects in Sect. 5.1 and subsequently give suggestion on tools which can help to structure and manage a Data Science project. This summary is the main contribution of this paper. The conclusions can be found, together with future research directions, in Sect. 6.

2   Methodology-Resulting Structures

In this section, we want to review suggestions on how to structure the project artifacts drawn from popular knowledge discovery and data mining methodologies and process models. An excellent survey on data mining process models is provided by Kurgan and Musialek [19]. Here, we only concentrate on the actual project structures, their content, and related suggestions.

Knowledge Discovery in Databases (KDD) was described in the seminal paper by Piatetsky-Shapiro and Frawley [21] and later extended into a more generic Knowledge Discovery and Data Mining (KDDM) methodology [22]. The Knowledge Discovery (KD) process is defined with the stages of selection, preprocessing, transformation, data mining and interpretation/evaluation [17]. It is worth noticing that even though the original work defined database as "logically integrated collection of data maintained in one or more files", it focuses strongly on relational databases. Thus, although, the process is very broadly defined, it gives no hints on the directory structure. It also does not stress the need of efficient communication unlike the later approaches.

CRISP-DM (Cross-industry Standard Process for Data Mining) [15] drafted initial in an EC-funded project and is currently driven forward by a consortium of mainly industrial partners. It is "based on the practical, real-world experience of how people conduct data mining projects" [15]. The process is comprised of six main phases: business understanding, data understanding, data preparation, modeling, evaluation and deployment. The process does not define a particular directory structure for a project. Nevertheless, we found an example of project called py-crisp [18] that organizes directory structure according to the phases of the project. It makes sense as the CRISP-DM defines outputs and deliverable of each step. From our perspective one important is, for example, the data description from the data preparation step. It describes what data are used and from what sources they are coming. Still it is not clear if the structure built in this way is really useful as the place for raw data, secondary data or analyzing software is not well defined. It is questionable if the management process used to produce the result is really relevant and need to be built into the structure of the project deliverable. We will elaborate on this subject more later in this paper.

An alternative to CRISP-DM is Team Data Science Process (TDSP) proposed by Microsoft [13]. TDSP also describes a life cycle for a data science project, and it is defined in following phases: business understanding, data acquisition and understanding, modeling, deployment, and customer acceptance. The process gives a lots of suggestions on how to execute particular steps and what outputs shall they produce. Already during the business understanding step, the locations of potentially relevant raw data should be fixed and later scripts for moving the data into the analytics environment should be written. The data acquisition and understanding part should among the others produce data quality report. Importantly, the TDSP defines a standardized project structure based on the provided template [12]. The three major components of this structure are programs (in `Code` directory), documents (`Docs`), and data (`Sample_Data`). The code directory should store artifacts for the defined project phases, i.e., data acquisition or modeling. On the lower hierarchy level, the template suggest to create separate directories for each experiment done in the modeling phase, where both optimal method and parameters are sought after. The data are divided into raw, processed, and modeling part and should not include large-size data but rather small sets of example data for verification and scripts to obtain the actual data. The reasoning behind is, among the others, the fact that the structure should be kept in a version control system. Such systems are usually not designed for storing large files. The TDSP and its directory structure is motivated by making projects compatible with computing facilities offered by Microsoft Azure Cloud platform [9]. The implicit assumptions captured in the structure help in interoperating between services. Similarly a well defined structure can help an organization or research community to achieve better exchange of working projects between distributed services.

A less formalized methodology to execute and structure data science project is Cookiecutter Data Science [16]. It is a logical, and standardized yet not binding project structure. The project is interesting for at least three reasons. Firstly, it motivates and describes an advanced directory structure for projects. The structure comprises of a `data` directory which is further divided into self-explanatory `raw`, `interim`, and `processed` folders. Furthermore, there are dedicated locations for documents, references, and reports. The code used is dived between sub-directories in `src` according to its functions, e.g., scripts for downloading data, scripts for building features, or make visualizations. The authors also foresee a place for notebooks (like Jupyter [11], Apache Zeppelin [2], etc.), which are expected to capture the exploratory phase of the project. Second reason why the project is interesting are the "opinions" collected in the project documentation.

They suggest, for instance, to make data immutable (and never make any manual changes to them), or to understand data analysis as directed acyclic graph, where transitions between subsequent steps are implemented in scripts. Lastly, the project make some good practical suggestion on tools that can be used, like `cookiecutter` [4] or `make` [7]. We devote a separate section to this problem (Sect. 5.1). A comparative analysis of different data science methodologies can be found in Table 1.

Table 1. Comparative analysis of methodology-resulting structures

Feature	Methodology			
	KDD 2	CRISP-DM	TDSP	Cookiecutter DS
Source	Academia	Industry	Vendor	Mixed
Formalization	High	High	Moderate	Adjustable
Main concern	Data mining	Data mining	Data science	Data science
Directory structure	None	Unclear	Defined	Flexible
Project phases	Defined	Defined	Defined	Not defined

3   Community Established Structures

Noble quick guide to organizing computational biology projects [20], describes "one good strategy for carrying out computational experiments." The paper suggest to use logical structure at the top level of a project and chronological one on the lower level. The most important parts of the logical structure are `data` and `results` directories storing fixed data sets and processing outcomes respectively. Furthermore, the author suggest to use `src` directory for scripts and programs, and put documentation in `doc`. On the lower hierarchy level, to organize data and results chronological ordering is used. Each new (sub) experiment in the project is started in a directory named after current date. Beside a dedicated documentation, it is advised to record each conducted step in `README` files. The project should also contain a driver script that execute experiments and processing automatically. To some extend, such a script is also a piece of documentation. The author strongly advocates to use version control system as a first line of backup, historical record of performed changes, and a way to facilitate collaboration.

Wilson et al. list "good enough practices in scientific computing" [27]. The paper has a similar goal to ours, and is worth reading as it captures practical experiences of the authors in fields of data management, software engineering, collaboration, and project organization. Proposed organization includes separate directories for raw data and metadata, source code, documents, and all compiled programs and scripts. The directory should also include a `README` file describing the project. All intermediate files created during the cleanup and integration of the data should be stored in the `/results/` directory, rather than together with the raw data. It also suggests to split the processing and cleansing into separate phases with their intermediate files. Such an approach allows for restarting only part of the procedure if required. Authors postulate to avoid proprietary data formats and use "tidy data" [26], i.e., to make each column in a file a variable and make each row an observation. The authors stress to keep track of changes. This can be achieved in a manual process, in which each time a significant change is done the whole project directory is copied. Alternatively, version control systems can be used.

4 Data Sharing Platforms

The platform myExperiment [24] enables sharing data and scientific workflows mostly from biological and medical sector. It is possible to upload so called packs, i.e., collections of files. The platform does not prescribe any particular structure for such collection. It enables, however, a description of the collection with help of an emerging standard OAI-ORE (Open Archives Initiative: Object Reuse and Exchange) [10]. Regardless of the structure adopted in a project it makes sense to document it.

Kaggle [8] is another online platform aiming at sharing of data sets and analysis tools and insights. It is not very formalized in terms of project structures but suggest to use data formats like CSV, and JSON or relational databases depending on the nature of the data. Sharing of analysis is possible by sharing kernels and notebooks. Such kernels run in the official Docker [1] images of the platform which include all the required libraries. The data are injected into running Docker containers in the /input/ directory. Docker is a light-weight virtualization solution which can be used to transport running programs with all their dependencies across many platforms. It was already postulated to use Docker to facilitate reproducible research [25] for many reasons. One of them is the possibility to clearly separate the processing from the data. It enables reuse of the methods on different data and also selection of best suited storage for data. Docker images can also be versioned to reflect on the process of their evolution. A clear lesson learned from this example is the separation of data and code but also a benefit of establishing a standardized running environment for the analyzes.

5 Discussion

In this section, we summarize the insights and point out aspects that should be part of an informed decision with regard to project structure. Later we also list some tools that can be used to define and implement such a structure.

5.1 High-Level Aspects

Data Science projects are, or at least should be, a place where software development meets data management. Thus, it makes sense to stick to the respective best practices in these fields. In particular, the software developed should apply logical code structuring, naming, and separation of concerns. Furthermore, the code should be stored in a version control system to facilitate change tracking, backup, and collaboration. The increasing popularity of notebooks in Data Science (Jupyter, Zeppelin, etc.) pose some additional challenges in this regard. Such notebooks are a mixture of code, visualization, and documentation. They are not getting well along with version control systems (as it is hard to track the changes in the mixture of code and markup). Most of the sources that we cited suggest to use notebooks (they are great vehicle for the ideas and results) but try to strip them of most of the data processing steps and outsource these in the separate programs and scripts. Such an approach allows for quickly check the results and visualizations whilst enabling the researchers to dig deeper into the processing if required. Lastly, for the developed code, it makes sense to provide ways of dependency management and easy execution of the code. This can be achieved in many ways. Staring from a *driver script*, i.e., one script that execute the whole processing, building tools like make [7] or cmake [3] to describe the processing workflow, or virtualization solutions like Docker [1] to share working running environment.

Data Science project of at least moderate complexity comprise of many sub-experiments. In our survey we identified three major ways of structuring the projects: functional, along the phases of the project life-cycle, and chronological. Some sources suggested to create directories for each of such a sub-experiment and name them according to starting dates. We believe that it makes sense to capture such single steps (even if they do not contribute to the final result), but using date as names might not be the best option. Especially when version control system is used, the starting dates can be obtained anyway. Perhaps more meaningful names, reflecting the intermediate hypotheses are a better option.

In the data management dimension common practice is to provide metadata (i.e. description of data sources and structures). Almost all reviewed sources suggest not to store raw data in the version control system along the code, but rather use data-oriented resources and provide scripts for the data retrieval. For the sake of reproducibility it is necessary not to make any manual changes in the raw data but rather use scripts for data cleansing, filtering, and processing.

With respect to data, it is advisable to use open formats like CSV, or JSON. It is usually possible, at least, for the intermediate processing steps. Also making the raw data tidy makes sense and can facilitate its reuse. This advice is not always followed in both academia and industry. The reasons are many fold. Industry often relies on proprietary tools imposing proprietary formats. Similarly, some research fields have established own formats that are used in the data repositories. Still for the intermediate processing steps done within the data science project it is advisable to use open formats.

Although, we see a merit in using defined methodology when implementing data science projects, we also see some shortcomings of orienting project structure strongly towards the phases of the methodology. The central point of this paper is to discuss means of making the results of the project understandable.

But is it really relevant what methodology was used to arrive at the final conclusions? Should they be formulated in such a way that it is understandable even without knowing the particular methodology? On the other hand, neglecting the strictly defined phases may lead to situations in which ad-hoc hypothesis are created to retrofit the obtained results. It is problematic in both academia and industry, as it leads to not publishing negative results (e.g., initial hypothesis was not confirmed) and can lead to producing something else than the customer required. Building the results structure according to life-cycle of the project is problematic when the results are shared outside of the organization, i.e., with people who do not know the methodology and had to first understand the process before diving into the results. The usage of version control system advocated above, make the history of the project visible and thus can to some extend prevent the retrofitting of hypothesis to the results obtained. Nevertheless, we suggest to get familiar with one of the methodologies like CRISP-DM or TDSP as it helps to put the Data Science project into a long-term perspective, and provide useful hints on its management.

5.2   Tooling

Regardless of the structure agreed upon in a project, a company, or research group, it makes a lot of sense to describe it in a more or less formal way. An obvious way is to write a document but changes in the structure must be later reflected in the documentation which is sometimes problematic. Another way of describing an existing structure is to use the already mentioned OAI-ORE. It seems to be a little bit too formal for many applications and not really human readable. A nice balance between the usability and ease of use is stroke by the `cookiecutter` tool [4]. We mentioned this tool together with the Cookiecutter Data Science project structure [16] but it is in fact a stand-alone tool to formalize and share any structure. There are numerous of existing templates which can be used, or adjusted to particular needs. Technically, the template is a JOSN-based description of the structure, upon creation of a templated project, some user input (like project name) might be required. The templates can be stored in a repository (e.g., GitHub).

Another important part of an artifact produced in a data science project is the entry point script or driver script. That is one script intended to start data retrieval, cleansing, and actual analysis. The idea is to provide the recipient of the project with a means to just redo the analysis. But the meaning of such a script is deeper, it is a formal (computer-understandable) way of describing the workflow of the performed analysis. It corresponds very well with the reasoning to understand the processing as acyclic graph. There are many tools to achieve that, some are generic, some are specific for the programming language or framework selected. On the more generic side, there are tools like `make` [7] and `cmake` [3]. Initially, conceived for building more complicated software projects. Yet they fit very well with data science project [14]. Abstractly speaking, `Makefile` describes parts of the artifact and ways of producing them. For instance it is possible to define that a particular piece of raw data should be present in the raw data directory and define a means to put it there if it is not present. Unlike the project structure itself, structuring or at least orienting the workflow description along the lines of the life-cycle from the adopted methodology seems to make sense. It is an additional meta description of the process.

An important part of making research from a data science project reproducible is to provide not only ways of sharing code but providing also description of the runtime environment. There are solutions for that which are specific to programming language used. Another way is sharing the actual running environment with help of virtualization technology. In this regard, there is a range of choices: virtual machine images, cloud infrastructures, and Docker. The last option is probably the easiest to work with. With Docker it is possible to define an image describing how a basis system should be extended and configured to run particular program. The Docker Hub [6] enables to share such images and also version them to capture the evolution of ideas and analysis. A valid option is to use existing Docker images like the one provided by Kaggle also in local projects. It already includes a lot of libraries and due to its popularity it can be seen as a de facto standard Data Science environment.

6   Conclusion

In this paper we compiled best practices on how to structure Data Science projects. They came from three different sources, thus reflect different approaches to the problem proliferated in both academia and industry. We summarize some common aspects like capturing the evolution of the project, separation of raw and intermediate data, or formal description of the processing workflow. The main contributions of this work boils down to following aspects. Firstly, we intended to make people and organization aware of the needs and benefits of common Data Science projects. Secondly, we provide a survey and links to popular sources of information on this subject. Finally, a list of recommendations in this regard is presented. The intended use of this contribution is a support in decision making and implementation of data science project structures. We claim that Data Science project structure should be viewed as one of the communication artifacts that can enable and support the exchange of the research findings in an understandable way.

In our future research we plan to use this work to define a common, formalized structure for Data Science projects conducted in our affiliating facility.

References

1. Docker, May 2017. https://www.docker.com/
2. Apache Zeppelin, May 2018. https://zeppelin.apache.org/
3. CMake, May 2018. https://cmake.org/
4. Cookiecutter, May 2018. https://github.com/audreyr/cookiecutter
5. DataOne: Best practices in data management, May 2018. https://www.dataone.org/all-best-practices

6. Docker Hub, May 2018. https://hub.docker.com/
7. GNU make, May 2018. https://www.gnu.org/software/make/
8. Kaggle, May 2018. https://www.kaggle.com/
9. Microsoft azure cloud computing platform and services, May 2018. https://azure. microsoft.com/
10. Open Archives Initiative: Object Reuse and Exchange, May 2018. http://www. openarchives.org/ore/1.0/toc
11. Project Jupyter, May 2018. https://jupyter.org/
12. TDSP project template, May 2018. https://github.com/Azure/Azure-TDSP-ProjectTemplate
13. TDSP: Team data science process, May 2018. https://docs.microsoft.com/en-us/ azure/machine-learning/team-data-science-process/overview
14. Butler, P.: Make for data scientists, May 2018. http://blog.kaggle.com/2012/10/ 15/make-for-data-scientists/
15. Chapman, P., Clinton, J., Kerber, R., Khabaza, T., Reinartz, T., Shearer, C., Wirth, R.: CRISP-DM 1.0: Step-by-step data mining guide, May 2018. ftp://ftp. software.ibm.com/software/analytics/spss/support/Modeler/Documentation/14/ UserManual/CRISP-DM.pdf
16. DriveData: Cookiecutter data science, May 2018. https://drivendata.github.io/ cookiecutter-data-science/
17. Fayyad, U., Piatetsky-Shapiro, G., Smyth, P.: Knowledge discovery and data mining: towards a unifying framework. In: Proceedings of the 2nd International Conference on Knowledge Discovery and Data Mining, pp. 82–88 (1996)
18. Jackson, M.: py-crisp, May 2018. https://github.com/ruffyleaf/py-crisp
19. Kurgan, L.A., Musilek, P.: A survey of knowledge discovery and data mining process models. Knowl. Eng. Rev. **21**(1), 1–24 (2006)
20. Noble, W.S.: A quick guide to organizing computational biology projects. PLOS Comput. Biol. **5**(7), 1–5 (2009). https://doi.org/10.1371/journal.pcbi.1000424
21. Piatetsky-Shapiro, G., Frawley, W.J. (eds.): Knowledge Discovery in Databases. AAAI/MIT Press, Cambridge (1991)
22. Reinartz, T.: Stages of the discovery process. In: Klosgrn, W., Zylkon, J. (eds.) Handbook of Data Mining and Knowledge Discovery, pp. 185–192. Oxford University Press, Inc., Oxford (2002)
23. Reitz, K., Schlusser, T.: The Hitchhiker's Guide to Python: Best Practices for Development (2016). ISBN: 978-1-49193-317-6
24. Roure, D.D., Goble, C., Stevens, R.: The design and realisation of the myExperiment virtual research environment for social sharing of workflows. Future Gener. Comput. Syst. **25**(5), 561–567 (2009)
25. Rybicki, J., von St. Vieth, B.: Reproducible evaluation of semantic storage options. In: Proceedings of the 3rd IARIA International Conference on Big Data, Small Data, Linked Data and Open Data (ALLDATA 2017), pp. 26–29, April 2017. ISBN: 978-1-61208-552-4, ISSN: 2519-8386
26. Wickham, H.: Tidy data. J. Stat. Softw. **59**(10), 1–23 (2014). https://www. jstatsoft.org/v059/i10
27. Wilson, G., Bryan, J., Cranston, K., Kitzes, J., Nederbragt, L., Teal, T.K.: Good enough practices in scientific computing. PLOS Comput. Biol. **13**(6), 1–20 (2017). https://doi.org/10.1371/journal.pcbi.1005510

Project Management Practices in Polish Nonprofit Organisations

Radoslaw Czahajda[(⊠)]

Faculty of Computer Science and Management, Wroclaw University of Science
and Technology, ul. Ignacego Łukasiewicza 5, 50-371 Wrocław, Poland
radoslaw.czahajda@pwr.edu.pl

Abstract. This paper presents a summary of the level of adoption of different
project management tools and approaches in Polish Nonprofit Organisations. It
is based on the responses from 133 polish NGOs in March 2018. The results are
compared with similar work on International Development projects to define if
the level of adoption of PM practices is different in polish nonprofit sector and if
there is a difference in the dependence of project management maturity on the
internal project success. Various aspects of project management throughout
entire project cycle were evaluated: project initiation, project definition, project
implementation and project evaluation. The research concludes with several
outcomes. Polish NGOs adapted project management practices to the lower
extent than International Development projects. Yet, different tools influence the
project success comparably. There is a lack of knowledge about project management among polish NGOs, which might be an important reason behind their
low performance on international arena. Paper brings also a broad perspective
on the level of adoption of different project management tools and practices that
was not evaluated before.

Keywords: Project management adoption · Project success
Nonprofit organisations

1 Introduction

1.1 State of Noprofit Sector in Poland

In 2015 Klon-Jawor association conducted a wide research on a sample of 4 000
nonprofit organisations in Poland. The scope of research included finances, human
capital and external relations. The following report included several issues in the sector
of non-profit organisations. It concluded with many important features of non-profit
organisations, including incapability of fundraising, lack of human capital, competitive
attitude towards other NGOs, diminishing potential of media and promotion [1]
CIVICUS Civil Society Index ranked Poland 1,1 on a 3 points scale in NGOs structure
field, pointing on low social activity, low representation of NGOs in society as main
drawbacks influencing the score [2]. Strategy of development of polish civic society for
years 2009–2015 defined plenty of weaknesses, including lack of strategic planning,
low organisational culture, low financial condition, lack of organised consulting for
NGOs, low quality of NGO work in comparison with potential of Poland [3].

© Springer Nature Switzerland AG 2019
Z. Wilimowska et al. (Eds.): ISAT 2018, AISC 854, pp. 358–370, 2019.
https://doi.org/10.1007/978-3-319-99993-7_32

According to Newsweek [4], Polish NGOs on average have a yearly budget of 48 421 EUR. That's lower than Hungary (74 494), Sweden (93 017), Portugal (150 046), United Kingdom (305 499) and United States of America (1 693 579). Polish non-profit organisations as well as entire sector are in weak condition and there is low understanding for the causes of it.

1.2 Research in Nonprofit Sector in Poland

The main reason to focus on specific case of Nonprofit organisations is because this sector is different from public and private sector and therefore requires scientific attention [5–8]. Yet, only few researchers considered polish non-profit sector in their work in recent history and there are still plenty of blind spots with project management practices being one of them. In Poland an extensive research on NGOs was made in the fields of marketing by Iwankiewicz-Rak [8], strategic management by Domanski [5] and Bogacz- Wojtanowska [7]. There is hardly any evidence on Project Management tools and methodologies adoption and their influence on organisation and project success.

1.3 Project Management Practices in International Development Projects

In 2015 Golini [9] conducted an extensive research among International Development Projects to identify the use of project management tools and approaches and its influence on project success. International Development projects are large in budget and aiming to support the developing countries [10]. Particular focus areas of these projects include training and education, housing, health aid, disease prevention, protection of basic human rights [11].

Using cluster analysis, Golini have differentiated 4 clusters of ID projects depending on the level of implementation of project management tools and to evaluate their performance. The clusters were created if an average score of 2.9 was exceeded in specific set of tools, which represented at least 25% of the projects in the cluster using the tool. Figure 1 presents the results of this study.

After defining the clusters Golini used statistical analysis to evaluate the influence of the project management tools adoption on internal project performance based on Ika's work (complying with the budget, complying with expected time, complying with expected quality [12]) and external project performance based on work on many authors (obtaining long-term project impact, stakeholder/partner involvement, owner-ship extension of the project to the local community, monitoring and reporting to the stakeholders, economic sustainability after the end of the project, satisfaction of the local community [13–17]. The performance was evaluated by project coordinators based on the 1–5 Likert scale, where 1 represented "Very Low" and 5 represented "Very High". As a result, tools from cluster 2 have strong positive impact on internal project performance, whereas tools from cluster 3 have a strong positive impact on external project performance. Other does not have a significant influence on project performance.

Practice	Cluster 1	Cluster 2	Cluster 3	Cluster 4	Average
Progress reports	2.94	4.63	4.67	4.95	4.47
Logical Framework	3.19	4.40	4.23	4.73	4.22
Cost accounting	2.08	4.22	4.40	4.85	4.11
GANTT diagram or project schedule	2.19	3.22	3.96	4.37	3.59
Risk analysis/management	1.92	3.15	3.82	4.42	3.49
Communication plan	2.35	2.65	3.86	4.55	3.46
Organizational chart or OBS	1.54	2.62	3.73	4.58	3.28
Milestone planning	1.94	2.46	3.66	4.50	3.26
Stakeholder matrix	1.94	1.97	3.51	4.30	3.03
Scope management	1.46	2.04	3.06	4.28	2.79
Contingency allocation	1.73	2.05	2.99	3.92	2.73
Responsibility assignment matrix (RAM)	1.38	1.66	3.15	4.60	2.77
Work breakdown structure (WBS)	1.46	1.88	2.89	4.37	2.68
Critical path method (CPM)	1.40	1.64	2.73	3.88	2.46
Issue log	1.29	1.47	2.70	3.85	2.38
Earned value management system (EVMS)	1.17	1.20	2.07	3.70	2.00
Number of NGO	46	92	150	60	
(%)	(13.2%)	(26.4%)	(43.1%)	(17.2%)	

Fig. 1. Clusters of PM tools adoption [9]

This was the only wide research in the field of implementing project management tools and approaches in nonprofit organisations conducted so far. Researchers defined this gap in scientific discovery and pointed out its importance [18]. This gap was a main inspiration to design and conduct this study.

2   Methodology

2.1   Research Questions

After a desk research on the topic of non-profit sector in Poland, a several research questions were formed for this research. Two of them are covered in this paper.

1. RQ1. What is the level of adoption of project management practices in polish NGOs and how does it relate to the case of International Development projects?

   Since Golini's work brought a lot of understanding to the state and issues of project management in International Development Projects, similar question were to be asked in polish sector to know if the results could be extrapolated or if the two are different.

2. RQ2. What is the influence project management tools adoption on the project success among polish NGOs?

   In order to understand state of project management in Polish NGOs wider, other fields of PM were incorporated in the study, from qualifications of respondents, through the decision about conducting the project, project scope definition, team composition, project implementation, evaluation, to project success.

2.2 Data Collection

In March 2018 a questionnaire was distributed among non-profit organisations in Poland through targeted marketing in social media. 133 responses were collected. The responses were equally distributed between foundations (51%) and associations (49%). There was slightly higher amount of organisations working in education and upbringing and culture and arts among others, but all the major branches were represented (Table 1). High amount of answers from both locally oriented and globally oriented organisations was achieved.

Table 1. Sample definition n = 133

Main focus area		Area of operation	
Education and upbringing	23%	Entire country	27%
Other	21%	City above 200 000 inhabitants	17%
Culture and arts	18%	International	13%
Sports, tourism, recreation, hobby	12%	Voivodership	12%
Local development	12%	Commune	11%
Social services, social care	8%	City below 50 000 inhabitants	8%
Healthcare	6%	Village	5%
		City below 200 000 inhabitants	4%

The majority of organisations participating in the survey were small and young which is indicated by the difference between median and average of amount of members, organisation yearly budget and age of the organisation (Table 2).

Table 2. Key sample features n = 133

	Members	Yearly budget (PLN)	Age (years)
Average	12,53	35 925,20	10,17
Median	6	57 500,00	4,00
Std dev	20,09	513 684,50	18,14

2.3 Questionnaire Construction

Questionnaire used in this research consisted of 87 open and closed questions. Some of them were adapted from Golini and other past project management researches to have a common comparison baseline. The survey was divided in 10 chapters ordered similarly to project management processes.

2.4 Data Analysis

In his research Golini was interviewing highly experienced project managers with many projects in their portfolio. Therefore he was asking about adoption of project management tools in all the projects that project managers were in charge of. Since local organisations in Poland did not have such experience, a survey was asking about specific project. The average level of adoption of specific project management tool could have been later easily adopted to 1–5 scale from Golini.

3 Project Management Approaches

The vast majority of organisations (82%) that participated in the research did not use any project management methodologies and approaches (Table 3). In an open question organisations declared that the major reason behind not using them is lack of knowledge and skills (80%). Moreover, 14% of organisations not using any methodologies declared they don't feel a need to use them. Other answers were "I don't know"(3%), Lack of financial reasons (2%), lack of project team members (1%).

Table 3. Which project management methodologies/frameworks were used in the project? n = 133

Methodology/framework	Percent of answers
None	82%
Prince2	2%
I don't know	2%
PMBoK	2%
Waterfall	4%
IPMA	2%
Agile PM/Scrum	2%

4 Project Initiation Phase

4.1 Needs Analysis

Organisations spread between different decision factors to start the project (Table 4). 16% of organisations (22 cases) decided to prepare a project based on the requirements from grant institution, ceding therefore the responsibility of assessing the needs to the grant institution. 49% of the organisations (65 cases) did not run a needs analysis before starting the project, either basing the decision on the coordinator instinct (24 cases), or on the previous project teams (41 cases). Only 26% of the organisations decided to run a needs analysis among project stakeholders (34 cases), or developed an MVP to test project assumptions.

Table 4. What was the main decision factor to start the project? n = 123

Decision factor	Percent of answers
Tradition – next edition of the project	33%
Conducting needs analysis among project stakeholders.	28%
Project proposal by project coordinator	20%
Project written to meet grant competition requirements	18%
Developing minimum viable product and testing it	2%

4.2 Goal Setting

In the survey, respondents were asked to write the precise goal of the project as it was stated in the project definition meeting. The author analysed the results to define how many of them defined any measure of success in the goal itself. Only 5% of the organisations have written goals that were measurable, so a precise level of accomplishing them could be calculated.

4.3 Project Definition Tools

The most popular project definition tool used by NGOs taking part in the survey was SWOT analysis, used by 31% of the organisations that participated in the survey. Only 13% use Gantt chart as a scheduling tool (Table 5). 70% of organisations use none to one project definition tools, which they explain similarly to not using project management methods – 75% of organisations due to lack of knowledge and 25% due to lack of perceived need.

Table 5. Which tools were used in project definition period? N = 133

Tool	Percent of answers
None	31%
SWOT analysis	27%
Gantt chart	13%
Risk analysis	10%
SMART goals	9%
Stakeholder analysis	5%
Work breakdown structure	4%
I don't know	2%
Framework logic	1%

In order to compare the results with research from Golini, data was adapted to represent the level of implementation of different project management tools on the scale 1–5, where 1 represents 0% of projects using the tool and 5 representing 100% of them. Results are presented in Table 6. In comparison with International Development projects polish non-profit organisations implement project management tools on very low level. In the collected sample the representation did not allow to make a cluster analysis and further investigate characteristics and results of organisations adopting project management tools.

Table 6. Comparison of PM practices adoption between ID projects and study results.

Project management practice	Average ID	Average PL
Progress reports	4,47	N/A
Logical framework	4,22	2,03
Cost accounting	4,11	**4,08**
GANTT diagram or project schedule	3,59	2,51
Risk analysis/management	3,49	2,39
Communication plan	3,46	N/A
Organizational chart or OBS	3,28	N/A
Milestone planning	3,26	**3,20**
Stakeholder matrix	3,03	2,18
Scope management	2,79	N/A
Contingency allocation	2,73	N/A
Responsibility assignment matrix (RAM)	2,77	2,12
Work breakdown structure (WBS)	2,68	2,15
Critical path method (CPM)	2,46	2,12
Issue log	2,38	N/A
Earned value management system (EVMS)	2	2,03
SWOT analysis	N/A	**3,08**
SMART goals	N/A	2,36

4.4   Project Team Recruitment

22% of organisations declared that they included every volunteer that was interested to join the team. 45% of the organisations declared doing a competence mapping and recruitment of candidates that fit the competencies profile. 31% of the organisations used the same team as in previous editions of the project. An interesting practice of including team composition in grant requirement was revealed in this question (1%).

5   Project Initiation Practices

Participants were asked about different practices that were undertaken during project initiation phase (Table 7). The majority of organisations declared they prepared a schedule (82% stated 5 or 4) as well as a budget (79%). A bit less conducted a needs analysis (74%) and included all the team members in project definition (69%). Based on this declaration one can assume a majority of the organisations implement these practices in their project management.

Table 7. On the scale 1–5 please evaluate how much do you agree with statements below.

Before the project started…	5	4	3	2	1	Average
A precise schedule was made	62	47	4	14	6	4,09
All the stakeholders role and needs were evaluated	47	51	12	16	7	3,86
A budget was prepared	69	36	9	11	8	4,11
Every team member had influence on the project definition	41	51	13	19	9	3,72

* 5- I fully agree 4 – I rather agree 3 – I don't know/I don't have an opinion 2 – I rather disagree 1 – I fully disagree

6 Project Implementation

6.1 Meetings

44% of the organisations that participated in the study had meetings once a week or more frequently. 9% of organisations did not organise meetings at all and 28% of organisations had meetings only once a month or less often.

In general, NGOs in Poland did not send reports to meeting attendees because of not preparing such. 42% of organisations declare some of the team members did not attend project meetings. 19% of the respondents had technical difficulties during online meetings (Table 8).

Table 8. How much do you agree with following statements about the meetings organised in the project?

Statements about the meetings	5	4	3	2	1	Average
Every attendee knew the meeting agenda	34	55	17	17	10	3,65
Every attendee had an influence on the meeting agenda	44	58	11	10	10	3,87
Entire project team attended all the meetings	22	47	9	33	22	3,11
Meetings and discussions were run fluently	37	63	14	9	10	3,81
There was one meeting facilitator and everyone knew who takes this role	44	45	21	14	9	3,76
Precise minutes from every meeting were taken	28	49	13	29	14	3,36
Every attendee received detailed reports from the meetings	22	36	15	32	28	2,93
Online meetings were held without any problems	18	46	44	13	12	3,34

* 5- I fully agree 4 – I rather agree 3 – I don't know/I don't have an opinion 2 – I rather disagree 1 – I fully disagree

6.2 Progress Control

The main progress control tool used by non-profit organisations that attended the study is milestones (30% of respondents). 38% of organisations did not use any tools to measure progress in their project. Even though 35% of organisations declared using

Priority Requirement List, since most of the organisations did not use agile project management tools – respondents probably declared using it by their own understanding (Table 9). A general conclusion that the non-profit organisations do not use different project management tools is because they do not know them (78%) and secondly, they don't feel a need to use them (22%).

Table 9. Which progress control tools were used during project implementation? (It was possible to select several answers) n = 133

Progress control tool	Percent of answers
None	38%
Priority Requirement List	35%
Milestones	30%
Surveys	3%
Responsibility Assignment Matrix	3%
Earned Value Management System	1%
I don't know	1%
Key Performance Indicators	1%

Project Evaluation

Vast majority of non-profit organisation that attended the study declares every stakeholder of project was satisfied with the results (Table 10). Interestingly, several organisations declared project achieved success despite high dissatisfaction of its stakeholders, including project beneficiaries. In other part of survey only 55% of organisations declared asking project stakeholders about their feedback, while over 90% declare they were satisfied with the project. The dissimilarity may suggest many organisations believe their stakeholders were satisfied even though they have no real proof on it.

Table 10. How much were different project stakeholders satisfied with the project?

Stakeholder	5	4	3	2	1	Success**	Average
Project coordinator	51	59	8	10	4	3	4,08
Sponsors	30	43	53	2	4	3	3,70
Org. managers	49	67	7	5	5	4	4,13
Beneficiaries	56	62	7	2	6	5	4,20
Other partners	34	46	46	4	3	2	3,78

* 5- Fully satisfied 4 – rather satisfied 3 – I don't know/I don't have an opinion 2 – Rather unsatisfied 1 – Fully unsatisfied
** Declared success among projects evaluated with 1 in stakeholder satisfaction.

In 62% of organisations team members do not prepare reports summarising their work, but despite this fact, 72% of organisations believe the next project team will have easy access to how the project was conducted (Table 11). This may indicate some issues with knowledge management in project portfolio.

Table 11. Which of the statements describe the project closure?

Statements about project closure	5	4	3	2	1	Avg
All the materials used during the project were archived and available for other projects	51	47	13	16	6	3,91
Every team member prepared a report summarising their work	7	23	20	39	44	2,32
The opinion of all the project stakeholders was verified	18	55	21	27	12	3,30
All the project stakeholders were satisfied with it	25	65	33	9	1	3,78
Team of next edition of the project will have easy access to how this project was conducted	47	49	19	11	7	3,89

* 5- I fully agree 4 – I rather agree 3 – I don't know/I don't have an opinion 2 – I rather disagree 1 – I fully disagree

6.3   Project Success

89% of the organisations declared their project ended up with success. Interestingly, many organisations declared project success despite not reaching one of project success criteria (Table 12). 15 out of 16 organisations exceeding budget believed they achieved success. 12 out of 13 organisations that exceeded project schedule believed they achieved project success and 12 out of 16 organisations that decreased project scope also declared achieving project success. In final summary, 66% of the organisations did succeed in all three internal project success criteria. There were 23% of organisations that declared project success despite not reaching one of project success criteria.

Table 12. Project success criteria and perceived success among respondents.

	Number of projects	Perceived success
Exceeding budget	16	15
Exceeding schedule	13	12
Reducing scope	16	12

Using Chi-Square test, interdependence between usage of particular project management tools and sets of tools with internal project success, defined as succeeding in each of the three project success criteria (instead of perceived project success) was evaluated (Table 13). SWOT Analysis, cost accounting and using any planning tools had strong, positive correlation with internal project success, whereas using SMART goals and beneficiary analysis had a positive correlation with project internal success. No correlation was determined for set of project control tools, nor risk analysis, Gantt chart, Work Breakdown Structure, Critical Path Method, Priority Requirement List, milestones and Responsibility Assignment Matrix. The results are similar with Golini, both having cost accounting as one of important tools influencing internal project success. However, Golini clustered them together with Gantt chart and Risk Management, which did not influence the project success in this study. When considering SWOT analysis as a kind of risk analysis and setting smart goals as the simplest scheduling method, all three are also represented in the study. The reason why Gantt chart nor risk analysis proven might be due to low adaptation of these tools in polish NGOs (Gantt chart used by 14% of surveyed organisations and risk analysis used by 25% of them). What is important, for polish NGOs preparing a plan is influencing project success more than controlling the progress (p = 0,01 against p = 0,29).

Table 13. Correlation of project management tools and internal project success.

Tool	p value	Correlation
Cost accounting	0,003351	Strong positive
SMART goals	0,088916	Positive
SWOT analysis	0,011278	Strong positive
Risk analysis	0,200403	No correlation
Beneficiary analysis	0,059188	Positive
Gantt chart	0,142466	No correlation
Work Breakdown Structure	0,594693	No correlation
Critical Path Method	0,822176	No correlation
Priority Requirement List	0,10189	No correlation
Milestones	0,76195	No correlation
Responsibility Assignment Matrix	0,182915	No correlation
Progress control tools	0,28803	No correlation
Planning tools	0,014173	Strong positive

7   Limitations

Participants of the survey were offered an insight to the results as well as a possibility to attend dedicated training aiming to increase their performance after filling the survey. This allowed to receive sufficient amount of data, but might not have been enough motivation for the organisations that are performing well.

8   Conclusions and Further Research

The research discovered several flaws in project management of polish non-profit organisations. Firstly, there is a low understanding of project success. Organisations are satisfied when they will accomplish at least a part of project scope, without analysing its influence on the targeted social problem. This complicates research on project success factors, as majority of the respondents declare reaching project success. Secondly, vast majority of organisations does not use any project management methodologies and approaches mainly due to lack of knowledge about them. Thirdly, organisations does not include all the stakeholders in project definition nor in project evaluation. Such attitude would not be possible in corporate sector as it is regulated by the customers and demand. Finally, in many cases organisations does not use many good project management practices; they do not meet often enough, they does not control project progress, they do not set measurable goals because of lack of knowledge about them.

There is a need for measures of how did the organisation influence the problem they are tackling as they measure their success mainly by beneficiaries satisfaction, level of accomplishment of project scope or amount of participants in their activities. None of these measures relates to purpose non-profit organisations exist for.

References

1. Adamiak, P., Charycka, B., Gumkowska, M.: Kondycja sektora organizacji pozarządowych w Polsce 2015 - RAPORT Z BADAŃ Stowarzyszenie Klon/Jawor, Warszawa (2016)
2. Indeks Społeczeństwa obywatelskiego 2007, p. 8. Stowarzyszenie Klon/Jawor, Warszawa (2008)
3. Strategia wspierania rozwoju społeczeństwa obywatelskiego na lata 2009–2015, Załącznik do uchwały nr 240/2008 Rady Ministrów z dnia 4 listopada 2008 r
4. http://www.newsweek.pl/polska/dzialalnosc-charytatywna-polakow-dobroczynnosc-filantropia-newsweek-pl,artykuly,283938,1.html. Accessed 10 Apr 2018
5. Domański, J.: Zarządzanie strategiczne organizacjami non profit w Polsce. Oficyna a Wolters Kluwer business, Warszawa (2010)
6. Courtney, R.: Strategic Management for Voluntary Nonprofit Organizations, pp. 47–50. Routledge, London (2002)
7. Bogacz, E.: Wojtanowska, Zarządzanie organizacjami pozarządowymi na przykładzie stowarzyszeń krakowskich, pp. 32–37. Wydawnictwo Uniwersytetu Jagielloń-skiego, Kraków (2006)
8. Iwankiewicz-Rak, B.: Marketing organizacji niedochodowych, Marketing Usług, red. A. Styś, PWE, Warszawa, p. 182 (2003)
9. Golini, R., Kalchschmidt, M., Landoni, P.: Adoption of project management practices: the impact on international development projects of non-governmental organizations. Int. J. Proj. Manag. (2015). https://doi.org/10.1016/j.ijproman.2014.09.006
10. Ahsan, K., Gunawan, I.: Analysis of cost and schedule performance of international development projects. Int. J. Proj. Manag. 28, 68–78 (2010)
11. Youker, R.: The nature of international development projects. In: PMI Conference, Baltimore (2003)
12. Ika, L.A., Diallo, A., Thuillier, D.: Critical success factors for World Bank projects: an empirical investigation. Int. J. Proj. Manag. 30, 105–116 (2012)
13. Zeller, M., Meyer, R.L.: The Triangle of Microfinance: Financial Sustainability, Outreach, and Impact. Johns Hopkins University Press, Baltimore (2002)
14. Hermano, V., Lopez-Paredes, A., Martin-Cruz, N., Pajares, J.: How to manage international development (ID) projects successfully. Is the PMD Pro1 Guide going to the right direction? Int. J. Proj. Manag. 31, 22–30 (2013)
15. Bracht, N., Finnegan, J.R., Rissel, C., Weisbrod, R., Gleason, J., Corbett, J., Veblen-Mortenson, S.: Community ownership and program continuation following a health demonstration project. Health Educ. Res. 9, 243–255 (1994)

370   R. Czahajda

16. Bryde, D.J.: Modelling project management performance. Int. J. Qual. Reliab. Manag. **20**, 229–254 (2003)
17. Edwards, M., Hulme, D.: Too close for comfort? The impact of official aid on nongovernmental organizations. World Dev. **24**, 961–973 (1996)
18. Ahlemann, F., Teuteberg, F., Vogelsang, K.: Project management standards—diffusion and application in Germany and Switzerland. Int. J. Proj. Manag. **27**, 292–303 (2009)

Profiling User's Personality Using Colours: Connecting BFI-44 Personality Traits and Plutchik's Wheel of Emotions

Katarzyna Kabzińska, Magdalena Wieloch, Dominik Filipiak,
and Agata Filipowska[✉]

Department of Information Systems, Faculty of Informatics and Electronic Economy,
Poznań University of Economics and Business, Poznań, Poland
agata.filipowska@ue.poznan.pl

Abstract. Personalisation and profiling are of interest of every entity that deals with clients, users, etc. Profiling may concern interests, preferences, demographics, etc. This study is concerned with the issue of user's colour preferences emerging from his personality. Connections between outcomes of the survey on the BFI-44 Personality Traits and 8 colours inspired by representations of Plutchik's Wheel of Emotions were analysed based on responses of 144 respondents. The results were analysed by application of the Apriori algorithm for different personality traits and compared with outcomes of a previous analysis applying linear regression on this data.

Keywords: User profile · User characteristics · BFI-44 · Colours

1 Introduction

Deriving personality of users based on their actions is not a new topic of research. This phenomenon was investigated e.g. in telecommunication [5] and education [7]. Such experiments need to be based on two sources of data: questionnaire and domain data from users targeted by the questionnaire – data must be collected from two sources and concern the same users. There are many different personality tests used. BFI-44 is the best known and the most frequently used [13]. However, what is rarely researched is a connection of a personality with person's colour preferences. Zuckerman–Kuhlman Personality Questionnaire [24] or Eysenks [9] personality tests have been used to measure personality traits and colour preferences – participants had to grade pictures with specific plain colours. These techniques deserve more attention – only several articles that use these questionnaires were published.

In this paper, we want to examine whether a colour is correlated with certain personality traits of people. The aim of our research is to analyse users' personalities and their colour preferences using the Big Five Inventory, also known as BFI-44 [13]. We divide colours into 8 different classes, which is roughly inspired

© Springer Nature Switzerland AG 2019
Z. Wilimowska et al. (Eds.): ISAT 2018, AISC 854, pp. 371–380, 2019.
https://doi.org/10.1007/978-3-319-99993-7_33

the Wheel of Emotions [20]. The paper is structured as follows: the next section presents a short review of the existing body of knowledge. Section 3 presents how the dataset was collected and prepared for further research. Findings are presented in Sects. 4 and 5. The former examines survey results, whereas the latter focuses on patterns found in the data. This is further compared with outcomes of other experiment using the same data, presented in [27]. The paper is summarised in Sect. 6.

2   Related Work

Image analysis is recently gaining on importance thanks to the deep learning methods, especially convolutional neural networks. An important area of research concerns studying pictures appearing on online social media (OSM). Following Souza et al. [23], one can enlist the following topics connected to OSM pictures analysis: engagement [3,28], self-presentation in social media [8,22], age and gender prediction from visual content [12], and recognising basic personality types from photos basing on the BFI-44 questionnaire [6]. OSM data analysis enables to study many aspects related to a personality of a user. Some researchers sought to understand phenomena such as loneliness or sadness. Pittman [19] established a link between social media use and offline loneliness. Students who are active online usually feel less lonely in the real life. By the examination of norms of expressing emotions in OSM, Lup et al. [15] claim that there is a relation between Instagram usage and depression symptoms. Waterloo et al. [26] have carried an extensive study on sadness, anger, disappointment, worry, joy, and pride. These studies proved that positive expressions are perceived better by users across all OSM platforms. Some scholars claim that there are differences in a way of expressing emotions between men and women. In other papers, colours of photos taken in two cities in a certain period of time were compared [10]. Additionally, a number of scholars investigate visual sentiment analysis [4].

The Big Five Inventory (BFI-44) is a personality test (based on the Big Five Model) and consists of 44 questions that measure the Big Five traits such as Extraversion/Introversion, Agreeableness/Antagonism, Conscientiousness/Lack of Direction, Neuroticism/Emotional Stability, and Openness/Closeness to Experience [13]. Following Ahrdnt et al. [2], these five personality features can be defined as follows: *Extraversion* is related to interactions with other people and gaining the energy from them, contrary to being more independent (e.g. action-oriented, outgoing and energetic behaviour vs. inward, solitary and reserved behaviour). *Agreeableness* stems from being trustful, helpful and optimistic, contrary to being antagonistic and sceptical (e.g. cooperative, friendly and compassionate behaviour vs. detached, analytical and antagonistic behaviour). *Conscientiousness* is connected to the level of self-discipline and acting dutifully, contrary to spontaneity (e.g. efficient, organised and planned behaviour vs. careless, easy-going and spontaneous behaviour), *Neuroticism* reflects the inability in dealing with stress, contrary to emotional stability and confidence, addressed the level of emotional reaction to events (e.g. nervous, sensitive and pessimistic behaviour vs. emotionally stable, secure and confident behaviour). *Openness* relies on creativity (e.g. curious, inventive and emotional behaviour vs. consistent, cautious and conservative behaviour).

There is no common agreement among scholars on solving the problem of dividing colours and associating them with particular emotions. For example, Liu et al. focused on red, green and blue [14], whereas Ferwerda et al. added yellow, orange, and violet to this list [6]. Other scholars investigated particular image low-level features, such as chrominance [18]. Plutchik's Wheel of Emotions [20] categorises 24 emotions (8 basic emotions with 3 intensity levels). For illustrative purposes, the wheel is often presented with 8 contrasting colours (each of them has 4 hues, 3 of them with a meaning). However, the intention was not to made connections between colours and emotions, as he focused only on the latter. As a consequence, one can find a number of interpretations of this colour schema in the literature. Sorting colours itself is a non-trivial task and considers a number of algorithms and colour spaces, of which none is appealing for a human eye. As there are many approaches to divide the spectrum of colours, this research was inspired by colours from the popular interpretations of Plutchik's Wheel of Emotions, since this palette contains blue, dark blue, green, dark green, orange, yellow, pink, and red. This work connects the BFI-44 personality traits and preferences for the aforementioned colours.

3 Experiment Setting

Our experiment is based on a questionnaire, which is a combination of the standard BFI-44 test and Flickr images. The latter are used to examine links between user's personality and his colour preferences.

Dataset Preparation. Flickr is one of the most popular OSM platforms. It is image oriented, what makes it particularly useful in our experiment. Flickr has a convenient API, which facilitates preparing a sample. Following approach presented in previous research [11,16], a random photographer was chosen using the *#photography* hashtag. Using a snowball sampling [3], users followed by that photographer were selected (329 users plus the photographer). For each selected Flickr profile, the most recent 100 pictures (or less if a user had fewer photos) have been chosen using Flickr API. This resulted in acquiring 32,056 photos in total. For every photo, a number of pixels of every specific colour has been encoded using the RGB palette. Values in the RGB scale were converted to HSV model, as an extension of previous ranges in order to enhance their perception by human eyes and therefore simplify the detection of more hues of colours. The downloaded images were divided into 9 categories (8 categories of colours inspired by palette from the Wheel of Emotions and a category for those colours that are not included in the wheel). Having assigned the categories of colours, a number of pixels corresponding to the one of 9 categories was assigned to each image. Following [17,21], an assumption that the colour that has the largest number of pixels is the dominant colour of the photo determines the category of it was made. As the dominant colour in a picture may be not obvious for a human eye, the acceptance threshold was set for 70% of pixels in the photo that have to constitute the dominant colour (following [25]). The resulting set was then manually checked to delete pictures with inscriptions, faces and vibrant symbols that can create biases in further research. From each of the 8 categories of photos, 4 randomly chosen pictures were selected (representing 4 different hues from the wheel) what resulted in a sample of 32 photos.

Questionnaire. Questions from BFI-44 were divided into groups that correspond to personality traits (each one has two opposite features, such as extraversion/introversion) from the Big Five model. These questions were associated with pictures divided into 8 categories. The scale for grading pictures was the same as in the questions from BFI-44: 1 – disagree strongly, 2 – disagree a little, 3 – neither agree nor disagree, 4 – agree a little, 5 – agree strongly. A questionnaire was created and then posted on the Internet.

Scoring Formula. The scoring process to assess the user personality traits was carried out as follows (Eq. 1):

1. At first, the number of questions for each of the five features was determined. Then, the number of points (P) for every trait was summed up.
2. For every question, the minimal number of points was equal to one. Therefore, the minimal number of points for each trait equals the number of the questions about it (min).
3. The middle of the scale is the number of the questions about the trait multiplied by three as it was the middle of the scale (M).
4. Maximum (max) is the minimum multiplied by five which is the largest number of points that could be marked by a participant of the survey.

$$F = \left(50 + \frac{100}{\max - \min} \left(\sum P - M\right)\right) [\%] \tag{1}$$

The result of the calculations is the percentage of a given trait (F) for a given user. If it is smaller than 50%, then it is treated as a value of the opposite trait by subtracting 100% minus the given value. The same process was performed in the case of scoring photos in order to determine how often a photo was liked.

4   Questionnaire Findings

144 responses to the questionnaire (of which there were 92 women and 52 men) were collected. The results of the survey are presented in Table 1 and Fig. 1. Respondents were divided into 20 groups, for each trait separately and depending on the score. The respondents based on the score were classified as follows: extreme (score above 75%), minimal (the opposite trait, up to 25%), and and two moderate groups for people who scored more than 25% up to 50% and more than 50% up to 75% accordingly. For each trait, the moderate (between) group was the largest and there are not many people who present extreme types. There was also no person who was classified into the group of Closeness to experience

Table 1. Percentage of participants by trait. Source: own elaboration

	Extraversion	Agreeableness	Conscientiousness	Neuroticism	Openness
[0, 25]	8.33	3.47	0.69	11.11	0.00
(25, 50]	36.81	22.92	34.03	36.81	13.19
(50, 75]	44.44	58.33	52.08	36.11	66.67
(75, 100]	10.42	15.28	13.19	15.97	20.14

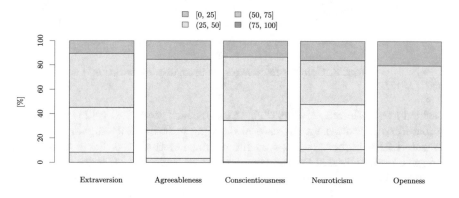

Fig. 1. Percentage of participants by trait. Source: own elaboration

(the opposite of the Openness trait) and there was only one person classified in the Lack of direction group (the opposite trait of Conscientiousness).

Figure 2 presents colour preferences of the respondents. The vast majority of surveyed liked blue photos. They graded it 66% on average and the median was close to 69%. Dark green, the second most liked colour, had the mean score equal to 56% and the median about 59%. Drawing conclusions, dark green also have a big impact on the most of the people. On contrary, respondents did not like yellow photos, as they scored 42% on average and the median at the level of 44%. The rest of the colours was rated around 50%.

Introverts graded most of the colours lower than extraverts and moderate people. For example, blue was marked by extraverts around 70%, moderate people graded it about 66% and introverts at the level of 60% on average. Extraversion or Introversion cannot be determined, as the differences between these two groups are too small to draw conclusions from them. None of the colours is significantly more important comparing extraverts to introverts. People who were classified as antagonistic rated dark blue higher than these with extreme high Agreeableness trait. It indicates that dark blue is a factor that differentiates agreeable and antagonistic people, because antagonistic people rated that colour about 16% higher. With regard to the analysed preferences, this colour is more significant for antagonistic people, so it may be likely that they are more prone to like this colour. The rest of the colours was graded lower or the same as in the Agreeableness group, except for dark blue and red. Regarding people who were

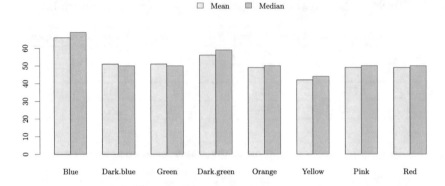

Fig. 2. Colour preferences. Source: own elaboration

classified as emotionally stable, their scores in dark green and yellow were higher compared to neurotics. In the case of dark green it was 25% higher and yellow was rated 21% higher. It leads to the conclusion that it is likely that people who are emotionally stable like more dark green and yellow colours. As only one person in the sample turned out to be a person with Lack of direction, there are not sufficient data to find differences between such distinct sets. None of the respondents was also classified into the Closeness to experience group, therefore a comparison of these groups can not be conducted.

5   Establishing Links Between BFI-44 and Colours

The conducted analysis involved two methods: linear regression and association finding. Firstly, we used a linear regression model in order to find explanation of personality traits with colour preferences. The second approach entails rule mining with the Apriori algorithm. In this paper we present details for the association rule mining, whereas the detailed description of the linear regression models may be found in [27].

Association rules mining was performed in R using the Apriori algorithm [1]. As an entry requirement for rule forming, the value of support parameter was set to 0.1 and confidence to 0.75 (which means that candidates for rules with lower values of at least one of these parameters had been discarded). Rule mining resulted in 28 association rules. Each of them can be perceived as an implication (transactions T) between colours and traits, which formally can be written as a set of items $I = \{i_1, i_2, \ldots, i_n\}$. We define a rule as $X \Rightarrow Y$, where $X, Y \subseteq I$. Since the algorithm does not operate on continuous values, we discretise both: traits and colour preferences using quartiles. For each trait or colour, 1st quartile contains values from 0 to 25%, 2nd quartile has a range of (25,50], 3rd quartile has values in the range of (50,75] and 4th quartile contains values from 75 to 100%. The outcomes of analysis are presented in Table 2 and Fig. 3. It should be noted, that the analysis was performed altogether for all traits, and also for each trait separately.

Table 2. Top 10 rules with the highest confidence. Source: own elaboration

LHS	RHS	Support	Confid.	Lift
{Blue = (50,75], Dark.green = (50,75], Yellow = (25,50]}	{Openness = (50,75]}	0.1250	0.9000	1.3500
{Blue = (50,75], Yellow = (25,50], Pink = (25,50]}	{Openness = (50,75]}	0.1181	0.8947	1.3421
{Blue = (50,75], Dark.blue = (25,50], Dark.green = (50,75]}	{Openness = (50,75]}	0.1181	0.8947	1.3421
{Green = (50,75], Yellow = (50,75]}	{Openness = (50,75]}	0.1042	0.8824	1.3235
{Green = (25,50], Pink = (25,50]}	{Openness = (50,75]}	0.1389	0.8696	1.3043
{Red = [0,25]}	{Openness = (50,75]}	0.1736	0.8621	1.2931
{Blue = (50,75], Orange = (50,75]}	{Openness = (50,75]}	0.1250	0.8571	1.2857
{Orange = (25,50], Red = (25,50]}	{Cons. = (50,75]}	0.1181	0.8500	1.6320
{Orange = (25,50],Red = (25,50]}	{Agreeabl. = (50,75]}	0.1181	0.8500	1.4571
{Blue = (50,75], Dark.blue = (25,50], Orange = (25,50]}	{Openness = (50,75]}	0.1111	0.8421	1.2632

Orange is important for open, extravert, conscious and agreeable people, whereas red was chosen by agreeable and conscientious people. Dark Blue is a factor that distinguishes extraverts, agreeable, conscientious and open people. There are also some interesting rules generated to compare preferences in case of two or more personality traits. For example, if someone is an extravert from the 3rd quartile (from medium to high), likes dark green and rather does not like red or yellow. It also means that this person has Openness value also from the 3rd quartile. Agreeable people from the 3rd quartile, who do not especially like orange and dark blue or red, have Conscientiousness value from the 3rd quartile. This rule works also for conscientious people from the 3rd quartile. If they did not like red and orange or dark blue, then they are agreeable people from the 3rd quartile. Also if someone is open (from medium to high) and does not like yellow and pink or dark blue, then this person perhaps likes blue. These results may be further compared with outcomes of our previous research. There are some colours for regression model and Apriori algorithm that turned out to be significant in both approaches. Table 3 sums up our findings. For Extraversion, it is the orange colour. Orange is also mutual for Agreeableness and Openness. Red is important for Conscientiousness. Green goes with Neuroticism. In the case of Openness, blue and orange are mutual.

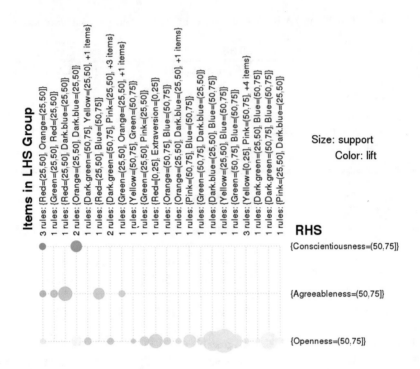

Fig. 3. Influence of colours on a personality trait (rules with confidence exceeding 0.75). Source: own elaboration in R

Table 3. Comparison of two methods' results. Source: own elaboration

	Extraversion	Agreeableness	Conscientiousness	Neuroticism	Openness
Linear model	Blue Orange	Blue Dark Green Orange	Blue Dark Green Red	Blue Green Red	Blue Orange Red
Rule Mining	Orange Dark Blue Dark Green	Blue Dark Blue Green Red Orange	Orange Red Dark Blue Dark green Pink	Yellow Green	Blue Dark Blue Green Dark Green Red Orange Pink Yellow

6   Conclusions

This paper presents how colours can be connected with personality traits. A survey using questionnaire based on Big Five Inventory 44 and a sample of photos classified using 8 colours inspired by Wheel of Emotions was carried out. The goal was to indicate colour preferences using a relation between colours and the strength of the trait, comparing the results of scoring pictures with information about criterion of differentiating people with opposite traits. The research shown that there are some colours that differentiate traits, and that the relation may be seen independently from a method applied for analysis (e.g. Orange in case of Extraversion, Red for Conscientiousness or Green for Neuroticism). Colours related to specific personality traits can lay foundations for creating a profile of user's preferences in OSM. In our case, the data was collected from Flickr, but it could emerge from different data sources. In case of future work, a larger data sample might be considered. One may also test different machine learning approaches for user profiling. A sample from the different parts of the world might be considered in order to spot cultural differences. Results from this study can be helpful for marketing, due to revealed ties between colours and traits.

References

1. Agrawal, R., et al.: Fast algorithms for mining association rules. In: Proceedings of the 20th International Conference Very Large Data Bases, VLDB, vol. 1215, pp. 487–499 (1994)
2. Ahrndt, S., Aria, A., Fähndrich, J., Albayrak, S.: Ants in the ocean: modulating agents with personality for planning with humans. In: European Conference on Multi-Agent Systems, pp. 3–18. Springer, Cham (2014)
3. Bakhshi, S., Shamma, D.A., Gilbert, E.: Faces engage us: photos with faces attract more likes and comments on instagram. In: Proceedings of the 32nd Annual ACM Conference on Human Factors in Computing Systems, pp. 965–974. ACM (2014)
4. Borth, D., Ji, R., Chen, T., Breuel, T., Chang, S.F.: Large-scale visual sentiment ontology and detectors using adjective noun pairs. In: Proceedings of the 21st ACM International Conference on Multimedia, pp. 223–232. ACM (2013)
5. Chittaranjan, G., Blom, J., Gatica-Perez, D.: Who's who with big-five: analyzing and classifying personality traits with smartphones. In: ISWC, pp. 29–36. IEEE Computer Society (2011)
6. Ferwerda, B., Schedl, M., Tkalcic, M.: Predicting personality traits with instagram pictures. In: Proceedings of the 3rd Workshop on Emotions and Personality in Personalized Systems, pp. 7–10. ACM (2015)
7. Geramian, S.M., Mashayekhi, S., Ninggal, M.T.B.H.: The relationship between personality traits of international students and academic achievement. Procedia Soc. Behav. Sci. **46**, 4374–4379 (2012). 4th World Conference on Educational Sciences (WCES-2012) 2012 Barcelona, Spain
8. Geurin-Eagleman, A.N., Burch, L.M.: Communicating via photographs: a gendered analysis of olympic athletes visual self-presentation on instagram. Sport Manag. Rev. **19**(2), 133–145 (2016)
9. Ghorawat, D., Madan, R.: Correlation between personality types and color shade preference. Int. J. Indian Psychol. **1**(04) (2014)
10. Hochman, N., Schwartz, R.: Visualizing instagram: tracing cultural visual rhythms. In: Proceedings of the Workshop on Social Media Visualization (SocMedVis) in Conjunction with the Sixth International AAAI Conference on Weblogs and Social Media (ICWSM 2012), pp. 6–9 (2012)

11. Jamil, N., et al.: Automatic image annotation using color k-means clustering. In: International Visual Informatics Conference, pp. 645–652. Springer, Heidelberg (2009)
12. Jang, J.Y., Han, K., Shih, P.C., Lee, D.: Generation like: comparative characteristics in Instagram. In: Proceedings of the 33rd Annual ACM Conference on Human Factors in Computing Systems, pp. 4039–4042. ACM (2015)
13. John, O.P., Srivastava, S.: The big five trait taxonomy: history, measurement, and theoretical perspectives. Handb. Pers. Theor. Res. 2(1999), 102–138 (1999)
14. Liu, L., Preotiuc-Pietro, D., Samani, Z.R., Moghaddam, M.E., Ungar, L.H.: Analyzing personality through social media profile picture choice. In: ICWSM, pp. 211–220 (2016)
15. Lup, K., Trub, L., Rosenthal, L.: Instagram# instasad?: exploring associations among instagram use, depressive symptoms, negative social comparison, and strangers followed. Cyberpsychol. Behav. Soc. Netw. 18(5), 247–252 (2015)
16. Machajdik, J., Hanbury, A.: Affective image classification using features inspired by psychology and art theory. In: Proceedings of the 18th ACM International Conference on Multimedia, pp. 83–92. ACM (2010)
17. Min, R., Cheng, H.: Effective image retrieval using dominant color descriptor and fuzzy support vector machine. Pattern Recogn. 42(1), 147–157 (2009)
18. Pazda, A.D.: Colorful personalities: investigating the relationship between chroma, person perception, and personality traits. Ph.D. thesis, University of Rochester (2015)
19. Pittman, M.: Creating, consuming, and connecting: examining the relationship between social media engagement and loneliness. J. Soc. Media Soc. 4(1) (2015)
20. Plutchik, R.: Emotion: A Psychoevolutionary Synthesis. Harpercollins College Division, New York (1980)
21. Potluri, T., Nitta, G.: Content based video retrieval using dominant color of the truncated blocks of frame. J. Theor. Appl. Inf. Technol. 85(2), 165 (2016)
22. Smith, L.R., Sanderson, J.: I'm going to instagram it! an analysis of athlete self-presentation on instagram. J. Broadcast. Electron. Media 59(2), 342–358 (2015)
23. Souza, F., de Las Casas, D., Flores, V., Youn, S., Cha, M., Quercia, D., Almeida, V.: Dawn of the selfie era: the whos, wheres, and hows of selfies on instagram. In: Proceedings of the 2015 ACM on Conference on Online Social Networks, pp. 221–231. ACM (2015)
24. Tao, B., Xu, S., Pan, X., Gao, Q., Wang, W.: Personality trait correlates of color preference in schizophrenia. Translational Neurosci. 6(1), 174–178 (2015)
25. Wang, P., Zhang, D., Zeng, G., Wang, J.: Contextual dominant color name extraction for web image search. In: 2012 IEEE International Conference on Multimedia and Expo Workshops (ICMEW), pp. 319–324. IEEE (2012)
26. Waterloo, S.F., Baumgartner, S.E., Peter, J., Valkenburg, P.M.: Norms of online expressions of emotion: comparing Facebook, Twitter, Instagram, and Whatsapp. New Media & Society (2017), https://doi.org/10.1177/1461444817707349
27. Wieloch, M., Kabzińska, K., Filipiak, D., Filipowska, A.: Profiling user colour preferences with BFI-44 personality traits. Lecture Notes in Business Information Processing (2018). (submitted, accepted for publication)
28. Zhong, C., Chan, H.w., Karamshu, D., Lee, D., Sastry, N.: Wearing many (social) hats: how different are your different social network personae? (2017). arXiv preprint arXiv:1703.04791

Research Project Planning Based on SCRUM Framework and Type-2 Fuzzy Numbers

Agata Klaus-Rosińska[(⊠)], Jan Schneider, and Vivian Bulla

Faculty of Computer Science and Management, Wrocław University of Science and Technology, ul. Ignacego Łukasiewicza 5, 50-371 Wrocław, Poland
agata.klaus-rosinska@pwr.edu.pl

Abstract. The aim of the paper is to present the way of research project planning. The solution is based on chosen agile algorithm - SCRUM with using type-2 fuzzy numbers. The case study of real research project was taken into account to show an exemplary usage of suggested solution.

Keywords: Research project · Greedy algorithm · Type-2 fuzzy numbers

1 Introduction

One of the characteristics of today's society is that people want more, better and faster products and services. These circumstances allow the growth of Research and Development area (R&D) where *"Research refers to the process of discovery of a product or service - related knowledge"* [15] and *"Development pertains to the application of this knowledge to create or improve products or services to meet market needs"* [15]. Research and Development is not only happening in the business but also in the educational sector like in universities. The benefits might not be as obvious as in the business world but the results are an important foundation of knowledge that make further innovations possible [9].

The main question is how to manage research projects, which one - traditional or agile approach should be used? The practice is more related to traditional one (e.g. regulations of European Union connected with research projects application forms). However, in the literature can be found also items which suggest to use agile methodology, e.g. [11, 14]. The authors of the article consider that suggestions to be right. This paper was written to demonstrate a theoretical approach of changing a research project planning at a university from traditional approach to agile one. **The science gap here is using type-2 fuzzy numbers to project management** (see Sect. 2). The authors suggest to take into account in research project planning SCRUM framework (agile approach) and type-2 fuzzy numbers (See Sect. 4).

© Springer Nature Switzerland AG 2019
Z. Wilimowska et al. (Eds.): ISAT 2018, AISC 854, pp. 381–391, 2019.
https://doi.org/10.1007/978-3-319-99993-7_34

2 Type-2 Fuzzy Numbers in Project Management

Type-1 fuzzy numbers have been extensively used in project management (e.g. for model stakeholder views and preferences, risk, uncertainty, and incomplete information) since many years. The Scopus search returns more than 600 papers referring to this subject, which means that fuzziness is an important aspect.

The applications found in literature practically use only type-1 fuzzy numbers, which do not offer the modeling possibilities type-2 fuzzy numbers have. Type-1 fuzzy numbers can express either the views or estimation of one stakeholder or of a homogeneous group of stakeholders, but cannot represent often conflicting or even contradictory views of groups of stakeholders with various influence or power degrees, which are different in type and cannot be aggregated in a direct way [2, 5, 6, 8]. *"Type-1 fuzzy logic by its (essentially) crisp nature is limited in modeling decision making as there is no uncertainty in the membership function"* [13]. Nowadays it is generally clear that numerous groups of stakeholders have to be considered and their views or objectives may be conflicting or even contradictory. Project success cannot be achieved or even defined without taking all of these persons or very different groups of persons into account (e.g. [1]). **Type-2 fuzzy numbers make up for this deficiency in modeling** [2, 8].

Above statements are necessary for research project, which are the projects with many types of stakeholders. Considering research projects, which are implemented in universities, stakeholders can be as follow [10]: Project manager, Project team, Administrative units supporting research project implementation (central level), Administrative units supporting research project implementation (departmental level), The authorities of the university/school which implementing research project (central level), The authorities of the university/school which implementing research project (departmental level), Institutions funding research projects, Potential recipients of the project results, Cooperating institutions (collaborators), Advisory (consultative) institutions. **Each of the group can have different opinion and power degree related to project management processes.**

3 The Case Study - Main Information About Chosen Research Project

The case study is based on a project undertaken by a project team at the Wroclaw University of Science and Technology. The research project title is *"Identification of success and failure factors of research projects, with a special emphasis on projects at universities and higher school in countries belonging to European Union and being in various development stages, using the example of Poland and France"*. The general aim was to increase the effectiveness and efficiency of research projects executed at universities and higher schools. The funding period was initially from the 17th of March 2015 until 16th of March 2016 and the projects budget was 182 585 PLN. The National Science Center supported the project with the stated amount.

To achieve the previously named main objective, six partial goals were needed. First of all, stakeholders of executed research projects at universities or higher school had to be identified and their attitudes, opinions, competences towards project management established. Furthermore, it was important to define individual definitions of a "successful research project", but also problems and dangers in various research stages. Another important partial goal was the identification of procedures, methods, techniques and approaches in research project management at universities and higher schools. Furthermore, research types at universities and higher schools had to be clarified. Finally, it was important that a comparative analysis of the obtained results for both countries took place.

4   The Case Study - How to Plan the Research Project with Using Agile (SCRUM) Approach and Type-2 Fuzzy Numbers?

In Sect. 4 the transformation from the traditional project planning (used for application process) to the agile project planning approach (SCRUM) is being undertaken. The approach uses type-2 fuzzy numbers.

4.1   Teamwork

Teamwork is something many people take for granted, but it is hard to build a team that is motivated and works well together. That point is very important in agile project management (APM) and there are several thoughts on how to accomplish a good team composition. Moreover, it is important to build team correctly right at the beginning with the focus on the needed outcome. First of all, team members have to have an updated mindset [12], which means that the team should be excited and motivated to work on the project [7]. It also means that everyone should know where the project is going and everyone should want to achieve the goals with the help of the suggested methods. Secondly, it is important to build a team that is balanced but also divers. A well-rounded team is composed of people that are experienced, have the technical know-how and workers who have personal - strengths like flexibility, creativity and willingness to collaborate. Focusing on both types is a strong step towards a great project outcome [12]. The last step of having a successful team is to build a supportive environment.

4.2   Formalities First

After building a team it is necessary to assure everyone understands the for the project chosen management approaches. There is a lot of literature about APM and SCRUM, but it unfortunately differs on the basics sometimes. To not have future disagreements on basic characteristics, it should be made sure that everyone is "on the same page" and understands the formalities. Another important point is to ensure the whole project team understands that APM is about being productive and efficient [16]. This assumption will lead to a manageable Product Backlog. During the first meeting, the

SCRUM roles (Product Owner, SCRUM Master and Team) can be assigned as well. The Product Owners task for the official first Sprint meeting would be in theory to prepare a list of requirements and their prioritization that he wants to have accomplished by the end of the project. Considered that the research project is held at a university, the team roles are most likely not as strictly defined as in e.g. IT-projects, which means that the Product Owner and SCRUM Master do not only execute their actual roles, but also work on the fulfillment of tasks during the sprints. These roles are most likely be taken by the same person during this specific project. This thought matches the observation of Klaus-Rosinska and Skowron [11] who named that kind of Product Owner an "*intern client with involvement in project*". Moreover, the team is also involved in preparing requirements for the Product Backlog.

4.3   How to SCRUM Research Projects

Right at the beginning the Sprint Planning Meeting, which represents the start of the project, has to be carried out. Here, the whole research team meets and theoretically discusses the prioritized list of requirements that lead to the overall goal. Since it is a research project, Product Owner and SCRUM Master are most likely the same person and the Team is involved in creating User Stories. Additionally, third group of stakeholders can be taken into account, e.g. representatives of administrative units supporting research project implementation (on departmental level).

The whole list might not be finished during the first meeting and has to be worked on. The overall goal of the project which was the increase of the effectiveness and efficiency of research projects executed at universities must be in the center of the discussion at all times.

Table 1 is a proposed "Epic-Product Backlog".

Table 1. Epic - Product Backlog.

ID	EPIC
1	As a project member, I want to identify stakeholders of a research project executed at a university or a higher school in both countries, so that I can find out their attitudes/opinions/competences with respect to project management
2	As a project member, I want to contact stakeholders in both countries, so that I can find out their attitudes/opinions/competences with respect to project management
3	As a project member, I want to analyze literature, so that I can identify procedures, methods, techniques and approaches used in research projects management at universities and higher schools
4	As a project member, I want to identify problems and dangers of research project management at universities and higher schools, so that I can analyze their input on success and failure factors of research projects
5	As a project member, I want to identify research types which were worked on at universities and higher schools, so that I will be able to classify research projects easier
6	As a project member, I want to compare the outcome of both countries, so that I can find out differences

Epic - Product Backlog is a list of Epics, that are large User Stories, which are going to be broken down into smaller tasks in the Product Backlog. Each Epic is written in the widely-used form of a User Story. After the big categories of User Stories are established, it is time to break these down in smaller ones which can be prioritized in a Product Backlog. For this article the authors decided to present only Epic-Product Backlog, so the Epic-User Stories will be the base for Sprint planning.

The team needs to evaluate the list of User Stories with the consideration of available technology, skills and capabilities of the members. The project members and other stakeholders in this project had very broad qualifications, which meant that each team member was able to work on each User Story.

4.4   Research Project Planning - Greedy Algorithm with Fuzzy Numbers in Suggested Approach

The idea of research project planning is to use the SCRUM framework and the greedy algorithm, which will have influence on the Sprint. The Greedy algorithm is supposed to find the order of Epic-User Stories, which would go to the different Sprints. How each User Story will be evaluate? With using two criterion "Effort points" and "Priority points".

One way to do this is to look at the respective proportions of the subjectively ascribed effort points (in the further referred to as weights w_i) and priority points (in the further referred to as values v_i). In our case the effort points of an epicare roughly equivalent to the product of (number of members) and (days). A scale from 1 to 5, where 1 is not important and 5 very important, has been used for the "Prioritization".

The decision as to the order in which the respective epics are to be executed is then very aptly taken by applying the 0-1 knapsack model to be solved by a simple greedy algorithm:

The 0-1 knapsack model
The 0-1 knapsack problem is of course formulated as maximizing the objective function:

$$x_1 \cdot v_1 + x_2 \cdot v_2 + \ldots x_n \cdot v_n \to \max \tag{1}$$

subject to the single constraint

$$x_1 \cdot w_1 + x_2 \cdot w_2 + \ldots x_n \cdot w_n < = \boldsymbol{B}. \tag{2}$$

With $x_i \in \{0, 1\}$ and the coefficients v_i, w_i (values and weights) as well as the upper bound B being real numbers in the classical setting, and type-2 fuzzy sets, in the setting of this paper.

A Simple Greedy Heuristic for the Knapsack Problem

A popular heuristic for the classic 0-1 knapsack problem is defined by first forming the proportions:

$$p_i = \frac{w_i}{v_i} \tag{3}$$

of values and weights, and then placing those fractions in ascending order, and subsequently "filling up the knapsack" with the corresponding weights $w_{i'}$ until the barrier B is met.

A Certain Class of Type 2-Fuzzy Numbers (Sets, Intervals)

With both priority points and effort points it is difficult to give point estimates. Therefore, to ensure maximum expertise and experience in the process of estimation authors have decided to engage three different teams of experts, each of which again consists of three individual experts. Each individual expert in the team is now asked to give an interval estimate of the expected duration of each of the six EPICS (Fig. 1).

EPIC 1	Expert A	Expert B	Expert C
Team 1	[2, 8]	[3.5, 10]	[4, 9]
Team 2	[4.5, 6.5]	[1, 12]	[5, 7]
Team 3	[3, 9]	[2.5, 8.5]	[1.5, 6]

Fig. 1. Table of team and individual interval estimates

In this paper authors propose to model the effort points as certain type-2 fuzzy sets, following a two-step procedure which was developed in 2015 by Wagner et al. Havens from the University of Nottingham [2].

1. The procedure first aggregates the three team intern individual expert interval estimates into a single compound type-1 fuzzy interval.
2. In the second step the three type-1 fuzzy intervals are aggregated into a single type-2 fuzzy set.

Step 1. The procedure of moving from a number of classic intervals to single type-1 fuzzy interval. Authors refer the reader to stop for the mathematical technicalities and instead demonstrate the procedure graphically by example of the first team's interval estimates.

For each of the team's three individual estimates a fuzzy type-2 interval $(E_i, i = 1...3)$ is now constructed, by ascribing the highest level of importance (1) to the intersection of all individual interval estimates, that is to the interval as to which all experts agree.

Next a level of importance of 2/3 is ascribed to where at least two experts agree. And finally 1/3 to interval values included in at least one of the experts estimate (Fig. 2):

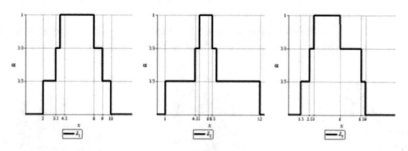

Fig. 2. Three type-1 fuzzy intervals generated by the nine classic intervals from Fig. 1.

The next Fig. 3 shows all three type-1 fuzzy intervals in one:

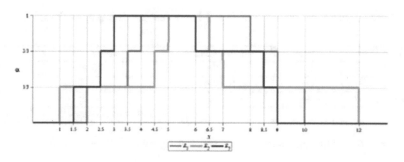

Fig. 3. The three type-1 fuzzy intervals of Fig. 2 in one graph

Step 2: From type-1 fuzzy intervals to general type-2 fuzzy sets. Authors now proceed as in step 1 and look for the intersections of the epigraph above to receive a nested sequence of type 1 intervals, F_1, F_2, F_3 as illustrated in Fig. 4:

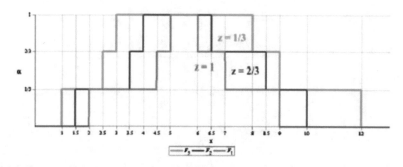

Fig. 4. Three nested type 1 intervals generated by intersections of the three fuzzy intervals of Fig. 3

Authors now proceed as in step 1 and ascribe a level of importance of 1 to the innermost area which represents agreement of all aggregated team opinions, (represented by the red graph in Fig. 4), an importance level of 2/3 for two agreeing estimates (blue color), and finally a level of 1/3 for the outermost region, represented by the green line F_3.

The arising two-dimensional object is a type-2 fuzzy set, which may be represented by a three-dimensional graph as shown below in Figs. 5 and 6).

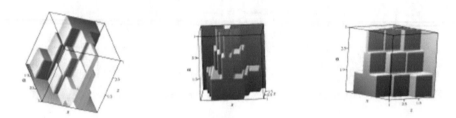

Fig. 5. The type-2 fuzzy set generated by the three type-1 fuzzy intervals of Fig. 4

The type-2 fuzzy set depicted in Fig. 5 has a representation as 9 classic intervals:

EPIC 1	z-level = 1/3	z-level = 2/3	z-level = 1
α-level = 1/3	[1, 12]	[1.5, 10]	[2, 9]
α-level = 2/3	[2.5, 9]	[3.5, 8.5]	[4.5, 7]
α-level = 1	[3, 8]	[4, 6.5]	[5, 6]
			Possibilistic mean value: 5.722.

Fig. 6. Table representation of the type-2 fuzzy set corresponding to Epic 1

What remains to be defined, in order to give meaning to Eqs. (1) and (2) is an arithmetic and a method of comparison $\leq$ for type-2 fuzzy sets of the specific kind defined in [2]. For both the arithmetic and comparison method we rely on the classic work of Goetschel and Voxman [3, 4]. Figure 7 shows the arithmetic of type-1 fuzzy intervals.

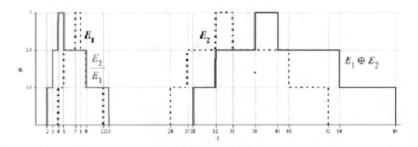

Fig. 7. The sum $\mathbf{E}_1^* \oplus \mathbf{E}_2^*$ (blue) and the fraction $\mathbf{E}_1^*/\mathbf{E}_2^*$ (red) of two step type fuzzy intervals $\mathbf{E}_1^*, \mathbf{E}_2^*$.

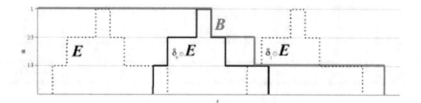

Fig. 8. E^* is trivially within the upper bound B^*, while $\delta_0 E^*$ is still just within, and in some sense maximal, whereas $\delta_1 E^*$ is outside the bound, although the intersection $\delta_1 E^* \cap B^*$ is not empty.

A graphic illustration of the upper bound to be satisfied in (2) is given in Fig. 8.

Finally, authors use the *possibilistic mean value* $E_{\mathcal{P}}$ as defined in [4] as a means of a representation of the proportions $\frac{w_i}{v_i}$. in real numbers ($\mathbb{R}$).

This is a so-called ranking function on the set of fuzzy sets. The *possibilistic mean value* corresponding to Epic 1 is $\frac{5.722}{5} = 1.444$.

4.5   Research Project Planning - Numerical Example

Returning to our case study, beneath are figures (Figs. 9 and 10) representing second EPIC.

EPIC 2	Expert A	Expert B	Expert C
Team 1	[3, 5]	[3.5, 6.5]	[4, 7]
Team 2	[2, 3]	[2.25, 4.25]	[3.25, 4.75]
Team 3	[3.25, 4.75]	[2.5, 3.5]	[1.5, 2.75]

Fig. 9. Table of team and individual interval estimates EPIC 2

EPIC 2	z-level = 1/3	z-level = 2/3	z-level = 1
α-level = 1/3	[1.5, 7]	[2, 4.75]	[3, 4.75]
α-level = 2/3	[2.25, 6.5]	[2.5, 4.25]	[3.5, 3.5]
α-level = 1	[3.25, 5]	[3.25, 3]	[4, 2.75]
			Possibilistic mean value: 3.708.

Fig. 10. Table representation of the type-2 fuzzy set corresponding to EPIC 2

Because of the pages limitations, the authors decided not to put tables related to other EPICS (for the record, it was identified six epics, look Table 1). However, Fig. 11 presents proportions for whole identified EPICS.

	EPIC 1	EPIC 2	EPIC 3	EPIC 4	EPIC 5	EPIC 6
$E_P(w_i)$	5.722	3.708	4.375	2.875	2.111	4.042
v_i	5	4	4	3	2	1
$E_P\left(\frac{w_i}{v_i}\right)$	1.144	0.742	1.094	0.9583	1.056	0.5052

Fig. 11. The proportions $\frac{w_i}{v_i}$

which gives us the following order:

$$EPIC\,1 > EPIC\,3 > EPIC\,5 > EPIC\,4 > EPIC\,2 > EPIC\,6 \qquad (4)$$

It is clear that epics 1, 3, 4 and 5 are not far apart, meaning that the order is not very stable.

It is enough that a single expert, for instance expert C in team 3, EPIC 5 changes his interval estimate from [0.5, 1.75] to [2.5, 7.5], thus yielding a *possibilistic mean value* of $E_P(EPIC\,5) = 2.333$ and subsequently $E_P\left(\frac{w_i}{v_i}\right) = 1.1665$, to change the order to

$$EPIC\,5 > EPIC\,1 > EPIC\,3 > EPIC\,4 > EPIC\,2 > EPIC\,6 \qquad (5)$$

And in this order User Stories should go for the Sprints. What should be determine here is the capacity of the Sprints. The capacity will indicate which User Stories are included in the first and in the next Sprints.

5   Conclusions and Further Works

The approach proposed by the authors for planning research projects is a "new approach", considering the use of 2-type fuzzy numbers.

It is simple in use but requires the establishment of a number of important issues, namely:

(1) what kind of Product Backlog adopt? In the presented case study, it was created on a large scale of generality (Epic-Product Backlog), due to the desire to simplify mathematical calculations. In fact, suggestion is to break down Epics-User Stories into more detailed items.

(2) When adopting more detailed solutions in point 1, the computational machine becomes more complicated. It would be worth considering how to make it easier for users to grasp the proposed approach.

(3) It should be considered how to determine the capacity of individual Sprints. Considering that in suggested approach to determine the order of User Stories the proportions $\frac{w_i}{v_i}$. is proposed.

References

1. Davis, K.: Different stakeholder groups and their perceptions of project success. Int. J. Project Manag. **32**(2), 189–201 (2014)
2. Garibaldi, J.M., Anderson, D.T., Wagner, C., Miller, S., Havens, T.C.: From interval-valued data to general type-2 fuzzy sets. IEEE Trans. Fuzzy Syst. **23**(2), 248–269 (2015)
3. Goetschel Jr., R., Voxman, W.: Topological properties of fuzzy numbers. Fuzzy Sets Syst. **10**(1–3), 87–99 (1983)
4. Goetschel Jr., R., Voxman, W.: Elementary fuzzy calculus. Fuzzy Sets Syst. **18**(1), 31–43 (1986)
5. Gong, Y., Hu, N., Zhang, J., Liu, G., Deng, J.: Multi-attribute group decision making method based on geometric Bonferroni mean operator of trapezoidal interval type-2 fuzzy numbers. Comput. Ind. Eng. **81**, 167–176 (2015)
6. Han, Z., Wang, J., Zhang, H., Luo, X.: Group multi-criteria decision making method with triangular type-2 fuzzy numbers. Int. J. Fuzzy Syst. **18**(4), 673–684 (2016)
7. Holmberg, O.: Six Steps for Implementing Agile across the Organization (2013). https://www.agileconnection.com/article/six-steps-implementing-agile-across-organization?page=0%2C. Accessed 25 May 2017
8. Ilieva, G.: Group decision analysis with interval type-2 fuzzy numbers. Cybernet. Inf. Technol. **17**(1), 31–44 (2017)
9. Johnson, P.: The importance of academic research (2012). http://www.news.uwa.edu.au/201203194542/vice-chancellor/importance-academic-research. Accessed 1 May 2017
10. Klaus-Rosińska, A., Kuchta, D.: The success and failure of research projects according to the opinions of various stakeholders. In: Gómez Chova, L., López Martínez, A., Candel Torres, I. (eds.) 9th International Conference of Education, Research and Innovation ICERI2016 Conference Proceedings, IATED Academy, pp. 1947–1956 (2016)
11. Klaus-Rosińska, A., Skowron, D.: Zastosowanie metodyk zwinnych do zarządzania projektami badawczymi szkół wyższych. In: Nauka i praktyka w zarządzaniu projektami: Project Management Excellence Forum, Tempo, Wrocław (2012)
12. Montony, A.M.: 3 key steps to build an agile team (2014). https://www.cprime.com/2014/04/3-key-steps-to-building-an-agile-team/. Accessed 25 May 2017
13. Runkler, T., Coupland, S., John, R.: Interval type-2 fuzzy decision making. Int. J. Approx. Reason. **80**, 217–224 (2017)
14. Spałek, S., Zdonek, D.: Zwinne podejście projektowe a projekty badawcze. Zeszyty Naukowe Politechniki Śląskiej, Seria: Organizacja i Zarządzanie **64**(1894), 242–249 (2013)
15. Wei, C.-C.: R&D Management Body of Knowledge, pp. 11–23. American Project Management Institute, Newtown Township (2012)
16. Weisbach, R.: How I led 6 R&D groups through an agile transition. https://techbeacon.com/how-i-led-6-rd-groups-through-agile-transition. Accessed 1 June 2017

Fuzzy Analytic Hierarchy Process for Multi-criteria Academic Successor Selection

Hamidah Jantan[1($\boxtimes$)], Yau'mee Hayati Mohamad Yusof[2], and Siti Nurul Hayatie Ishak[1]

[1] Faculty of Computer and Mathematical Sciences,
Universiti Teknologi MARA (UiTM),
21080 Kuala Terengganu, Terengganu, Malaysia
{hamidahjtn,sitinurul}@tganu.uitm.edu.my
[2] Faculty of Business and Management, Universiti Teknologi MARA (UiTM),
23000 Dungun, Terengganu, Malaysia
Yaume555@tganu.uitm.edu.my

Abstract. The selection of leader for academic position in Higher Learning Institution (HLI) involves Multi Criteria Decision Making (MCDM) process. The decision making becomes complicated once it deal with the multiple candidates, multiple conflicting criteria and imprecise parameters. In addition, the uncertainty and vagueness of the experts' opinion is considered as the prominent characteristics of the problem. This paper proposed an academic multi-criteria succession selection approach using Fuzzy Analytic Hierarchy Process (FAHP). This study consists of three phase's i.e. academic multi-criteria model development, data collection, successor selection using FAHP and result analysis. The dataset was collected from several assessors in selected HLI based on the proposed multi-criteria model for academic leader. The aims of this study is to determine the best candidate for academic position based on the higher weight obtained by the candidate. The potential academic successor was obtained after analyzing different dataset for the same candidate that evaluated by different assessors. In future, this study attempts to optimize the result in selection process by incorporating with soft computing method such bio-inspired method such as Particle Swarm Optimization (PSO), Ant Colony Optimization (ACO), Artificial Bee Colony (ABC) and etc.

Keywords: Multi-Criteria Decision-Making (MCDM)
Fuzzy Analytic Hierarchy Process (FAHP) · Academic leader
Successor selection

1 Introduction

Successor selection for any career position is an important task in Human Resources Management (HRM) department that requires adequate selection criteria for difference positions. Succession planning is a subset of talent management activity to satisfying organization requirements for optimally effective organization and to maintain

Z. Wilimowska et al. (Eds.): ISAT 2018, AISC 854, pp. 392–404, 2019.
https://doi.org/10.1007/978-3-319-99993-7_35

organization continuity though successor selection process [1, 2]. The successor selection involves the process of searching the best candidate as successor in selection decision making process. This successor selection for specific position process mainly involves the evaluation of different criteria and various candidate attributes which dealing with vagueness decision judgement, it can be considered as a multiple criteria decision making (MCDM) problem for uncertainty problems [3].

Decision making proses is intensively performed in daily activities involve multiple evaluation criteria for selecting the best decision that needed for future planning. MCDM models normally consist of a finite set of alternatives among which a decision-maker has to rank and decide. One main goal of MCDM is to aid decision making in integrating objective measurements with value judgments that are not based on individuals' opinion but on collective group ideas [4]. The list of some popular MCDM methods which have been frequently used by researchers to solve some real-world multiple criteria problems such as Analytic Hierarchy Process (AHP), Analytic Network Process (ANP), TOPSIS and etc. [5, 6]. One of the most prevalent MCDM techniques is the Analytic Hierarchy Process (AHP). AHP is a structured technique for helping people deal with complex decision that helps decision makers that suitable to their needs. However, due to vagueness and uncertainty in assigning priorities for the selected decision, fuzzy set theory based AHP approach is a potential approach to handle this issue [7]. In Fuzzy AHP (FAHP), the triangular fuzzy numbers (TFNs) and linguistic variables are used to achieve better performance especially when dealing with the vague input data and uncertainty decision judgment. Due to that reason, this study attempt to use FAHP approach in academic successor selection for specific academic position.

This paper introduces in following manner: related work regarding multi-criteria decision making (MCDM), Fuzzy Analytic Hierarchy Process (FAHP). The second section discussed the research method and the third section is described result and discussions; the last section is conclusion and future work.

2 Related Work

2.1 Multi-Criteria Decision Making (MCDM)

Multi-criteria decision making (MCDM) methods deal with the process of making decisions in the presence of multiple objectives. A decision-maker (DM) is required to choose among quantifiable or non-quantifiable and multiple criteria. The main goal of MCDM is to aid DMs in integrating objective measurements with value judgments that are not based on individuals' opinion but on collective group ideas. The objectives are usually conflicting and therefore, the solution is highly dependent on the preferences of the decision-maker and must be a compromise.

Multi-criteria decision making environment exist in many areas where multiple conflicting criteria raise and need to be evaluated. MCDM models normally consist of a finite set of alternatives among which a decision-maker has to rank and decide. Often a finite set of criteria need also to be weighted according to their relative importance. The evaluation of multi-criteria decision making (MCDM) problem in the presence of many

Table 1. MCDM methods and application areas

Application areas	Method
Selection of University Academic Staff	Fuzzy Analytic Hierarchy Process (FAHP) [9]
Selection of a New Hub Airport	SAW (Simple Additive Weighting), TOPSIS (Technique for Order Preference by Similarity to the Ideal Solution) and AHP (Analytic Hierarchy Process) [10]
System Platforms for Mobile Payment	(Elimination and Choice Translating Reality - English) (ELECTRE) [11]
Banking Performance Evaluation	SAW, TOPSIS, and VIKOR [12]
Mobile Banking Adoption	AHP (Analytic Hierarchy Process) [13]
Finding The Right Personnel in Academic Institutions	Simple Additive Weighing (SAW), Weighted Product Method (WPM), Analytical Hierarchy Process (AHP), TOPSIS (Technique for Order Preference by Similarity to Ideal Solution) [14]
Academic Staff Selection	Fuzzy (Elimination and Choice Translating Reality - English) (ELECTRE) [15]
Supplier Selection Problem	Fuzzy Analytic Hierarchy Process [3]

criteria and sub-criteria. MCDM methods have been applied to different applications and find the best solution to choose the best alternative. Some of widely used MCDM methods include Analytic Hierarchy Process (AHP), Analytic Network Process (ANP), TOPSIS and etc. [8]. The following Table 1 is the list of some MCDM methods which have been frequently used by researchers to solve some real-world multiple criteria problems.

2.2   Fuzzy Analytic Hierarchy Process (FAHP)

Fuzzy Analytic Hierarchy Process (FAHP) method is a systematic approach to the alternative selection and justification problem by using the concepts of fuzzy set theory and hierarchical structure analysis [5]. The FAHP technique can be viewed as an advanced analytical method developed from the traditional Analytic Hierarchy Process (AHP) to handle the uncertainties inherent in the AHP method. AHP is a useful mathematical method for solving MCDM problems, where a choice has to be made from a number of alternatives based on their relative importance. The AHP techniques form a framework for decisions that use a one-way hierarchical relation with respect to the decision layers. The hierarchy is constructed in the middle level(s), with decision alternatives at the bottom, as shown in (see Fig. 1).

In FAHP method, the pair-wise comparisons in the judgment matrix are fuzzy numbers and use fuzzy arithmetic and fuzzy aggregation operators, the procedure calculates a sequence of weight vectors that will be used to choose main attribute. Triangular fuzzy numbers (TFNs) were introduced into the conventional AHP in order to enhance the degree of judgment of decision maker. There is an extended literature on multi-criteria decision making that use FAHP that show in Table 2.

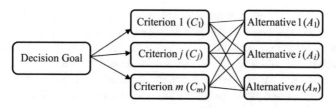

Fig. 1. Hierarchy for a typical three-level MCDM problem [16].

Table 2. Selection process using FAHP approach.

Application	Issues
Selection of University Academic staff [9]	Academic staff selection contains uncertainties which pose another problem, since the AHP lacks the ability to deal with imprecise and subjective judgment in its pair-wise comparison process
Academic Staff Selection [16]	Unsatisfactory selections may occur with assessment and evaluation tools, such as written or oral exams and tests, which may not be based upon any certain criteria and/or weight
Evaluation of students' projects [17]	Traditional evaluation methods is difficult to carry out paired comparison among projects when the number of projects increases; it takes a lot of time to carry out numerical procedures for each project; and there can occur probable errors in the evaluation process
Landfill location selection [18]	The conventional methods for landfill location selection are insufficient in dealing with the vague or imprecise nature of linguistic assessment
Supplier selection [3]	Decision becomes complicated in case of multiple suppliers, multiple conflicting criteria, and imprecise parameters

2.3   Successor Selection in Succession Planning

Succession planning is a subset of talent management that required optimally effective solution for an organization to maintain future continuity though successor selection process. Succession planning involved the process of identify key areas and positions; identify capabilities for key areas and positions; and identify interested employees and assess them against capabilities. Although succession planning has spelt many affirmative returns to organizations, several reviews have indicated that this corporate initiative lacks in its planning, implementing and managing. Due to these setbacks, this initiative, all too often revered by many successful organizations, may not be too popular by educational organization. However, several literatures [2, 19] have indicated hopes that this initiative can be a significant step in planning for effective pool of talented academicians for the purpose of managerial duties execution in the educational sector. The selection of teaching personnel is the process of choosing individuals that have the necessary up-to-date knowledge, research performance, and language skills and those who match the qualifications required to perform a defined job in the best way [14].

Recent studies show that, in the selection process, the teaching staff is assessed and evaluated based on written and oral exams, based on which the selection is made [16]. Unsatisfactory selections may occur with assessment and evaluation tools, such as written or oral exams and tests, which may not be based upon any certain criteria and/or weight. Succession planning provides continuity of leadership. Therefore, an organization needs to identify, develop, and select successors who are the right candidates with the right skills at the right time for leadership positions [20].

2.4 Multi-criteria Selection in Academic Leadership

There are several studies proposed the model of multi-criteria academic leadership for academic leader in selection process. Australian Higher Education was identified the capabilities that characterize effective academic leaders in a range of roles and has produced resources to develop and monitor these leadership capabilities [21]. This study proposed on the review of the higher education literature on leadership in the UK, US and Australia, notes that little research in higher education is concerned with the issue of effectiveness in leadership. Earlier studies of effective leaders in school education, a review of the limited literature on leadership effectiveness in higher education, benchmarking with overseas higher education leadership groups, and an analysis of existing position descriptions and input from the project's National Steering Committee, identified 25 key indicators, each phrased as a specific form of achievement or outcome. These indicators were clustered into five discrete leadership effectiveness scales as shown in Table 3.

Table 3. Leadership effectiveness criteria scales and items [21].

Scale	Item
Personal and Interpersonal Outcomes	Achieving goals set for your own professional development
	Establishing a collegial working environment
	Formative involvement of external stakeholders in your work
	Having high levels of staff support
	Producing future learning and teaching leaders
Learning and Teaching Outcomes	Achieving high-quality graduate outcomes
	Enhanced representation of equity groups
	Improving student satisfaction ratings for learning and teaching
	Increased student retention rates
	Producing significant improvements in learning and teaching quality
	Winning learning and teaching awards and prizes

(*continued*)

Table 3. (*continued*)

Scale	Item
Recognition and Reputation	Achieving a high profile for your area of responsibility
	Achieving positive outcomes from external reviews of the area
	Being invited to present to key groups on learning and teaching
	Publishing refereed papers and reports on learning and teaching
	Receiving positive user feedback for your area of responsibility
Financial Performance	Achieving a positive financial outcome for your area of responsibility
	Meeting student load targets
	Securing competitive funds related to learning and teaching
	Winning resources for your area of responsibility
Effective Implementation	Bringing innovative policies and practices into action
	Delivering agreed tasks or projects on time and to specification
	Delivering successful team projects in learning and teaching
	Producing successful learning systems or infrastructures
	Successful implementation of new initiatives

The academic leadership capabilities identifying by the optimum focus for each academic leadership role in HCL such as Deputy Vice-Chancellor; Pro Vice-Chancellor; Dean; Associate Dean and etc. In other study, Strengthening Leadership Development Model was proposed by Malaysian as a part of the initiatives of the Government-Linked Companies (GLC) Transformation Programme that contains approaches to leadership development including academic leader that draw on global best practices and the experiences of companies in Malaysia [22]. The model was designed to help develop the human capital that will drive the transformation. Besides, this model also sets out a framework to assess and strengthen companywide leadership development. This leadership model will set out a specific set of leadership behaviors that together will create maximum value for, and embed the desired leadership culture in. Besides that, other research was conducted to propose the solution of academic successor selection in HLI for selected academic main criteria and sub-criteria during the candidates evaluation proses using MCDM approach [9, 16].

3   Research Method

This study consist of three phase's i.e. academic multi-criteria model development, data collection, successor selection using FAHP and result analysis. There are four level hierarchical process for successor selection that was devised as shown in Fig. 2. The first level indicates the goal of the process such as to find the best candidate for academic's position of Deputy Vice Chancellor (DVC). In order to identify the best candidate to this position, some candidates (Level 4) need to fulfill the main criteria (Level 2) and its sub-criteria (Level 3). Four candidates will be evaluated by two assessors. The main focus of this research study is to find the best candidate for academic position of Deputy Vice Counselor. The dataset was collected from several assessors in selected HLI based on the proposed multi-criteria model for academic leader.

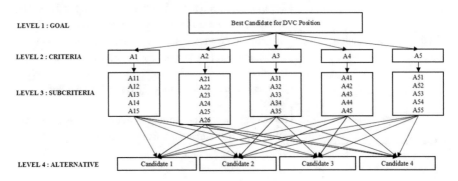

Fig. 2. Hierarchy for three-level MCDM problem.

Fuzzy Analytic Hierarchy Process (FAHP) was used in successor selection process which is consists of several steps (Fig. 3):

Step 1: Develop fuzzy pairwise metrics. The pairwise comparison of different criterion against the overall goal is conducted to construct the fuzzy comparison matrix and then priority value of each criterion with respect to the overall goal is calculated.

Step 2: Calculate fuzzy extent value.

Step 3: Determine and calculate degree of possibility.

Step 4: Determine minimum value of degree of possibility.

Step 5: Calculate normalized weight. The basic 1–5 steps of FAHP are repeated to determine the normalized weight of sub-criteria in response to criteria and that of alternatives with respect to sub-criteria. Then, calculate the final weight. Rank the final weight.

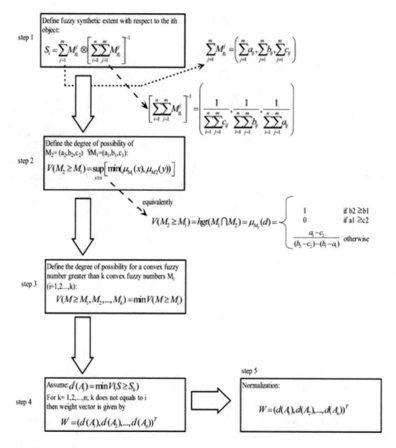

Fig. 3. The steps involve in Fuzzy Analytic Hierarchy Process (FAHP).

In result analysis phase, this study using two datasets for the similar candidates (four candidates) that evaluated by the different assessor in order to verifying the consistency of the selected potential candidate with the proposed academic multi-criteria model.

4   Results Discussion

Academic leadership criteria that are highlighted in previous studies was used in order proposed the model of multi-criteria academic leader as shown in Fig. 4. Thera are five main criteria and 25 sub-criteria are categorized into specific label. In this study, the parameters in the selection of academic staff are labelled with five main criteria which include Effective Implementation (A1), Learning and Teaching Outcomes (A2), Personal and Interpersonal Incomes (A3), Recognition and Reputation (A4) and Financial Performance (A5) that contain 5, 6, 5, 5 and 4 sub items respectively.

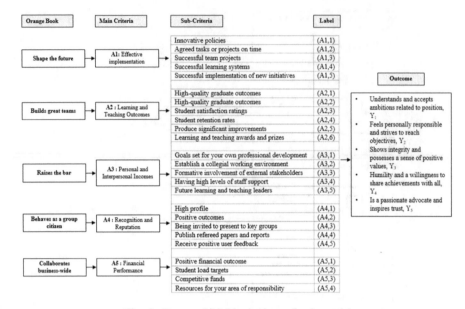

Fig. 4. Proposed Multi-criteria academic model

In experimental phase, FAHP method was applied for two datasets of the different assessors apply for the same candidate. As a case study, the academic position of DVC was used in order to identify the weight for all criteria produce by the FAHP step-by-step process (Fig. 3). Four potential candidates for this position which are C1, C2, C3 and C4 was evaluated by two assessors. A*n* and A(*n,m*) are represent the criteria used in academic successor selection depends on the selected academic position. The consistency measure in FAHP method that represents the consistency index was determined by the *n* x *n* reciprocal pair-wise comparison matrix with triangular fuzzy numbers, the calculation of fuzzy extend value; and the determination of degree of possibility and minimum degree of possibility. The final normalized weight is calculated to rank the

Table 4. Weight distribution for candidates - First accessor

Main criteria	Weight	Sub-criteria	Weight	C1	C2	C3	C4
A1	0.5027	A11	0.299	0.1877	0.5554	0.0693	0.1877
		A12	0.299	0.1877	0.5554	0.0693	0.1877
		A13	0.0515	0.1877	0.5554	0.0693	0.1877
		A14	0.0515	0.1739	0.4783	0.1739	0.1739
		A15	0.299	0.2365	0.5990	0.0822	0.0822
		A11	0.299	0.1877	0.5554	0.0693	0.1877
		A12	0.299	0.1877	0.5554	0.0693	0.1877

(*continued*)

Table 4. (*continued*)

Main criteria	Weight	Sub-criteria	Weight	C1	C2	C3	C4
A2	0.1514	A21	0.5382	0.0856	0.5961	0.0856	0.2328
		A22	0.1114	0.1877	0.5554	0.0693	0.1877
		A23	0.1114	0.1739	0.4783	0.1739	0.1739
		A24	0.0688	0.0856	0.5961	0.0856	0.2328
		A25	0.1114	0.2142	0.5833	0.1013	0.1013
		A26	0.0589	0.2328	0.5961	0.0856	0.0856
A3	0.2097	A31	0.1422	0.1666	0.5003	0.1666	0.1666
		A32	0.3928	0.1666	0.5003	0.1666	0.1666
		A33	0.0360	0.1739	0.4783	0.1739	0.1739
		A34	0.3928	0.2142	0.5833	0.1013	0.1013
		A35	0.0361	0.1666	0.5003	0.1666	0.1666
A4	0.0497	A41	0.1028	0.1666	0.5003	0.1666	0.1666
		A42	0.5271	0.2142	0.5833	0.1013	0.1013
		A43	0.1028	0.2142	0.5833	0.1013	0.1013
		A44	0.0364	0.2142	0.5833	0.1013	0.1013
A5	0.0865	A51	0.3759	0.2142	0.5833	0.1013	0.1013
		A52	0.3759	0.1666	0.5003	0.1666	0.1666
		A53	0.1241	0.2328	0.5961	0.0856	0.0856
			Final weight	0.1879	0.5594	0.0993	0.1534
			Rank	2	1	4	3

Table 5. Weight distribution for candidates - Second accessor.

Main criteria	Weight	Sub-criteria	Weight	C1	C2	C3	C4
A1	0.5027	A11	0.299	0.1539	0.5384	0.1539	0.1539
		A12	0.299	0.4282	0.1015	0.0421	0.4282
		A13	0.0515	0.1539	0.5384	0.1539	0.1539
		A14	0.0515	0.3143	0.3143	0.0571	0.3143
		A15	0.299	0.1539	0.1539	0.1539	0.5384
		A11	0.299	0.1539	0.5384	0.1539	0.1539
		A12	0.299	0.4282	0.1015	0.0421	0.4282
A2	0.1514	A21	0.5382	0.0909	0.4091	0.4091	0.0909
		A22	0.1114	0.0909	0.4091	0.0909	0.4091
		A23	0.1114	0.1539	0.5384	0.1539	0.1539
		A24	0.0688	0.0571	0.3143	0.3143	0.3143
		A25	0.1114	0.1539	0.1539	0.1539	0.5384
		A26	0.0589	0.5384	0.1539	0.1539	0.1539

(*continued*)

Table 5. (*continued*)

Main criteria	Weight	Sub-criteria	Weight	C1	C2	C3	C4
A3	0.2097	A31	0.1422	0.1539	0.1539	0.1539	0.5384
		A32	0.3928	0.0909	0.4091	0.0909	0.4091
		A33	0.0360	0.0000	0.0000	0.0000	0.0000
		A34	0.3928	0.3143	0.3143	0.0571	0.3143
		A35	0.0361	0.1539	0.1539	0.5384	0.1539
A4	0.0497	A41	0.1028	0.1539	0.5384	0.1539	0.1539
		A42	0.5271	0.1539	0.1539	0.1539	0.5384
		A43	0.1028	0.3143	0.3143	0.0571	0.3143
		A44	0.0364	0.1539	0.1539	0.1539	0.5384
A5	0.0865	A51	0.3759	0.0909	0.4091	0.0909	0.4091
		A52	0.3759	0.0571	0.3143	0.3143	0.3143
		A53	0.1241	0.0000	0.0000	0.0000	0.0000
			Final weight	0.1929	0.2955	0.1422	0.3290
			Rank	3	2	4	1

candidates. The highest value of final weight indicates the first rank or the potential candidate for that position. The final weight result will be used to rank the four candidates as shown in Tables 4 and 5 respectively.

The result in Table 4 show the potential candidate for DVC position which is candidate no 2 (C2) with highest final weight 0.5594 and then followed by other candidates C1, C3 and C4. Besides that, the higher weight for main criteria is A1 that representing the Effective Implementation criteria focus on some aspect of managerial skill that needed for an academic leader.

Table 5 shows that C4 is the best candidate for DVC position among four candidates that has 0.3290 as their final weight. Meanwhile, C2 is placed as second potential candidate with 0.2955 final weight. Besides that, the higher weight for main criteria from second assessor also A1 that representing the Effective Implementation criteria. The consistency result can be analyzed based on the results of rank obtained using different sample data given by two assessors. The results show that the rank for each candidate is not the same. But C2 seems to have potential to be the best successor of DVC position based on the ranks of two sample datasets that candidate C2 obtain first and second rank.

5 Conclusion

Decision makers always facing in complex environment to assign a successor. Succession planning is critical to mission success and creates an effective process for recognizing, developing, and retaining top leadership. There are always come out with a new effective approach to achieve accuracy and consistency in succession planning. The aim of this study was to provide an adequate MCDM on academic position selection in HLI. Current approach are based on assumptions and some contains uncertainties. In order to handle uncertainty in order to identify the best candidate for selected position, some criteria have been measured using fuzzy multi criteria selection model as proposed in this study. In future, this study attempts to optimize the result in selection process by incorporating with soft computing method such bio-inspired method such as Particle Swarm Optimization (PSO), Ant Colony Optimization (ACO), Artificial Bee Colony (ABC) and etc.

References

1. Cannon, J.A., McGee, R.: Talent Management and Succession Planning. Chartered Institute of Personnel and Development, London (2011)
2. Abdullah, Z., Samah, S.A.A., Jusoff, K., Isa, P.M.: Succession planning in Malaysia institution of higher education. Int. Educ. Stud. **2**, 129–132 (2009)
3. Ayhan, M.B.: A fuzzy AHP approach for supplier selection problem: a case study in a Gearmotor company. Int. J. Manag. Value Supply Chains **4**, 11–23 (2013)
4. Hosseini-Nasab, H.: An application of fuzzy numbers in quantitative strategic planning method with MCDM. In: International Conference on Industrial Engineering and Operations Management, pp. 555–562 (2012)
5. Aruldoss, M., Lakshmi, T.M., Venkatesan, V.P.: A survey on multi criteria decision making methods and its applications. Am. J. Inf. Syst. **1**, 31–43 (2013)
6. Zardari, N.H., Ahmed, K., Shirazi, S.M., Yusop, Z.B.: Literature review. In: Weighting Methods and their Effects on Multi-Criteria Decision Making Model Outcomes in Water Resources Management, pp. 7–67 (2015)
7. Kahraman, C., Cebeci, U., Ulukan, Z.: Multi-criteria supplier selection using fuzzy AHP. Logist. Inf. Manag. **16**, 382–394 (2003)
8. Li, W., Yu, S., Pei, H., Zhao, C., Tian, B.: A hybrid approach based on fuzzy AHP and 2-tuple fuzzy linguistic method for evaluation in-flight service quality. J. Air Transp. Manag. **60**, 49–64 (2017)
9. Asuquo, S.E., Onuodu, F.E.: A fuzzy AHP model for selection of university academic staff. Int. J. Comput. Appl. **141**, 19–26 (2016)
10. Janic, M., Reggiani, A.: An application of the multiple criteria decision making (MCDM) analysis to the selection of a new Hub Airport. Eur. J. Transp. Infrastruct. Res. EJTIR **2**, 113–141 (2002)
11. Tehrani, M.A., Amidian, A.A., Muhammadi, J., Rabiee, H.: A survey of system platforms for mobile payment. In: 2010 International Conference on Management of e-Commerce and e-Government, pp. 376–381 (2010)
12. Wu, H.-Y., Tzeng, G.-H., Chen, Y.-H.: A fuzzy MCDM approach for evaluating banking performance based on Balanced Scorecard. Expert Syst. Appl. **36**, 10135–10147 (2009)
13. Komlan, G., Koffi, D., Kingsford, K.M.: MCDM technique to evaluating mobile banking adoption in the Togolese banking industry based on the perceived value: perceived benefit and perceived sacrifice factors. Int. J. Data Min. Knowl. Manag. Process (IJDKP) **6**, 37–56 (2016)
14. Kumar, D.S., Radhika, S., Suman, K.N.S.: MCDM methods for finding the right personnel in academic institutions. Int. J. u- and e- Serv. Sci. Technol. **6**, 133–144 (2013)
15. Rouyendegh, B.D., Erkan, T.E.: an application of the fuzzy ELECTRE method for academic staff selection. Hum. Factors Ergon. Manuf. Serv. Ind. **23**, 1–9 (2012)
16. Rouyendegh, B.D., Erkan, T.E.: Selection of academic staff using the fuzzy analytic hierarchy process (FAHP): a pilot study. Tech. Gaz. **19**(4), 923–929 (2012)

17. Çebi, A., Karal, H.: An application of fuzzy analytic hierarchy process (FAHP) for evaluating students' project. Acad. J. Educ. Res. Rev. **12**, 120–132 (2017)
18. Hanine, M., Boutkhoum, O., Tikniouine, A., Agouti, T.: Comparison of fuzzy AHP and fuzzy TODIM methods for landfill location selection. SpringerPlus **5**, 501 (2016)
19. Luna, G.: Planning for an American higher education leadership crisis: the succession issue for administrators. Int. Leadersh. J. **4**, 56–79 (2012)
20. Seniwoliba, A.J.: Succession planning: preparing the next generation workforce for the University for Development Studies. Res. J. Educ. Stu. Rev. **1**, 1–10 (2015)
21. Scott, G., Coates, H., Anderson, M.: Learning Leaders in Times of Change: Academic Leadership Capabilities for Australian Higher Education. ACER, Australian Council for Educational Research, Australian Learning & Teaching Council, University of Western Sydney, Sydney (2008)
22. Strengthening Leadership Development. The Putrajaya Committee on GLC High Performance (PCG) (2006)

The Identification of the Pro-ecological Factors Influencing a Decision to Become a Prosumer

Edyta Ropuszyńska-Surma and Magdalena Węglarz$^{(\boxtimes)}$

Faculty of Computer Science and Management, Wroclaw University of Science and Technology, 50-370 Wroclaw, Poland
{edyta.ropuszynska-surma,
magdalena.weglarz}@pwr.edu.pl

Abstract. The aims of the article are (1) the identification of factors influencing the households' willingness to install renewable energy sources (RES) – and to became a prosumer, (2) the estimation which variables have the stronger influence on probability of the RES installation and which have the weaker influence. The prosumers were defined as households who have the RES installation, produce electricity on their own needs, and any surplus is given back to the grid. They have electricity cost balance once a year. There were taken into consideration the factors connected with pro-ecological and pro-effectiveness behaviors such as: sorting rubbish, the utilization of electro-rubbish, switching off the lights, using LEDs, installing and using energy-saving home appliances and so on. The input data comes from the survey research conducted in November and December 2015 within a project "Modelling prosumers' behaviour on the energy market" funded by the National Centre of Science, by the grant no. 2013/11/B/HS4/01070. It was observed that significant relationships exist between a household's decisions about the RES installation and (1) using LEDs, (2) installing energy-saving home appliances (A+++), (3) sorting rubbish, (4) washing or ironing at particular times, (5) using eco-options in washing machine, (6) switching off electrical equipment (e.g. computer) if nobody uses. The significance of qualitative variables, which influence a household's decisions about the RES installation, was verified by the results of logit model.

Keywords: Prosumer · Logit model · Pro-ecological households' behaviour

1 Introduction

Energy innovations (e.g. increase in the efficiency of micro-installations generating electricity, including renewable energy sources (RES)) and development of IT technologies in energy, e.g. smart metering, changes the logistic chain model of electric energy. The current model of energy flow from the power generator to the passive recipient begins to change in the direction of electricity flow from many dispersed generation sources (including RES installed directly in households) to the so-called active electricity consumers. The change of the attitude of the electricity recipient from passive to active is understood as: increasing the energy consciousness of the recipient and the resulting conscious management of electricity consumption, so as to maximize their benefits, e.g. limiting energy consumption during periods of energy rush, in which

© Springer Nature Switzerland AG 2019
Z. Wilimowska et al. (Eds.): ISAT 2018, AISC 854, pp. 405–416, 2019.
https://doi.org/10.1007/978-3-319-99993-7_36

energy prices are higher, and increasing consumption during periods of reduced daily electricity demand from the electricity grid (in the so-called bottom of the load curve). Such recipients' behavior can and are stimulated by the use of the so-called Demand Side Management (DSM). In addition, the activity of the recipient may take a more advanced form, because it may become so-called prosumer. The concept of the prosumer was introduced to futurist literature in the 1980s by Toffler [13], noting that in the so-called "third wave" of economic development, consumers will operate themselves in accordance with the "do-it-yourself" principle. Later, this concept was adopted in the energy industry [6] and the term "prosumer" began characterize an entity that is both a producer and a consumer of energy. It should also be noted that the change of the existing logistic model in the energy sector was possible due to political and economic changes (liberalization of electricity production and trading) and legislative changes.

The act on renewable energy sources [14], by virtue of the amendment of 2016, defines that the prosumer is only a natural person not conducting economic activity, and the premise for launching microinstallation is to generate energy for own needs and returning its surplus to the distribution network [15]. Details of the legal analysis of the situation of prosumers in Poland are presented in the paper by Ropuszyńska-Surma and Węglarz [7]. This article broadens the concept of a prosumer and includes households that already have RES installed, as well as those who are interested in installing renewable energy and producing electricity for their own needs and want to transfer the surplus to the grid to be able to download it from the grid during the period of greater demand in relation to own production. Such balance settlement is possible under Polish law [15]. The process of changing passive energy consumers to prosumers is slow, because it depends on many factors, such as: knowledge and ecological awareness, household income, age, family model and lifestyle, or place of residence.

The purpose of the article is to identify factors affecting households' decisions regarding the installation of RES and to determine which variables have a stronger impact and which have a lower impact on the likelihood of RES installation. The following alternative H1 hypothesis was proposed in the paper: *Installation of RES depends on pro-ecological and pro-efficiency behaviors*. The pro-ecological and pro-efficiency attitudes include: sorting rubbish (garbage), the utilization of electro-rubbish, battery utilization, switching off the lights, washing or ironing at particular times, switching off chargers after charging the device, unplugging devices that are on standby, using eco-options in a washing machine, boiling water under the lid, switching off electrical equipment (e.g. computer) if nobody uses, using energy-saving lights, using LED lights bulbs, installing energy-saving household equipment (A+++).

Earlier work [10] analyzed demographic, social and economic factors. The results of the obtained model confirmed the hypotheses about the existence of dependence between the installation of RES and: the household decision-maker's gender, age and education, knowledge of electricity tariff, number of people in the household, type of inhabited building (single or multi-family), residential area and status on the labour market.

Research on pro-ecological behaviors of energy consumers (including households) and factors conditioning their installation of RES is conducted by various research centers around the world, for example Scarpa and Willis, using logit model for the majority of the British households [11]. Some authors draw attention to the relationship between pro-ecological habits of recipients and a greater willingness to pay for energy coming from renewable sources [2], or demographic and social factors relation with pro-ecological behavior [1, 12].

2 Logit Model

2.1 Research Sample

The article was prepared on the basis of the results of a research project entitled "Modelling prosumers' behaviour on the energy market" financed by the National Science Center, granted on the basis of Decision No. DEC-2013/11/B/HS4/01070. As part of the project, in November and December 2015, a questionnaire survey was conducted using a telephone method on a sample of 2000 households from Lower Silesia. The survey had 34 questions. A more detailed description of the survey was presented in the works of Ropuszyńska-Surma and Węglarz [7–9]. 949 questionnaires were qualified for the statistical analysis, because of the rejection of questionnaires with incomplete answers.

2.2 Variable Characteristic

As mentioned in the introduction, the aim of the article is to determine which variables have a greater impact and which have a lower impact on the likelihood of installing RES by a household. The endogenous variable presents the result of a rational economic decision of the household. Future prosumers are understood as households that are interested in installing RES, so the endogenous variable has a zero-one form: a $y_i = 1$, when the household has installed RES or is considering its installation in the future, and $y_i = 0$, when there is no RES and household is not interested in its installation in the future.

Because the endogenous variable is a dichotomous variable, a logit model was used for the analysis. The aim of modelling the binomial variable is the forecast of the change in the probability of making a decision to install RES caused by a change in the value of one of the exogenous variables [4]. As explanatory variables: pro-ecological or pro-efficiency attitudes of households were adopted. These variables are multi-variant quality variables (take values from 0 to 5 or values from 0 to 3). At the same time, ranking does not allow to indicate differences in the odds ratios for individual categories of features. For this reason, all explanatory variables have been converted into 0–1 regressors. The original nominal variable, with k variants, was transformed into k-1 artificial variables, and one of the variants is not introduced to the model but is a reference point (the so-called reference group) for other artificial variables [3, 5]. The size of groups and the method of coding variables (e.g. S1) are shown in Tables 1 and 2.

Correspondingly, as the reference subgroups the authors assumed groups of households, which answered that the given behavior does not concern them, e.g. S0 means households, which consider that the problem of sorting waste does not concern them. Other reference subgroups were determined: UE0, UB0, L0, P0, Z0, UD0, E0, G0 and K0.

Table 1. The number of household subgroups.

Feature	Variable (reference group)	Number of observation	Percentage
1	2	3	4
Sorting rubbish	Does not apply (S0)	27	2.8
	Never (S1)	76	8.0
	Rarely (S2)	35	3.7
	Often (S3)	99	10.4
	Very often (S4)	42	4.4
	Always (S5)	670	70.6
The utilization of electro-rubbish	Does not apply (UE0)	113	11.9
	Never (UE1)	136	14.3
	Rarely (UE2)	71	7.5
	Often (UE3)	66	7.0
	Very often (UE4)	27	2.8
	Always (UE5)	536	56.5
Battery utilization	Does not apply (UB0)	59	6.2
	Never (UB1)	87	9.2
	Rarely (UB2)	69	7.3
	Often (UB3)	66	7.0
	Very often (UB4)	24	2.5
	Always (UB5)	644	67.9
Switching off the lights	Does not apply (l0)	4	0.4
	Never (l1)	26	2.7
	Rarely (l2)	40	4.2
	Often (l3)	138	14.5
	Very often (l4)	99	10.4
	Always (l5)	642	67.7
Washing or ironing at particular times	Does not apply (P0)	552	58.2
	Never (P1)	232	24.4
	Rarely (P2)	47	5.0
	Often (P3)	26	2.7
	Very often (P4)	25	2.6
	Always (P5)	67	7.1
Switching off chargers after charging the device	Does not apply (Z0)	16	1.7
	Never (Z1)	90	9.5
	Rarely (Z2)	78	8.2
	Often (Z3)	86	9.1
	Very often (Z4)	47	5.0
	Always (Z5)	632	66.6

(*continued*)

Table 1. (*continued*)

Feature	Variable (reference group)	Number of observation	Percentage
1	2	3	4
Unplugging devices that are on standby	Does not apply (UD0)	10	1.1
	Never (UD1)	463	48.8
	Rarely (UD2)	154	16.3
	Often (UD3)	61	6.4
	Very often (UD4)	17	1.8
	Always (UD5)	244	25.7
Using eco-options in a washing machine	Does not apply (E0)	257	27.1
	Never (E1)	187	19.7
	Rarely (E2)	63	6.6
	Often (E3)	103	10.9
	Very often (E4)	25	2.6
	Always (E5)	314	33.1
Boiling water under the lid	Does not apply (G0)	34	3.6
	Never (G1)	62	6.5
	Rarely (G2)	37	3.9
	Often (G3)	93	9.8
	Very often (G4)	38	4.0
	Always (G5)	685	72.2
Switching off electrical equipment (e.g. computer) if nobody uses	Does not apply (K0)	106	11.2
	Never (K1)	114	12.0
	Rarely (K2)	44	4.6
	Often (K3)	81	8.5
	Very often (K4)	33	3.5
	Always (K5)	571	60.2

However, in the case of installing devices, as reference subgroups the authors assumed groups of households, which responded that they do not have a given type of equipment, e.g.: A0 – households that do not have energy-saving light bulbs. The other B0 and C0 reference groups were defined in an analogous manner. Next, the dependence of the probability of installing RES on the variables listed in the tables was examined.

2.3   The Chance of Installation of RES

While constructing the regression equation of the logit model, all explanatory variables were included in it, which from the point of view of the formulated hypothesis may explain the behavior of the explained variable (y_i). For each feature, a separate logit model was built. Hence, eight models were obtained. Then, using the logit models, odds ratios were determined $OR = \exp\beta_i$ [4], which is interpreted as a relative chance

Table 2. The number of household subgroups.

Feature	Variable (reference group)	Number of observation	Percentage
1	2	3	4
Using energy-saving lights	I have not such devices (A0)	156	16.4
	I have, but there are a few (A1)	130	13.7
	I have a lot of these devices (A2)	355	37.4
	I have only such devices (A3)	308	32.5
Using LED lights bulbs	I have not such devices (B0)	331	34.9
	I have, but there are a few (B1)	210	22.1
	I have a lot of these devices (B2)	242	25.5
	I have only such devices (B3)	166	17.5
Installing energy-saving household equipment (A+++)	I have not such devices (C0)	489	51.5
	I have, but there are a few (C1)	163	17.2
	I have a lot of these devices (C2)	179	18.9
	I have only such devices (C3)	118	12.4

of occurrence of an event in a given subgroup in comparison with the reference group. The calculations were made using the *Gretl* program. The level of significance was 95%. Figures 1, 2, 3, 4, 5 and 6 show the odds ratio of making a decision to install RES through different subgroups of respondents, distinguished for each examined feature (see Tables 1 and 2). The distinguished number of subgroups is equal k-1 and they are so-called artificial variables. Moreover, the reference group is marked with a black line.

In the case of models with the exogenous variable L, Z, UD, G, A, all tested parameters were statistically insignificant, which means that the chance of deciding on installing RES by distinguished household subgroups is similar to the reference group. Therefore, the results for these models are not presented in the paper. In other models, some exogenous variables are not statistically significant ($p > 0.05$) and there was a suggestion to remove these variables from the model. However, this was not done because the purpose of building the model is to analyze the odds ratio of k-1 subgroups in relation to the determined subgroup called reference [5]. The sign of

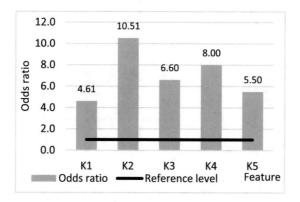

Fig. 1. Odds ratio for exogenous variable: on switching off electrical equipment if nobody uses.

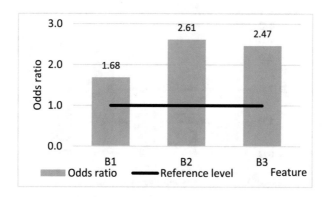

Fig. 2. Odds ratio for exogenous variable: using LED lights bulbs.

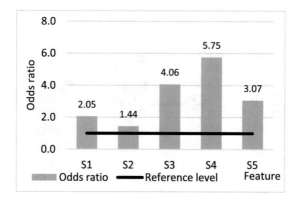

Fig. 3. A chance to make a decision about installing RES depending on sorting rubbish.

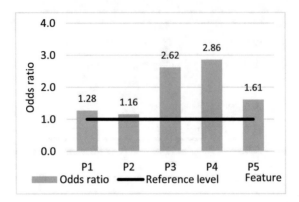

Fig. 4. Odds ratio for exogenous variable: washing or ironing at particular times.

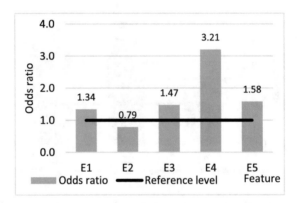

Fig. 5. Odds ratio for exogenous variable: using eco-options in a washing machine.

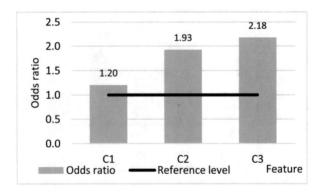

Fig. 6. Odds ratio for exogenous variable: using home appliances A+++.

estimating the parameter (β_i) standing next to the exogenous variable in the logit model determines the direction of changes in the endogenous variable [4], i.e. if the parameter value (β_i) is:

1. greater than zero, it should be interpreted that the chance of installing RES by a given subgroup is higher than in the reference group. However, to determine how many times it is greater, one should calculate the value of OR,
2. below zero, it should be interpreted that the chance of installing RES is lower than in the reference group.

Exogenous Variables having the Greatest Impact on RES Installations
The results of estimation of logit model parameters for variables: switching off electrical equipment if nobody uses (K) and using LED lights bulbs (B) are shown below. In models (1) and (2) all parameters are statistically significant.

$$logit(P) = -2.26 + 1.53 \cdot K1 + 2.35 \cdot K2 + 1.89 \cdot K3 + 2.08 \cdot K4 + 1.7 \cdot K5 \quad (1)$$

$$logit(P) = -1.19 + 0.52 \cdot B1 + 0.96 \cdot B2 + 0.9 \cdot B3 \quad (2)$$

Figures 1 and 2 show the odds ratio of the decision to install RES by different subgroups of respondents. The greatest chance of making a decision about installing RES is characterized by households that rarely turn off the computer when it is not in use. More precisely, an average of 10, 51 times higher is the chance of installing RES than in the reference subgroup. Also, in the other subgroups of households, the chance to make a decision on the installation of RES is much greater than in the case of a households, which believes that this problem does not concern them (see Fig. 1).

The chance of making a decision to install RES in comparison with the reference subgroup is much higher among households, which have mostly LED light bulbs (this chance is 2.61 times higher) or have only such bulbs (the chance is 2.47 times higher). The reference subgroup is constituted by households that do not have such light bulbs.

Exogenous Variables with a Medium Impact on RES Installations
For exogenous variables such as: sorting rubbish (S), washing or ironing at particular times (P), using eco-options in washing machine (E) and installing energy-saving household equipment (C), as a result of estimation of logit model parameters, the following results have been obtained:

$$logit(P) = -1.75 + 0.72 \cdot S1 + 0.36 \cdot S2 + 1.4 \cdot S3 + 1.75 \cdot S4 + 1.12 \cdot S5 \quad (3)$$

$$logit(P) = -0.81 + 0.24 \cdot P1 + 0.96 \cdot P2 + 1.05 \cdot P3 + 0.48 \cdot P4 + 0.15 \cdot P5 \quad (4)$$

$$logit(P) = -0.92 + 0.29 \cdot E1 - 0.24 \cdot E2 + 0.39 \cdot E3 + 1.17 \cdot E4 + 0.46 \cdot E5 \quad (5)$$

$$logit(P) = -0.91 + 0.18 \cdot C1 + 0.66 \cdot C2 + 0.78 \cdot C3 \quad (6)$$

In model (3) variables S1 and S2 are insignificant, which means that the chance of making a decision about installing RES by household that never segregated waste or did it is rarely is similar to the reference subgroup, i.e. to households to which problem of segregation does not apply.

The chance of making a decision on installing RES by households that very often sort rubbish is on average 5.75 times higher than the chance for households, to which problem of segregation does not apply (see Fig. 3).

In the model (4) the variables P1 and P2 are insignificant, which means that the chance of making a decision on installing RES by households that never wash or iron at a particular time (night tariff) or did it is rarely is similar to the reference subgroup, which is constituted by the households, that do not have such a possibility.

The greatest chance of making a decision about installing RES (see Fig. 4) is characterized by households who often and very often wash or iron using the second tariff (i.e. at particular times). Chance is on average 2.86 times (P4) and 2.62 times (P3) higher compared to the reference group.

In the model (5), the following variables are insignificant: E1, E2 and E3, which means that the chance of making a decision on installing RES by households that never used eco-options in the washing machine, did it rarely or even often, is close to the subgroup reference value, i.e. to households that do not have this option in their washing machines.

The chance of making a decision on installing RES by households, which very often use the eco-option in washing machines (E), is on average 3.21 times higher than the chance for the reference subgroup to make this decision (see Fig. 5).

However, in the (6) model, only the C1 variable is insignificant. It means that the chance of making a decision on installing RES by households, which have some household appliances type A+++ is similar to the reference subgroup, i.e. to households, which do not have such equipment.

The chance of making a decision to install RES grows with the amount of A+++ household appliances owned by households (see Fig. 6). The chance of making this decision by households having only such devices (C3) is on average 2.18 times higher than the chance of the households not having such devices.

Exogenous Variables with a Negative Impact on RES Installations

For the exogenous variable utilization of electro-rubbish (UE) and battery utilization (UB), the following results were obtained as a result of the estimation of logit model parameters:

$$logit(P) = -0.41 - 0.81 \cdot UE1 - 0.82 \cdot UE2 - 0.21 \cdot UE3 + 0.19 \cdot UE4 - 0.11 \cdot UE5 \tag{7}$$

$$logit(P) = -0.03 - 1.24 \cdot UB1 - 1.01 \cdot UB2 - 0.53 \cdot UB3 - 0.13 \cdot UB4 - 0.59 \cdot UB5 \tag{8}$$

Almost all estimated parameters with explanatory variables are negative, which means that the probability of making decisions by particular groups is lower than for the reference subgroup. For both models, variables that characterized more frequent pro-ecological behavior (UE3, UE4 and UE5 and UB3, UB4) proved to be insignificant. The fact that the household never utilizes of electro-rubbish reduces the chance of making a decision to install RES by 55%, and if it does it rarely, it decreases the chance by 56%. However, in the case of battery utilization, the following dependencies were identified. If a household:

- does not utilize the batteries, the chance for a decision to install RES decreases by as much as 71%,
- rarely utilizes batteries, then the chance to make a decision decreases by 63%,
- always utilizes batteries, it also has a negative impact on the decision to install RES (the chance is reduced by 44%).

3   Conclusions

The goals of the article have been achieved. The results of the obtained models confirmed the hypotheses about the existence of a relationship between decision making by households about installing RES and pro-ecological and pro-effectiveness behaviors, such as: switching off electrical equipment (e.g. computer) if nobody uses (K), using LED lights bulbs (B), sorting rubbish (C), washing or ironing at particular times (P), using eco-options in a washing machine (E), installing energy-saving household equipment (C), the utilization of electro-rubbish (UE), battery utilization (UB). Most of them are positive. The exception is battery utilization (UB) and the utilization of electro-rubbish (EU). In these two cases, those who perform a given activity more often are less likely to become prosumers than the reference subgroup.

The results did not confirm the relationship between households deciding to install RES and switching off the lights, switching off the chargers after charging the device, unplugging devices that are on standby, boiling water under the lid and using energy-saving lights.

Identified dependencies may find application in the promotion of RES among more flexible groups of households and addressing prosumer programs to them. Therefore, market segmentation based on criteria corresponding to the examined features of pro-ecological behavior (see Tables 1 and 2) can increase the effectiveness of marketing campaigns.

References

1. Barr, S.: Factors influencing environmental attitudes and behaviors: a UK case study of household waste management. Environ. Behav. **39**, 435–473 (2007)
2. Diaz-Rainey, I., Ashton, J.K.: Profiling potential green electricity tariff adopters: green consumerism as an environmental policy tool? Bus. Strategy Environ. **20**, 456–470 (2011)
3. Górecki, B.R.: Ekonometria podstawy teorii i praktyki. Wydawnictwo Key Text, Warszawa (2010)
4. Gruszczyński, M. (ed.): Mikroekonometria. Modele i metody analizy danych indywidualnych. Oficyna Wolters Kluwer, Warszawa (2010)
5. Markowicz, I.: Statystyczna analiza żywotności firm. Wydawnictwo Naukowe Uniwersytetu Szczecińskiego, Szczecin (2012)

6. Parag, Y., Sovacool, B.K.: Electricity market design for the prosumer era. Nat. Energy **1**, 1–6 (2016). https://doi.org/10.1038/nenergy.2016.32
7. Ropuszyńska-Surma, E., Węglarz, M.: Bariery rozwoju energetyki rozproszonej. Przegląd Elektrotechniczny **2017**(4), 90–94 (2017)
8. Ropuszyńska-Surma, E., Węglarz, M.: Społeczna akceptacja dla OZE – perspektywa odbiorców (prosumentów). In: Proceedings of conference Rynek Energii Elektrycznej REE'17, Polityka i ekonomia, Vol. I, pp. 44–55. Kazimierz Dolny (2017)
9. Ropuszyńska-Surma, E., Węglarz, M.: The pro-economical behaviour of households and their knowledge about changes in the energy market. In: E3S Web of Conferences, vol. 14, Energy and Fuels 2016 (2017). https://doi.org/10.1051/e3sconf/20171401006
10. Ropuszyńska-Surma, E., Węglarz, M.: Identyfikacja czynników wpływających na przyszłych prosumentów. Studia i Prace WNEiZ US (2018). (in Publishing)
11. Scarpa, R., Willis, K.: Willingness-to-pay for renewable energy: primary and discretionary choice of British households' for micro-generation technologies. Energy Econ. **32**, 129–136 (2010)
12. Swami, V., Chamorro-Premuzic, T., Snelgar, R., Furnham, A.: Personality, individual differences, and demographic antecedents of self-reported household waste management behaviours. J. Environ. Psychol. **31**, 21–26 (2011)
13. Toffler, A.: Trzecia fala. Państwowy Instytut Wydawniczy, Warszawa (1997)
14. Ustawa z 20 lutego 2015 r., o odnawialnych źródłach energii. Dz.U. 2015, poz.478 z późn. zmianami
15. Ustawa o zmianie ustawy o odnawialnych źródłach energii oraz niektórych innych ustaw. Dz.U. poz. 925 z dnia 22 czerwca 2016 r

The Assessment of GDPR Readiness for Local Government Administration in Poland

Dominika Lisiak-Felicka[1](✉) ⓘ, Maciej Szmit[2] ⓘ,
and Anna Szmit[3] ⓘ

[1] Department of Computer Science in Economics,
University of Lodz, Lodz, Poland
dominika.lisiak@uni.lodz.pl
[2] Department of Computer Science, University of Lodz, Lodz, Poland
maciej.szmit@uni.lodz.pl
[3] Department of Management, Lodz University of Technology, Lodz, Poland
anna.szmit@p.lodz.pl

Abstract. The article presents the most important changes introduced by the Regulation (EU) 2016/679 of the European Parliament and of the Council of 27 April 2016 on the protection of natural persons with regard to the processing of personal data and on the free movement of such data, and repealing Directive 95/46/EC (General Data Protection Regulation). The processing of personal data takes place in various spheres of economic and social activity, including public administration. The article presents results of the Computer Aided Web Interview (CAWI) survey that has been conducted among local government administration offices in Poland between March and April 2018. The aim of the research was to determine the degree of preparation of local government administration to implement changes resulting from the GDPR, and to identify sources and forms of support and problems in the implementation of these changes. On the basis of the conducted survey, an assessment the General Data Protection Regulation readiness for local government administration was performed and presented.

Keywords: General data protection regulation (GDPR) · Data protection
Local government administration

1 Introduction

The Regulation (EU) 2016/679 of the European Parliament and of the Council of 27 April 2016 on the protection of natural persons with regard to the processing of personal data and on the free movement of such data, and repealing Directive 95/46/EC (General Data Protection Regulation) will take effect on the 25th of May, 2018 [9]. Entities will have to make considerable efforts to get their data protection system into compliance with the GDPR [16, 17].

It lays down rules relating to the protection of natural persons with regard to the processing of personal data and rules relating to the free movement of personal data and protects fundamental rights and freedoms of natural persons and in particular their right

Z. Wilimowska et al. (Eds.): ISAT 2018, AISC 854, pp. 417–426, 2019.
https://doi.org/10.1007/978-3-319-99993-7_37

to the protection of personal data [9]. GDPR defines the conditions that the entities that process personal data will have to implement. The most important of them are (see e.g. [2–6, 8, 11]):

- extended rights of the data subject,
- Data Protection Officer position,
- information obligation and consent to data processing,
- notification of a personal data breach obligation,
- administrative fines,
- records of processing activities,
- processor and responsibility,
- data protection by design and by default [7, 10],
- data protection impact assessment,
- limitations on profiling.

As a part of the research, it was planned to check if the local government administration offices in Poland are prepared to implement changes resulting from the GDPR. Data protection in these units are directly concerned with personal data protection of citizens. Offices are processing data on the basis of legal regulations and everyone of citizens transfers his/her personal data to the office.

The administrative division of Poland is based on three organizational levels [12]. The territory of Poland is divided into voivodeships (provinces, "województwo" in Polish); these are further divided into districts ("powiat" in Polish), and these in turn are divided into municipalities ("gmina" in Polish). Major cities have the status of both gmina and powiat [13–15]. There are currently: 16 voivodeships, 380 districts (including 66 cities with districts status), and 2,478 municipalities in Poland.

The organizational units whose aim is to provide assistance to municipality officers, districts heads and marshals in the tasks defined by the law of the state are as follows: municipality offices, districts offices and marshal offices.

2   Method

The aim of the research was to determine the degree of preparation of local government administration to implement changes resulting from the GDPR, and to identify sources and forms of support and problems in the implementation of these changes. The survey has been conducted using Computer Aided Web Interview (CAWI) method between March and April 2018.

The survey invitation was sent by email to all local government administration offices. It was explained that the obtained data would be used in an aggregated form only for the preparation of statistical summaries and analyses in scientific publications. The questionnaire was anonymous.

The questionnaire had been available on the web for a few weeks and 462 offices decided to participate in the survey: 6 offices at the voivodeship level (marshal offices), 66 at the districts level (district offices) and 390 at the municipalities level (municipal offices). Figure 1 presents their location structure. Table 1 presents the numbers of employees in the offices.

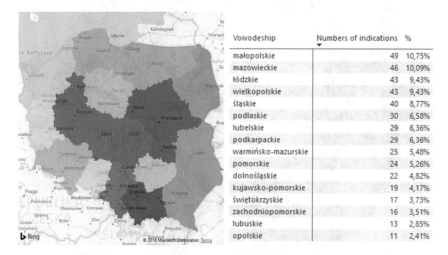

Voivodeship	Numbers of indications	%
małopolskie	49	10,75%
mazowieckie	46	10,09%
łódzkie	43	9,43%
wielkopolskie	43	9,43%
śląskie	40	8,77%
podlaskie	30	6,58%
lubelskie	29	6,36%
podkarpackie	29	6,36%
warmińsko-mazurskie	25	5,48%
pomorskie	24	5,26%
dolnośląskie	22	4,82%
kujawsko-pomorskie	19	4,17%
świętokrzyskie	17	3,73%
zachodniopomorskie	16	3,51%
lubuskie	13	2,85%
opolskie	11	2,41%

Fig. 1. The geographical location of offices participating in the survey. Due to the anonymous survey, the marshal offices were not asked about the location because of the possibility of identification (in each voivodship there is one marshal office). Source: own survey.

Table 1. Numbers of employees in the offices. Source: own survey.

Numbers of employees	Numbers of offices	%
up to 50 people	279	60,39%
51 to 100 people	104	22,51%
101 to 500 people	59	12,77%
501 to 1,000 people	7	1,52%
1,001 to 2,000 people	7	1,52%
2,001 to 3,000 people	4	0,87%

The structure of municipal offices responded to the survey were similar to the structure in Poland (chi-squared test p value = 0,044) – see Fig. 2.

Also the percent of different types of municipalities were similar in the sample and in the whole country – see Fig. 3.

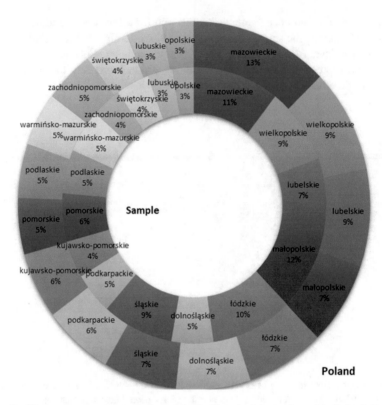

Fig. 2. The percentage structure of municipal offices in the sample and in Poland. Source: own survey.

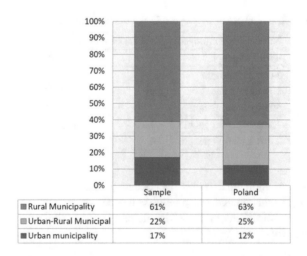

	Sample	Poland
■ Rural Municipality	61%	63%
▨ Urban-Rural Municipal	22%	25%
■ Urban municipality	17%	12%

Fig. 3. The percentage structure of types of municipal offices in the sample and in Poland. Source: own survey.

3 Results

Among 462 offices, only 96 (21%) formally defined implementation strategy for the GDPR (objectives, deadlines, responsible persons, procedures), but only 12 of them provided a link or attached a file that includes strategy specification. Other respondents did not want to include attachment or they declared that the strategy is in preparation.

Only 32 offices (7%) defined the indicators of readiness/maturity of the GDPR implementation, such as following (described by offices):

- % of trained employees, % of updated documentation, % of contracts/annexed contracts for entrusting data processing that complying with the requirements of the GDPR;
- completing tasks specified in the schedule on time;
- measures corresponding to the requirements of specific GDPR articles;
- checklists;
- monitoring of changes in regulations, participation in training;
- training, documentation preparation, security policy, records of processing activities, risk analysis,
- method of conducting the audit, DPO appointment;
- preparation of appropriate documentation, introduction of changes in the security level of IT systems and access control, development of training improving employees' competences;
- developing procedures in the required time;
- performance measures, in the context of the organization's objectives, time intervals, integrity and confidentiality, storage constraints, correctness, data minimization, purpose limitation, legal compliance, reliability and transparency;
- deadlines for projects, persons responsible for implementation, annotations on performance;
- threats, list of incidents;
- specified in the contract with the company.

The evaluation process of the GDPR implementation is conducted at the 139 offices (30%), in 92 cases by self-evaluation and the others by an external company. The evaluation tools are used in 32 offices. These includes:

- computer programs;
- information security risk analysis, data encryption programs, backup programs and fast recovery;
- forms, questionnaires, analyses;
- documentation review,
- training;
- security analysis;
- observation, interview, document analysis;
- self-control, functional control;
- planning, information gathering, analysis and evaluation, evaluation of the objectives implementation;
- analysis of resources and documents, observations, data sets;

- checking the legality, adequacy, purpose and scope of data processing; auditing, among others the area of data processing, access to the area; risk assessment; updating of data protection documentation, including authorizations; analysis of entrustment agreements; consents analysis; updating the content of the information obligation; analysis of data sets, etc.;
- verification card, information on the stand, application verification,
- ordinance;
- interviews with employees, case analysis, data sets;
- inventory of personal data processing processes, identification of risks and control mechanisms;
- risk management method.

Respondents were asked to assess the degree of office readiness for implementing changes resulting from the GDPR? (on a scale of 1 to 5, where 1 – no readiness, 5 – all GDPR requirements have been already implemented). Results are shown on Fig. 4.

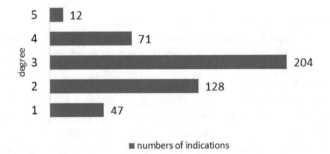

Fig. 4. Degrees of office readiness for implementing changes resulting from the GDPR. Source: own survey.

In the next question respondents were asked to specify the level of difficulty in implementing changes resulting from the GDPR to the office on a scale of 1 to 5 (1 – very easy, 5 – very difficult). Results are shown on Fig. 5. The most difficult for the respondents are: data protection impact assessment, data protection by design and by default and Data Protection Officer hiring. On the other hand a lot of respondents assessed DPO position as very simple change.

Among 462 offices, 216 (47%) declared that they use or plan to use the services of an external company to help in the implementation of requirements brought forward by the GDPR.

Figure 6 shows the answers on the question: Who is responsible for ensuring the security of personal data in the office?

The 18 respondents indicate also: data administrator (controller), contracted person, employees and external company, all office workers, data security administrator, municipality heads, unit manager and office manager.

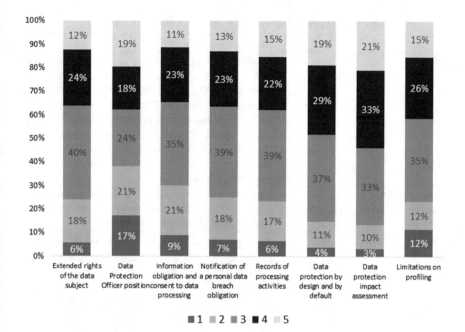

Fig. 5. The assessment of difficulty in implementing changes resulting from the GDPR. Source: own survey.

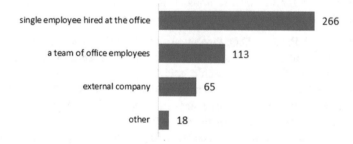

Fig. 6. Responsibility for ensuring the security of personal data in the office. Source: own survey.

Only 24 (5%) offices declared that there have been a case of personal data security breach during the last year (21 of them had one to 5 breaches and 3 offices – from 6 to 20 cases).

About one-third respondents receive support from higher organizations regarding the implementation of the GDPR. Table 2 presents types and sources of support.

Only 75 (16%) offices are conducting training on the changes resulting from the GDPR and all employees have been trained, 201 (44%) offices are conducting training but not all employees have been trained yet and 186 (40%) offices are not conducting such training.

Table 2. Types and sources of support. Source: own survey.

Type	Source
Training	conducted by for e.g. external company, voivodeship office, district office, the Poviats Association, Municipals Association, National Institute of Local Government, Regional Chamber of Accounts, Ministry of digitization, Foundation for the Development of Local Democracy, Regional Institute of Local Government and Administration, centers of education and self-government studies; regional projects
Information materials, brochure, interpretations, consultation, www	from the Inspector General for the Protection of Personal Data (GIODO in Polish), Association of Information Security Administrators
Conferences, webinars	GIODO, voivodeship office, higher organizations
Publications	from the Internet, specialized web pages devoted to GDPR, GIODO, LEX Wolters Kluwer, ABIExpert, Successpoint, legal guides on personal data protection, subscription to dedicated magazines
Consultations	with external companies, using the services of specialists in the personal data protection

Respondents were asked to indicate the biggest concern (in their opinions) in preparing the office for the GDPR implementation. They mentioned the following problems (grouped in legal, financial, organizational, essential and human aspects):

- legal – the Polish Personal Data Protection Act has not been adopted yet (during the article development process), the rules are unclear, the absence of specific legal acts, implementing regulations and specific guidelines;
- financial – the lack of sufficient financial resources for the GDPR implementation;
- organizational – the lack of time, excess of duties, lack of support from superior authorities, problem in finding qualified staff;
- essential - problems with the procedures implementation, preparation of documentation, risk analysis and assessment, implementation of tasks resulting from changes introduced in the GDPR, technical barriers, lack of tested solutions;
- human - resistance and reluctance against changes, lack of awareness of employees and management staff, lack of sufficient knowledge about GDPR.

4   Discussion and Conclusion

On the basis of the research it can be concluded, that the surveyed offices will have a huge problem with the implementation of changes resulting from the General Data Protection Regulation (GDPR) in the required time. A large group of respondents have not defined even the implementation strategy for the GDPR.

There is no doubt that all actions concerning GDPR implementation are taken too late. The readiness for implementing changes seems to be insufficient, especially taking into account the near deadline. At the same time responders' self-assessment of the readiness seems to be overstated. The degrees are more related to the choice of the middle element of scale instead of a reliable analysis of the situation at the office. Offices can only count on support from superior authorities in the field of training and information activities. The other alarming conclusion concerns immaturity of implementation approach itself. Only minor part of offices defined strategy, maturity measures or performed any evaluation of the process. This indicates a lack of a process approach.

As the biggest problems in GDPR implementation offices indicate: lack of Polish Personal Data Protection Act, unclear rules, the absence of specific legal acts, implementing regulations and specific guidelines. There is a fear that after the development and implementation of changes, the Act will introduce some additional rules/procedures and offices will have to adapt it again. Respondents listed also: lack of financial, time and human resources, problems with the procedures implementation and resistance and reluctance against changes by officers.

However, it should not be a surprise that the offices are not prepared to GDPR implementation, when even the Inspector General for the Protection of Personal Data declares that it is also not prepared for this Regulation [1].

References

1. Biekak-Jomaa, E.: Wdrożenie RODO w Polsce zagrożone. https://www.giodo.gov.pl/pl/1520281/10380. Accessed 20 Apr 2018
2. Ferrara, P., Spoto, F.: Static analysis for GDPR compliance. Paper presented at the CEUR Workshop Proceedings, vol. 2058 (2018)
3. Gellert, R.: Understanding the notion of risk in the General Data Protection Regulation. Comput. Law Secur. Rev. 34(2), 279–288 (2018)
4. Kolah, A., Foss, B.: Unlocking the power of data under the new EU General Data Protection Regulation. J. Direct Data Digit. Mark. Pract. 16(4), 270–274 (2015). https://doi.org/10.1057/dddmp.2015.20
5. Krystlik, J.: With GDPR, preparation is everything. Comput. Fraud Secur. 2017(6), 5–8 (2017)
6. Lisiak-Felicka, D., Nowak, P.: Wybrane aspekty zarządzania bezpieczeństwem informacji w podmiotach prowadzących działalność leczniczą. Przedsiębiorczość i Zarządzanie, Społeczna Akademia Nauk, Łódź-Warszawa (2018). (in print)
7. O'Connor, Y., Rowan, W., Lynch, L., Heavin, C.: Privacy by design: informed consent and internet of things for smart health. Procedia Comput. Sci. 113(2017), 653–658 (2017)
8. PWC: 10 najważniejszych zmian, które wprowadza RODO. https://www.pwc.pl/pl/artykuly/2017/10-najwazniejszych-zmian-ktore-wprowadza-rodo.html. Accessed 20 Mar 2018
9. Regulation (EU) 2016/679 of the European Parliament and of the Council of 27 April 2016 on the protection of natural persons with regard to the processing of personal data and on the free movement of such data, and repealing Directive 95/46/EC (General Data Protection Regulation)

10. Romanou, A.: The necessity of the implementation of Privacy by Design in sectors where data protection concerns arise. Comput. Law Secur. Rev. **34**(1), 99–110 (2018)
11. Tikkinen-Piri, C., Rohunen, A., Markkula, J.: EU General Data Protection regulation: changes and implications for personal data collecting companies. Comput. Law Secur. Rev. **34**(1), 134–153 (2018)
12. Ustawa z 24 lipca 1998 r. o wprowadzeniu zasadniczego trójstopniowego podziału terytorialnego państwa (Dz. U. z 1998 r. Nr 96, poz. 603)
13. Ustawa z 5 czerwca 1998 r. o samorządzie powiatowym (Dz. U. z 2001 Nr 142, poz. 1592, z późn. zm.)
14. Ustawa z 5 czerwca 1998 r. o samorządzie województwa (Dz. U. z 2001 r. Nr 142, poz. 1590 z późn. zm.)
15. Ustawa z 8 marca 1990 r. o samorządzie gminnym, (Dz. U. z 2001, nr 142, poz. 1591, z późn. zm.)
16. Voight, P., von dem Bussche, A.: The EU General Data Protection Regulation (GDPR). A Practical Guide. Springer International Publishing AG (2017). https://doi.org/10.1007/978-3-319-57959-7
17. Zerlang, J.: GDPR: a milestone in convergence for cyber-security and compliance. Netw. Secur. **2017**(6), 8–11 (2017)

Author Index

Printed in the United States
By Bookmasters